NEUROLOGY *of Music*

NEUROLOGY of Music

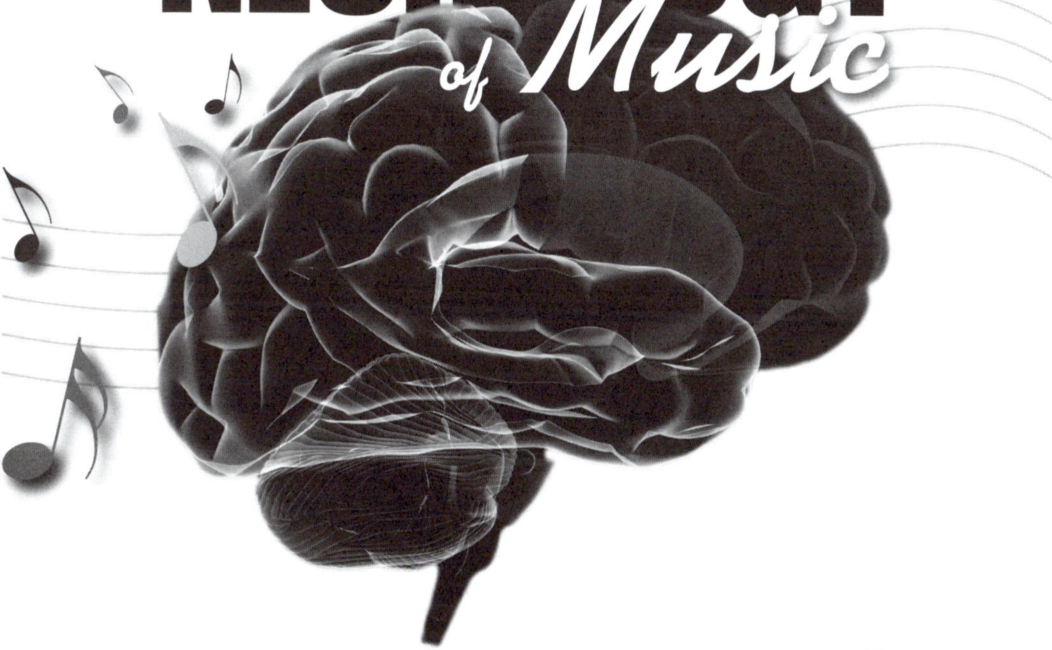

editor

F Clifford Rose

Founding Director, Academic Unit of Neuroscience
Charing Cross and Westminster (now Imperial College) Medical School

ICP Imperial College Press

Published by

Imperial College Press
57 Shelton Street
Covent Garden
London WC2H 9HE

Distributed by

World Scientific Publishing Co. Pte. Ltd.
5 Toh Tuck Link, Singapore 596224
USA office: 27 Warren Street, Suite 401-402, Hackensack, NJ 07601
UK office: 57 Shelton Street, Covent Garden, London WC2H 9HE

British Library Cataloguing-in-Publication Data
A catalogue record for this book is available from the British Library.

NEUROLOGY OF MUSIC

ISBN-13 978-1-84816-268-6
ISBN-10 1-84816-268-5

Typeset by Stallion Press
Email: enquiries@stallionpress.com

Printed in Singapore by World Scientific Printers.

Contents

Preface

In 1977, a two-day symposium was held at the Medical Society of London and was published as *Music and the Brain*, edited by Drs Macdonald Critchley and Ronald Henson.[1] As an avid attender of the meeting and reader of its proceedings, I have to report that there has been no further volumes on this subject from that society. And yet the subject has engaged the attention of researchers with new findings. This book is an attempt to fill in some of the gaps in our knowledge acquired during the past three decades, and is also derived from a symposium held at the Medical Society of London, thirty years later and sponsored by the Mansell Bequest of that society. To give some idea of the topics covered since the Mansell Bequest Sympoisa started in 1976, a list of the proceeding titles is given. Although pleading guilty to having been the organiser/convener/chairman of all these meetings and editor of the proceedings for the past 35 years, this book is to be my last endeavour for the Mansell Bequest. There have been recent volumes on the brain and music[2–4] so there can be little doubt that this subject will continue to excite the minds of all who enjoy listening to music — which means everyone.

Dr Frank Clifford Rose

[1] *Music and the Brain*, (Eds.) M. Critchley and R. A. Henson, 1977. William Heinemann, London.

[2] *The Cognitive Neuroscience of Music*, (Eds.) I. Perez and R. Zatorre, 2003. Oxford University Press, Oxford.

[3] *This is Your Brain on Music*, D. J. Levitin, 2007. Penguin, London.

[4] *Musicophilia*, Oliver Sachs, 2007. Picador, London.

The Mansell Bequest Symposia of the Medical Society of London

Dr Harry Mansell, a graduate of St. George's Hospital Medical School, died in 1941 at the age of 40 from Motor Neurone Disease. He had joined as a Fellow of the Medical Society of London in 1937, and when his widow died, her will included a Bequest to the Society to use the money to further neurological studies. After much discussion at council meetings and sub-committees in the early 1970s, it was agreed that the money should not be passed on to other societies, charities or institutes, but that the Medical Society of London would invest the capital (nearly £35,000 at that time) and use the interest accumulated for organising symposia. The symposia were not meant to compete in popularity with the ordinary meetings of the society, but open to Fellows of the society, they would promote an exchange of ideas between international neuroscience experts along the lines of symposia organised by the Ciba Foundation and would be based on topics not frequently dealt with by commercial publishers, and the proceedings published.

The first Mansell Bequest Symposium was, understandingly, on 'Motor Neurone Disease'. It lasted only one day since, at that time, there was little interest in the disease, a situation that has since changed markedly. The second Mansell Bequest Symposium was on 'Clinical Neuro-Immunology', which lasted two days and although the first of its kind, was followed by several others. The same applied to the third symposium, 'Clinical Neuro-Epidemiology', which was also the first in its field but again followed by many others. The full list of Mansell Bequest Symposia is on the following page:

Year	Title	Publisher
1977	Motor Neurone Disease	Pitman Medical
1979	Clinical Neuro-Immunology	Blackwell Scientific
1980	Clinical Neuro-Epidemiology	Pitman Medical
1981	Metabolic Disorders of the Nervous System	Pitman Medical
1982	Advances in Stroke Therapy	Raven Press
1984	Advances in Aphasiology	Raven Press
1985	Advances in Neuro-Oncology (*with W. S. Fields*)	Karger, Basle
1988	Physiological Aspects of Clinical Neuro-Ophthalmology (*with C. Kennard*)	Chapman & Hall Medical
1989	The Control of the Hypothalamo-Pituitary-Adrenocortical Axis	International Universities Press, Chicago
1993	Advances in Neuropharmacology	Smith Gordon
1995	Recent Advances in Tropical Neurology	Elsevier Science
1997	Towards Migraine 2000	Elsevier Science
1999	A Short History of Neurology: The British Contribution (1660–1910)	Butterworth Heinemann
2001	Twentieth Century Neurology: The British Contribution	Imperial College Press
2004	Neurology of the Arts: Painting, Music and Literature	Imperial College Press
2006	The Neurology of Painting	Elsevier
2010	Neurology of Music	Imperial College Press

List of Contributors

Johan Aarli
Department of Neurology
University of Bergen
Bergen
Norway

Eckart Altenmüller
Institute of Music Physiology and Musician's Medicine
Hannover University of Music and Drama
Hohenzollernstrasse 4730161
Hannover
Germany

Hansjörg Bäzner
Department of Neurology
University of Heidelberg
Universitätsklinik Mannheim 68135
Germany

Stavia Blunt
4 St Helen Terrace
Riverside
Richmond, Surrey
TW9 1NR
UK

Tomislav Breitenfeld
Department of Neurology
University Hospital Sisters of Charity
Vinogradska 29
HR 10000 Zagreb
Croatia

John Brust
College of Physicians and Surgeons
Harlem Hospital Center and Columbia University
506 Lennox Avenue
New York NY 10037
USA

Katherine Butler
London Hand Therapy
30 Cumberland Mansions
Seymour Place
London
W1H 5FT
UK

Ian Cross
Department of Biological Anthropology
and Centre for Music and Science
Cambridge University
West Road
Cambridge
CB3 9DP
UK

Stefan Evers
Department of Neurology
University of Münster
Albert Schweitzer Str 33, 48129
Münster
Germany

Amy B Graziano
Conservatory of Music
College of Performing Arts
Chapman University
1 University Drive
Orange CA 92866
USA

Jacky Hayward
Memory and Aging Center
Department of Neurology
University of California, San Francisco
350 Parnassus, Suite 905
San Francisco CA 94117
USA

Michael Hennerici
Department of Neurology
University of Heidelberg
Universitätsklinik Mannheim 68135
Germany

Julene K. Johnson
Memory and Aging Center
Department of Neurology
University of California, San Francisco
350 Parnassus, Suite 905
San Francisco CA 94117
USA

Steven Jones
Honorary Consultant
Department of Clinical Neurophysiology
National Hospital for Neurology and Neurosurgery
Queen Square, London
WC1N 3BG
UK

Marjorie Lorch
Reader in Brain and Language
Birkbeck College
43 Gordon Square
London
WC1H 0PD
UK

Lorenzo Lorusso
Neurology Department
Azienda Ospedaliera "Mellino Mellini"
25032 Chiari Viale Mazzini 6 — (BS)
Italy

Ivan Moseley
Music Department
Royal Holloway
University of London
Egham Hill
Egham
TW20 0EX
UK

Jock Murray
Dalhousie University
Halifax
Nova Scotia
B3H 4HZ
Canada

Klaus Podoll
Department of Psychology and Psychotherapy
Aachen RTWH University Clinic
Paulwelsstrasse 30
D-52074 Aachen
Germany

Alessandro Porro
Dipartimento di Specialita Chirurgiche
Scienze Radiologiche de Medico Forensi
Via Europa
11-25123 Bresica
Italy

Gottfried Schlaug
Department of Neurology
and Music and Neuroimaging Laboratory
Beth Israel Deaconess Medical Center
and Harvard Medical School
330 Brookline Avenue
Boston MA 02215
USA

Ola Selnes
Division of Cognitive Neuroscience
Johns Hopkins University School of Medicine
1620 McElderry Street
Baltimore MD 21287
USA

Geir Olve Skeie
Department of Neurology
Institute of Clinical Medicine
and Department of Biological and Medical Physchology
University of Bergen
Norway

Frank Stahnisch
Department of Social Studies of Medicine
McGill University
3647 Peel Street
Montreal
Quebec H3A 1X1
Canada

Mario Wiesendanger
Institute de Physiologie
Département de Médicine
Rue du Musée 5
CH-1700 Fribourg
Switzerland

Ian Winspur
The Hand Clinic
30 Devonshire Street
London
W1G 6PU
UK

George K. York III
Kaiser Permanente Stockton Medical Center
Säo Institute
21201 Ostrom Road
Fiddletown CA 65629
USA

Abbreviations

ADHD	Attention Deficit Hyperactivity Disorder
AEP	Auditory Evoked Potential
AS	Aphasia Severity
BAEP	Brainstem Auditory Evoked Potential
BPD	Bipolar Disorder
CBD	Corticobasal Degeneration
CBF	Cerebral Blood Flow
DA	Dopamine
DMN	Default Mode Network
DTP	Dipole Tracing Method
EEG	Electroencephalogram
FTLD	Fronto-Temporal Lobar Dementia
GABA	Gamma-Aminobutyric Acid
HT	Hydroxytryptamine
MIT	Melodic Intonation Therapy
MEG	Magnetoencephalography
MRI	Magnetic Resonance Imaging
MRS	Magnetic Resonance Spectroscopy
NA	Noradrenaline
OCD	Obsessive Compulsive Disorder
PD	Parkinson's Disease
PET	Positron Emission Tomography
TLE	Temporal Lobe Epilepsy
TS	Tourette's Syndrome
SPECT	Single Photon Emission Computed Tomography

Chapter 1

The Evolutionary Basis of Meaning in Music: Some Neurological and Neuroscientific Implications

Ian Cross

Introduction

When we explore music in neurological terms, the neurology that we find, and the neurology that we might expect to find, is influenced by what we mean by the term 'music'. In this chapter, I shall be arguing that what we currently know of music in neurological and neuroscientific terms is constrained by a conception of music that is narrowly shaped by historical and cultural notions of what constitutes 'music'. I shall suggest that music, rather than simply being a complex sonic pattern produced and received for aesthetic or hedonic ends, can better be interpreted as a communicative medium complementary to language that is deeply embedded in, and that may be foundational in respect of, the species-specific human capacity to manage complex social relationships (see Herrmann *et al.*, 2007).

In order to ground this argument, it may be helpful to consider two musical examples. The first consists of an audience listening to a performer in a Western concert hall. This constitutes an example of what might be called 'music-as-display', and is fairly typical of music as it tends to be conceived of within contemporary Western — and increasingly global — culture. This is an apparently quite conventional musical scenario; a performer performs, and an audience listens. In this situation, music may appear to be akin to chess, or perhaps gymnastics, a result of long training, motoric skill and muscular resilience (on the part of the performer, and sometimes, the composer), yet comprehensible by virtue of being rooted in general cognitive and neural processes. The underlying biology appears to be a biology of perception and action control, albeit one that may be shaped (on the part of the composer and performer) by specialised training and learning but that is rooted in general biological process, as is evidenced by the capacity of the audience to engage with the music by virtue of enculturative processes rather than formal learning. Most of what we know about the neuroscience and neurology of music derives from this type of musical situation — indeed, it derives largely from one side of this situation, that of the listener.

But now let us explore another type of musical situation, involving a group of people, none of whom have been formally musically trained, singing and dancing together, stamping their feet, slapping their bodies and hissing rhythmically, creating the illusion of a percussion accompaniment to the singing. The text that they sing veers between praise of Allah and praise of their local councilman for the efforts that he has put into establishing a local AIDS clinic. This is a brief description of a recent *Dakira* performance from northern Mozambique (from the work of my graduate student Lydia Slobodian), in which all members of the community are as likely to participate as performers as they are simply to spectate and listen. This second type of musical situation — engagement with music through interactive performance and appraisal — is probably more characteristic of music in the majority of world cultures (and certainly in all known traditional cultures) than is the type of situation where a performer performs and an audience simply listens (see, e.g., Blacking, 1976; 1995). The biology that underlies this collaborative performing situation requires to be understood not only as concerned with perception and action control, but also with social interaction and communication. Considered from this perspective, music appears likely to be rooted in neural circuitry and social processes that overlap, or share features with, those that constitute the matrix of language.

Language and Music as Universal and Discrete Human Capacities

Language is a universal human faculty; all humans (except those suffering from a pathological condition) are able to acquire the use of language; for all humans, the acquisition of language depends on engagement in linguistic interactions within a sensitive period (generally, from birth to around four to five years of age); and language is specifically human in that it is not acquired spontaneously by members of other species, and only acquired to a rudimentary level by our closest primate relatives (Pinker and Jackendoff, 2005; Hauser, Chomsky and Fitch, 2002; Nowak, Komarova and Niyogi, 2002).

Language seems to be a discrete capacity in that although it is implicated in most other domains of human behaviour, it appears differentiable from them on the basis of its complex and propositional intentionality; language is generally about something (an object, event, belief, or goal), and the use of language, unlike any other mode of human behaviour, enables the 'something' in question to be specified (see Pinker, 1994). The neural correlates of language in production and perception have been found to overlap significantly with those of other aspects of human action and perception (such as theory of mind, object perception, etc) as might be expected of such a multi-functional and ubiquitous mode of human interaction with the world, but some areas do appear to be specific to language, being largely lateralised in the left hemisphere and involving a network that incorporates Broca's area, known to be involved in the processing of hierarchical sequence structure (see Friederici *et al.*, 2006).

Is musicality also a universal and discrete human capacity? This question has been answered variously and differently, and answers depend on what is intended by the term 'music' in research that purports to address this question. At least in part, it can be suggested that some of these difficulties of operational definition arise from the different evidentiary bases on which we can make claims about language and about music. Questions of universality and discreteness can be asked about language largely because language research is grounded in a large body of comparative data gleaned from very many different languages, which enables cross-cultural regularities of linguistic behaviour and cognition to be identified (see, e.g., Comrie, 1989). Such a body of comparative data does not exist for music; in general, from cross-cultural perspectives, music does not appear to be identifiable on the basis of regularities of usage and of structure in the way that language is.

Conventionally, music has tended to be partitioned into pitch structure and rhythmic structure, and insofar as any cross-cultural comparative studies have been conducted, they have tended to focus on perception and cognition of music in terms of one or other of these structural aspects. However, within the cognitive science and neuroscience of music, the overwhelming majority of research has explored the cognitions and neural processes of Western listeners and performers in respect of Western music (largely, that of the Western common-practice period); hence scientific studies of music are not in any position to elucidate putatively universal features of music as a human capacity. Nevertheless, research within this tradition, focusing on Western musical cognitions and behaviours, has been brought to bear on the issue of whether musicality constitutes a discrete human capacity with identifiable and distinct behavioural and neural correlates.

Amusia as an Indicator of Domain-Specificity/Modularity

On the basis of a long-term research programme, reported in a substantial series of publications, Isabelle Peretz has suggested that the basis for music in cognition and neurophysiology is modular, and that there are specialised brain processes for music, in particular, musical pitch (Peretz and Coltheart, 2003). Much of that research programme has been concerned with the exploration of the phenomenon of *amusia* — the incapacity to process musical information coherently, first named as such by Knoblauch in the nineteenth century (Johnson and Graziano, 2003). Peretz and her collaborators have demonstrated that the condition may be either acquired or congenital (Ayotte, Peretz and Hyde, 2003; Peretz and Hyde, 2003); it involves a 'double dissociation' between speech and music (evidenced in the discovery that individuals may have an inability to deal with semantic and/or syntactic aspects of speech while preserving musical and intonational capacities, or *vice versa*).

For Peretz (2006, 12), such a finding 'implies the existence of anatomically and functionally segregated systems for music and speech', and leads her to suggest that, for musical cognition, (*ibid*., 8), 'components, especially those involved in pitch-based computations, rely on domain-specific mechanisms and specialised neural networks', and are hence modular. Hence (*ibid*., 9), 'encoding pitch in musical contexts appears to be a domain-specific ability that can be localised in the adult brain'. It is important to note that Peretz's research identifies the condition of amusia with an inability to process *pitch* in musical contexts; amusics typically perform in much the same way as 'normal' individuals in rhythm-based tasks; as she states (*ibid*., 14): 'To date, however, only abilities related to fine-grained processing of pitch appear to be uniquely engaged in music.'

Of course, as Peretz herself notes, the idea that a particular capacity is modular (or even domain-specific) does not imply that that capacity can be correlated with specific brain locations or neural networks. Indeed, the inability of amusics to process fine-grained pitch structures has not been localised to or identified with specific networks; it is not simply identifiable with deficits either in the peripheral auditory pathway or in the primary auditory cortex (superior temporal lobe). However, as Peretz and Hyde (2003) note, structures in the rostromedial prefrontal cortex that have been identified as operational in judgments of pitch relationships (described by Janata *et al*., 2002, 2169 as 'a nexus… for mediating interactions between sensory, cognitive, and affective information') seem to be good candidates for a prospective brain area where such processes might be carried out. More recently, Peretz (2006, 9) has suggested that a lack of connectivity to the inferior frontal gyrus may be responsible for congenital amusics' incapacity to deal with musical pitch in perception. It remains to be seen whether or not amusics demonstrate atypical activity (or lack of activity) in these areas when engaged with music.

Peretz may be right in identifying the human capacity to engage with music with the ability to process fine-grained pitch structures and relationships. However, it should be noted that all the research that points to this conclusion has been conducted using Western-encultured participants, and has overwhelmingly employed musical examples that conform to Western common-practice musical principles. Before any firm conclusions can be drawn, there is a need to extend the range of types of pitch discrimination capacities beyond those that have been explored, in particular, a need to explore pitch discrimination capacities and propensities in other cultures. Moreover, it also remains to be seen whether or not the brain structures hypothesised by Peretz as implicated in the processing of musical pitch are consistently implicated in representations of musical pitch for members of non-Western — particularly, traditional — cultures.

Peretz, particularly in her modular theory of musical processing in the mind, is happy to posit a clear distinction between processes that are domain-general and those that are wholly specific to music. However, this picture is complicated by recent findings reported in Patel *et al.* (2008) which indicate that some amusic individuals also have difficulty in processing speech intonation patterns, which might be taken to imply that in some cases amusia is not dissociable from the phonological (and pragmatic) aspects of speech and casts some doubt on the modularity hypothesis.

Action, Interaction and Rhythm as Adaptive Behaviour

For the most part, the cognitive science and neuroscience of music has tended to focus on individual responses to music, or on individual musical receptive behaviours, with little sense that music may involve action. However, a few studies have pointed towards the active dimensions of musical behaviour; a meta-review by Janata and Grafton (2003) demonstrated that even apparently 'passive' listening to music tended to elicit activation in premotor regions of the brain concerned with planning for action, leading them to suggest that engagement with music is best conceived of as a perception-action cycle that necessarily involves, if not overt action, at least covert planning for action as an essential constituent of the experience.

These findings do point in the direction of the notion that music, like language, is a mode of interacting with others. But most current neuroscientific methods severely limit the extent to which an experimental participant is in a position to engage in any type of interaction. Indeed, it is only very recently that behavioural studies have begun to explore music explicitly as interactive behaviour (see, e.g., Moran, 2007). However, one aspect of music appears likely to be particularly fruitful as a window into musical interactivity: that of musical rhythm. Thaut (for a summary, see Thaut, 2005) has explored the behavioural and neural correlates of musical rhythm in a number of adaptive tasks, requiring participants to tap in synchrony with sequences of pulses that deviate from complete temporal regularity to different degrees. He and his collaborators have found (Stephan *et al.*, 2002) that adaptive responses (where participants manage to align the periodicity and/or phase of their tapping with an external signal that continuously deviates slightly from temporal regularity) appear to involve self-monitoring, may be either consciously or non-consciously — automatically — performed, and that, whether or not responses are conscious or automatic, participants can either be aware of their actions or not. Specific brain regions identified as being activated in adaptive synchronisation tasks are primarily parts of the cerebellum, basal ganglia, thalamus and areas in the prefrontal cortex; these areas and networks are implicated in a wide range of motoric and temporal behaviours other than just musical, lending credence to the notion that the correlates of musical behaviours in the brain are as likely to rest on domain-general processes as on domain-specific ones, and be as concerned with action as perception.

While these studies have been highly informative, they have focused on music as evidenced in individual responses and action capacities. In an exploration of music as an interactive and social phenomenon, Himberg (forthcoming) has been conducting a series of behavioural experiments on interactive rhythmic behaviour, requiring pairs of participants to tap along with each other under a variety of experimental conditions. The results of these experiments have shown that participants strongly prefer to entrain with other humans rather than with a non-responsive pacing signal. To explore the factors that appear to lead to this preference, an experiment was conducted in which an 'averaged' playback of human tapping on computer was substituted for one participant without the knowledge of the other participant. Participants demonstrated less co-ordination with the recorded tapping than with 'live' human partners. In these experiments it appeared that participants were highly attuned to the presence of their tapping partners, being sensitive to the degree to which these partners engaged in processes of mutual co-adaptation of phase and period in their tapping to maintain perceived synchrony.

Music as Communicative Medium

To summarise, neuroscientific research has suggested that the human capacity to engage with music is rooted in the ability to process fine-grained pitch structures and relationships and is domain-specific; it has also suggested that some domain-general aspects of cognition and neural functioning are implicated in the processing of music; and it supports the notion that active aspects of musical behaviours are rooted in domain-general processes and structures. However, the scope of neuroscientific explorations of music is limited by the constraints of the methodologies currently available. Moreover, both cognitive science and neuroscience have focused almost entirely on the processes involved in individual Western musical cognitions — particularly, listening — with almost no research exploring other modes of engagement with music, or engagement with music in the contexts of non-Western cultures.

This raises the question of how music is being operationally defined within neuroscientific and cognitive research. In general, it appears to be being addressed as an auditory pattern of which the primary 'musical' features are constituted of complex pitch and rhythmic structures, and brain areas that have been proposed as specialised for music tend to be those associated with auditory perception and with information integration across different domains. Music appears to be conceived of — is certainly being explored as — individual responses to complex sonic pattern. While this conception of music may apply in situations such as the 'music-as-display' scenario, it does not seem adequate to address the complexities of music as an interactive medium, as in the second scenario sketched at the outset.

When compared with language, it appears difficult to determine what might be the necessary and sufficient conditions for identifying music as a generic human capacity. Language can be described in terms of phonology, morphology, syntax and semantics, and its portrayal in terms of the features that it exhibits in these structural domains lies at the heart of the study of linguistic practices across different linguistic cultures. Music appears to lack equivalent, established structural features in terms of which it can be described cross-culturally. While some have been proposed by ethnomusicologists seeking a basis to compare music across cultures (Lomax *et al.*, 1978), and others have been proposed by scientists wishing to establish a basis for exploring music as a generic human capacity (see, e.g., Carterette and Kendall, 1999), counterexamples can be found in at least some musical cultures; indeed, the notion that 'music' constitutes a mode of human interaction that is clearly distinct from language appears absent in some emic cultural conceptions (see, e.g., the account by Lewis (2009) of communication among the Mbendjele hunter-gatherers).

Rather than focusing on structural features in attempting to define music, it may be more fruitful to evaluate whether or not music fulfils common functions within and across cultures and to define music in pragmatic terms. The ethnomusicological literature reveals that music fulfils a wide range of diverse functions in different societies, in entertainment, ritual, healing and in the maintenance of social and natural order (see, e.g., Feld and Fox, 1994; Nettl, 2005). However, it can be suggested that one very generic feature that all these functions appear to share is the management of social relationships, particularly in situations of social uncertainty. Recently (Cross and Woodruff, 2009), I have suggested that music as a universal human trait might be best conceived of as a communicative medium optimally adapted to manage situations of social uncertainty by virtue of its semantic indeterminacy or 'floating intentionality' and by the affiliative nature of musical interactions. Music manages social uncertainty in part by presenting the characteristics of an 'honest signal' (see Számadó and Szathmáry, 2006). It also under-specifies goals in ways that permit individuals to interact even while holding to personal interpretations of goals and meanings that may actually be in conflict, and music's exploitation of the human capacity for entrainment increases the likelihood that participants will experience a sense of joint action. In contrast to language, which can be thought of as being optimised to facilitate goal-directed social interactions, music can be thought of as facilitating social interaction *per se*, being essentially phatic.

Music in all cultures seems to exhibit this semantic indeterminacy, or 'floating intentionality'. Elsewhere (Cross, 2007; 2008; 2009) I have suggested that this is likely to arise because music may draw on at least three different dimensions simultaneously that endow it with meaning for performer and listener, or for participants, and that each dimension can be considered to have a different time-depth in the course of human evolution. The most evolutionarily recent dimension can be termed *culturally-enactive*, and arises from the distinct types of meaning that can be attached to conceptions of music within particular cultures. An example might be the singing of *Abide With Me* at the FA Cup Final, an activity that would seem perverse in the context of a match between Glasgow Rangers and Glasgow Celtic, and positively incomprehensible in the context of a match between Boca Juniors and River Plate in Buenos Aires.

The second dimension is rooted in cross-cultural, vocal and gestural, commonalities of form in interpersonal communication. It could be referred to as prosodic, though the term *socio-intentional* better captures its applicability across all domains of human vocal and gestural communication. It is evident in the ways in which music embodies tonal shapes and rhythmic structures that appear homologous with those that are employed in linguistic interaction to convey particular pragmatic stances: that is, not *what* is said, but *how* what is said conveys or betrays the *attitude* or *intent* of the speaker towards what is said or towards their interlocutor. This second dimension can be thought of as having an evolutionary time-depth at least as great as that of communicative humanity and is likely to be species-specific and cross-culturally general.

The third dimension in terms of which music can be experienced as bearing meaning is species-general. It derives from the evolutionary binding between the form of an acoustical signal and the functions that signal may bear for sound-producers and sound-perceivers. The relationship is involuntary and reflexive, conditioned by its survival value for an organism (and species), and deeply embedded in the genomes of those species for whom sound carries survival-critical information. This dimension can be referred to as the *motivational-structural*, the acoustical structure of the signal impinging on, or being conditioned, by the motivational — affective — state of the organism that perceives or produces it. It is in respect of this dimension that music can be interpreted as constituting an honest signal. It is manifested in music's global structural characteristics, and can be interpreted as lying at the root of the effects that the musical signal has been shown to have on non-human species (Rickard *et al.*, 2005).

These three dimensions of meaning operate simultaneously in any musical experience, allowing for volatile meaning to migrate between dimensions and enabling participants in collective musical behaviour each to interpret the purpose and function of that behaviour to their own ends without impairing its collective integrity. And that integrity is likely to be maintained by the possibility of entrainment — of temporal alignment if not synchrony between participants' behaviours — which is likely to promote a sense of group affiliation. Taking the proposal that music is a communicative medium optimal in the management of situations of social uncertainty as an operational definition of music, it would suggest that, across cultures, core musical behaviours — and brain areas activated in music — are likely to be those implicated in situation-specific social interactions that are oriented towards the reduction of social uncertainty. Such a hypothesis, though self-evidently broad-brush, might be of use in explorations of the behavioural and neural correlates of musicality as a universal human trait.

Music and the Social Brain

The 'social mind' and the 'social brain' have become, in recent years, major foci for cognitive and neuroscientific investigation, in part precipitated by the realisation that the complexity of human sociality is one of the principal defining characteristics that differentiates us from other species (see, e.g., Tomasello *et al.*, 2005; Herrmann *et al.*, 2007). Behavioural research has focused on issues such as theory-of-mind (concerning the cognitive processes whereby inferences are made about others' states of mind), social reasoning, and decision-making, and has often adverted to evolutionary approaches in seeking explanations for the scope of, and the constraints on, human social behaviours (see chapters in Dunbar and Barrett, 2007). Neuroscientific research has begun to turn its attention to these and other aspects of human social capacities, and a helpful overview is provided in Adolphs (2003). He notes that together with limbic regions, the orbitofrontal cortex is clearly implicated in most aspects of social cognition, suggesting that (*ibid.*, 175) it might be 'specialised for social and moral judgements', although limbic regions, concerned with affect or emotion, are also central to social cognition.

Adolphs proposes that the social mind-brain can be conceived of in terms of generic reward/punishment or approach/withdrawal systems, though noting that while this may constitute a useful first approximation, the binary dichotomy that it suggests might be too coarse-grained for adequate behavioural and neural analysis. Nevertheless, in the context of Adolphs' proposal, music, as a communicative mode optimised for the management of social uncertainty, might be conceptualised as facilitating 'approach' or 'affiliative' behaviours, as opposed to 'avoidance' or 'withdrawal' behaviours. One might thus expect to find that engagement with music primarily involves behaviours, and motivates neural structures and networks that are involved in initiating and sustaining approach-affiliative behaviours. While the hypothesis appears almost self-fulfilling at the behavioural level, it would be difficult to test by means of current neuroscientific method as it would require that the focus of study be shifted to explore much more 'ecologically-valid' engagement with music than is currently achievable.

It would be critical to identify the loci and networks whereby music achieves these ends, and to explore the extent to which they are differentiable from those implicated in other domains of human behaviour. Some preliminary theories have been developed; for instance, Molnar-Szakacs and Overy (2006) present a theoretical account of the neural substrates that are associable with music construed as an active, interactive and affective behaviour. And of course none of the foregoing is intended to suggest that the results and foci of current research on the neuroscience of music should either be ignored or discounted. On the contrary, the notion that music should be explored primarily as a social and interactive medium seeks to provide an extended context for their conduct and interpretation by focalising the ways in which music as social act engages the pitch and rhythm systems identified by Peretz, Thaut and others in ways that are shared with, or are distinct from, other modes of human communicative interaction such as language.

References

1. Adolphs, R. (2003). Cognitive neuro-science of human social behaviour. *Nature Reviews Neuroscience* **4**(3): 165–178.
2. Ayotte, J., Peretz, I. and Hyde, K. L. (2001). Congenital amusia: A group study of adults afflicted with a music-specific disorder. *Brain* **125**: 238–251.
3. Blacking, J. (1976). *How Musical is Man?* London: Faber.
4. Blacking, J. (1995). *Music, Culture and Experience.* London: University of Chicago Press.
5. Carterette, E. C. and Kendall, R. A. (1999). Comparative music perception and cognition. In Deustch, D. (Ed.), *The Psychology of Music*, pp. 725–791. London: Academic Press.
6. Comrie, B. (1989). *Language Universals and Linguistic Typology* (2nd ed.). Chicago: University of Chicago Press.
7. Cross, I. (2007). Music and cognitive evolution. In Dunbar, R. I. M. and Barrett, L. (Eds.), *Handbook of Evolutionary Psychology*, pp. 649–667. Oxford: Oxford University Press.
8. Cross, I. (2008). Musicality and the human capacity for culture. *Musicae Scientiae*, Special Issue, 127–143.
9. Cross, I. (2009). The evolutionary nature of musical meaning. *Musicae Scientiae*, Special Issue, 179–200.
10. Cross, I. and Woodruff, G. E. (2008). Music as a communicative medium. In Botha, R. and Knight, C. (Eds.), *The Prehistory of Language*, Vol. 1, pp. 113–144. Oxford: Oxford University Press.

11. Dunbar, R. I. M. and Barrett, L. (Eds.). (2007). *Handbook of Evolutionary Psychology*. Oxford: Oxford University Press.
12. Feld, S. and Fox, A. A. (1994). Music and language. *Annual Review of Anthropology* **23**: 25–53.
13. Friederici, A. D., Bahlmann, J., Heim, S., Schubotz, R. I. and Anwander, A. (2006). The brain differentiates human and non-human grammars: functional localization and structural connectivity. *Proc Natl Acad Sci USA* **103**(7): 2458–2463.
14. Hauser, M., Chomsky, N. and Fitch, T. (2002). The faculty of language: What is it, who has it and how did it evolve? *Science* **298**(5598): 1569–1577.
15. Herrmann, E., Call, J., Hernández-Lloreda, M. V., Hare, B. and Tomasello, M. (2007). Humans have evolved skills of social cognition: the cultural intelligence hypothesis. *Science* **317**(5843): 1360–1366.
16. Himberg, T. (forthcoming). *Cognitive Foundations of Interaction in Musical Time*. Ph.D thesis, University of Cambridge, Cambridge, UK.
17. Janata, P., Birk, J. L., Van Horn, J. D., Leman, M., Tillmann, B. and Bharucha, J. J. (2002). The cortical topography of tonal structures underlying Western music. *Science* **298**(5601): 2167–2170.
18. Janata, P. and Grafton, S. (2003). Swinging in the brain: Shared neural substrates for behaviours related to sequencing and music. *Nature Neuroscience* **6**(7): 682–687.
19. Johnson, J. K. and Graziano, A. B. (2003). August Knoblauch and amusia: A nineteenth-century cognitive model of music. *Brain and Cognition* **51**(1): 102–114.
20. Lewis, J. (2009). As well as words: Congo Pygmy hunting, mimicry, and play. In Botha, R. and Knight, C. (Eds.), *The Cradle of Language* 2: 381–413. Oxford: Oxford University Press.
21. Lomax, A., Rudd, R., Grauer, V. A., Berkowitz, N., Hawes, B. L. and Kulig, C. (1978). *Cantometrics: An Approach to the Anthropology of Music: Audiocassettes and a Handbook*. Berkeley: University of California Extension Media Center.
22. Molnar-Szakacs, I. and Overy, K. (2006). Music and mirror neurons: From motion to 'e'motion. *Social Cognitive and Affective Neuroscience* **1**(3): 235–241.
23. Moran, N. S. (2007). *Measuring Musical Interaction*. Unpublished Ph.D, Open University, Milton Keynes.
24. Nettl, B. (2005). *The Study of Ethnomusicology: Thirty-one issues and concepts* (2nd ed.). Urbana and Chicago: University of Illinois Press.
25. Nowak, M. A., Komarova, N. L. and Niyogi, P. (2002). Computational and evolutionary aspects of language. *Nature* **417**: 611–617.
26. Patel, A. D., Wong, M., Foxton, J., Lochy, A. and Peretz, I. (2008). Speech intonation perception deficits in musical tone deafness (congenital amusia). *Music Perception* **25**(4): 357–368.

27. Peretz, I. (2006). The nature of music from a biological perspective. *Cognition* **100**(1): 1–32.
28. Peretz, I. and Coltheart, M. (2003). Modularity of music processing. *Nature Neuroscience* **6**(7): 688–691.
29. Peretz, I. and Hyde, K. L. (2003). What is specific to music processing? Insights from congenital amusia. *Trends in Cognitive Science* 7(8): 362–367.
30. Pinker, S. (1994). *The Language Instinct.* London: Allen Lane.
31. Pinker, S. and Jackendoff, R. (2005). The faculty of language: What's special about it? *Cognition* **95**(2): 201–236.
32. Rickard, N. S., Toukhsati, S. R. and Field, S. E. (2005). The effect of music on cognitive performance: Insight from neurobiological and animal studies. *Behavioral and Cognitive Neuroscience Review* **4**(4): 235–261.
33. Stephan, K. M., Thaut, M. H., Wunderlich, G., Schicks, W., Tian, B., Tellmann, L., *et al.* (2002). Conscious and subconscious sensorimotor synchronization — prefrontal cortex and the influence of awareness. *NeuroImage* **15**: 345–352.
34. Számadó, S. and Szathmáry, E. (2006). Selective scenarios for the emergence of natural language. *Trends in Ecology and Evolution* **21**(10): 555–561.
35. Thaut, M. H. (2005). Rhythm, human temporality, and brain function. In Miell, D, MacDonald, R. and Hargreaves, D. (Eds.), *Musical Communication*, pp. 171–191. Oxford: Oxford University Press.
36. Tomasello, T., Carpenter, M., Call, J., Behne, T. and Moll, H. (2005). Understanding and sharing intentions: The origins of cultural cognition. *Behavioral and Brain Sciences* **28**(5): 675–691.

Chapter 2

Historical Perspectives on the Study of Music in Neurology

Julene K. Johnson, Amy B. Graziano, and Jacky Hayward

The evolution or our knowledge about how the brain processes music holds an important place in the history of neurology. An examination of early neurology literature reveals that music abilities were occasionally examined in an attempt to better understand brain function; in particular, music was used as a tool to assess patients with aphasia. Early neurologists were fascinated by the observation that some patients with severe expressive aphasia were able to sing the text of songs. This observation became a theme in nineteenth-century neurology literature that was discussed by some of the most prominent neurologists at the time, including John Hughlings Jackson and Jean-Baptiste Bouillaud.

Introduction

Although music has been used as a tool to study the human brain for centuries, very little has been written about how an interest in studying music in neurologic patients emerged. An examination of early neurology literature reveals that music abilities were often examined in an attempt to better understand brain function. In particular, music was used as a tool to assess patients with aphasia. Early neurologists were fascinated by the observation that some patients with severe expressive aphasia were able to sing words of songs. The purpose of this chapter is to review early studies of music in neurology with a particular emphasis on how early ideas about the relationship between music and the brain developed.

Pre-Nineteenth-Century Descriptions of Music and Neurology

On a mute who can sing

In 1745, the Swedish historian Olof Dahlin published an article titled *Berättelse om en Dumbe, som kan siunga* (*On a Mute Who Can Sing*) in the Swedish Academy of Sciences journal.[1] He described a farmer's son who had a stroke that resulted in right-side paralysis and complete loss of speech. A priest, who had known the farmer's son for years, told Dahlin about the gentleman's noteworthy ability to sing and whistle hymns he had learned before the stroke. Although he never saw the patient, Dahlin was struck by the dissociation between the patient's loss of conversational speech and his preserved ability to sing words of a song. In his article, Dahlin provided more details about the patient's singing ability. The complete text of Dahlin's (1745) account reads:

> Jon Persson, a farmer's son from Ofvankihl in Juleta county in Södermanland, born in 1703, and raised in the usual, simple way to know his Christian religion and read, was in 1736, after having been married for three years, taken ill with an acute disease (rapid onset), from which he suffered a stroke that resulted in paralysis of his entire right side of the body and total speechlessness. After almost half a year of bed confinement, he began moving a bit; though when he walked, he limped and had his right arm in a sling. For two years after that, he was at a spa close to the Juleta parish residence, from which many had benefited. But he did not think he got better, except for more steadiness in walking and also somewhat better ability to say on many occasions, the important word "yes". One advantage, however, was that he had already been noticed and is the reason why this person now became famous; he can sing and whistle hymns, which he had learned before he became ill. And these so purely and explicitly like any other normal person: but one should notice that he in the beginning of the hymn had to be assisted a bit by another person who sings (as well). Even

> more, he is able with the same help to pronounce clearly certain prayers without singing: But in time and with a high pitched crying voice. However, this man is mute in daily life and needs to sign with the hands, to express his meaning, and who cannot pronounce one single word except for "yes".
>
> Somewhat stupid and little silly he has probably always been. Yet now, as before, his hearing and comprehension are not set back. For the rest, he is God-fearing, quiet and is moral.
>
> The priest in Juleta, Magistrate Johan Ihering, according to his own and assigned story, I report this (without having seen this person or hearing his singing) has now for eight years had him in the parish quarters. He tried in every way to determine if this has been some illusion by this man to better earn his food. But he found the case completely unfalsified. The inhabitants of Juleta parish of higher or lower rank tell the same thing. The varied and peculiar effects that a stroke may cause are all too well known: I want to add to this story nothing else other than the remark that those who stutter, even if they cannot manage ten words in a row without interruption when they speak, however, can sing without impairment and with sense.[1] (pp. 114–115) (translated by Arne Brun)

Dahlin's report identified several key observations that would be discussed in more detail over 100 years later. First, Dahlin noted that the speechless patient was able to verbalise words while singing or praying, that he was otherwise unable to verbalise in conversational speech. However, Dahlin also noted that the patient needed assistance in starting the hymn and sang with a, 'high-pitched, shouting voice', so it appears that his singing ability was not completely normal. Benton and Joynt[2] suggest that Peter Rommel (in 1683) was likely the first to describe the preserved ability to recite prayers and biblical verses in patients with severe motor aphasia. Dahlin's case report is important because it may be the earliest description of the ability to verbalise words in singing but not in conversational speech. This theme has remained an important topic in the study of music and neurology and will be discussed further in this chapter.

An abbot who could not recite psalms or sing

Approximately 30 years later, Johann August Philipp Gesner, a German physician, recorded another observation regarding the dissociation between conversational speech and singing in his well-known five-volume monograph, *Samlung von Beobachtungen aus der Arzneygelahrheit und Naturkunde.* Gesner wrote a chapter about aphasia (*Die Sprachamnesie*) in the second volume.[3] In this chapter, he described an abbot (case three of six) who was suddenly unable to speak and had difficulty understanding others when he was giving instructions to a peasant. Gesner also noted that the abbot was not able to read printed materials, recite longer Psalms, or sing songs. However, the abbot could recite a prayer. Unfortunately, no further information is given regarding the abbot's difficulty with singing songs. Gesner also noted that the abbot had other cognitive deficits, such as difficulty with calculations and a decline in judgment. However, he did not have difficulty recognising objects or have any sensory or motor deficits. Gesner's monograph is considered one of the first important writings on aphasia.[4,5] It is important to note that singing was included in this early monograph, although Gesner did not discuss how it might be related to reciting memorised texts.

John Hughlings Jackson and *Singing by Speechless Children*

John Hughlings Jackson continued the interest in music and aphasia and provided some of the first explanations as to why singing remained preserved in severe aphasia. Jackson, who is considered the founder of British neurology, had a strong interest in studying language and aphasia. He discussed singing abilities in several of his early papers about language. In 1866, Jackson[6] described a three-year-old boy with epilepsy who could only verbalise 'mam' and 'dad'. His mother reported that the boy could learn tunes that his father played on his flute, but he would not sing when asked. In this paper, Jackson also referred to another young boy with epilepsy who could not speak but could sing. Jackson, however, did not personally observe the singing abilities of either of these two children and, instead, relied on family report. In the same article, Jackson described a woman who could accurately sing a familiar melody using the sounds 'lor', 'deah' and 'me' instead of the song text. Jackson used these examples to show how the vocal and articulatory muscles were intact, despite severe difficulty in producing speech. Jackson (1866) noted:

> In order to develop her baggle, her husband said: 'Go and talk to the bird'. She went to the cage, which was hanging from the ceiling in one corner of the room, and, standing up, cried: 'Ah! Ah! O deah! deah! deah! Pittymy, pittymy. Lor, lor, lor' etc. She seemed quite delighted with her task, and varied her voice wonderfully, uttering one set of the gabble in one tone, and the others in other tones. At the same time she gesticulated incessantly, throwing her arms up and down, seeming to accompany her voice with a sort of dance of the arms.[6] (p. 175)

In another passage, Jackson (1866) stated:

> And dancing has all the world through been regarded as natural to an elevated state of mind. The woman would dance when a barrel organ was played in front of her house. Thus, then, she could use her laryngeal muscles not only to utter single sounds like 'ah! oh!' but also in the complex process of singing. Again, she could not only use her hands in simple gesticulations, but could use her legs in the more cultivated movements of dancing.[6] (p. 176)

Jackson's ideas about movement and singing were heavily influenced by Herbert Spencer's (1857) essay, *On the origin and function of music.*[7] Jackson linked the dance-like movement of the woman's arms to Spencer's idea that dancing was a rhythmic behavior that represented elevated emotion. Jackson also linked singing to Spencer's idea that singing is an expression of elevated emotion since it involves mental and muscular 'excitement'. Spencer (1857) stated, 'Every one of the alterations of voice which we have found to be a physiological result of pain or pleasure, is carried to its greatest extreme in vocal music'.[7] (p. 400). Spencer was a prominent English philosopher who wrote several essays about music and corresponded with Jackson for many years.[8,9]

Approximately ten years later, Jackson (1871) returned to the topic and published an article titled *Singing by speechless children.*[10] Here, he described two young boys who had minimal speech output but could sing songs. The first patient was a 10-year-old boy who, at the age of one, experienced a series of seizures that resulted in right-sided hemiplegia. Similar to his previous descriptions, the boy was only able to verbalise a few words (e.g., 'here', 'there', 'I won't') and never developed speech. However, the boy could sing several children's songs (i.e., *Not for Joe* and *Heads and Tails*). Jackson was again struck by the preserved ability to sing words despite very limited speech output. Jackson (1871) described a visit with the boy saying:

> He walks about the room, he takes up a book, puts it down; then goes to a chair, which he moves. He comes next to the fireplace, and points to a bust; then gives his mother a push, under which she staggers, and breaks out singing, 'Not for Joe,' 'Not for Joe.' He only sang these three words twice, and would not sing anything more. The word 'Joe' was uttered very clearly, the other two words indistinctly.[10] (p. 430–431)

The second patient was an eight-year-old boy who had epilepsy and mental retardation. The boy was only able to verbalise a few words (e.g., 'here' and 'Eleanor'), but was able to sing a melody, however, without words. Jackson examined this patient, but he was not able to convince the boy to sing in his presence. Jackson argued that speech and voice are distinct and that the preserved ability to sing provided evidence that, '...the larynx and its nerves are uninterfered with.'[10] (p. 431). Jackson felt that it was important to point out that loss of speech could occur without damage to the larynx or articulatory mechanisms.

Jackson also speculated about the parallels between aphasia and amusia, a syndrome that would be named by Knoblauch 17 years later in 1888. Jackson stated, 'A symptom corresponding to aphasia would be a loss of power to sing *tunes* previously acquired.'[10] (p. 431). In addition, he predicted, 'It is probable that a musician who had become speechless, and unable to read printed or written words, would easily read music...'[10] (p. 431). In this paper, Jackson also referenced several others who had made comments about music and neurology. For example, Langdon Down noted that some mentally handicapped children could not speak but could sing. He also remarked that Falret (1866)[11] differentiated aphasics who could sing with and without a text. Finally, he referenced Trousseau (1865)[12] who described a musician who could not read or write but could write a musical phrase to dictation. Jackson was, therefore, aware of other cases with the preserved ability to sing despite limited speech output.

Jackson continued to use the example of singing to argue the differences between emotional and intellectual speech. He published a series of three papers titled *On affections of speech from disease of the brain* from 1878–1879 that appeared in *Brain*.[13,14] He more firmly linked singing to emotional language and explained how singing and other automatic speech could remain preserved in severe aphasia. Jackson (1878) described:

> His vocal organs act apparently well; he may be able to sing. His emotional language is apparently unaffected. He smiles, laughs, frowns, and varies his voice properly. His recurring utterance comes out now in one tone and now in another, according as he is vexed, glad etc.; strictly we should say he sings his recurring utterance; variation of voice being rudimentary song (*Spencer*); he may be able to sing in the ordinary meaning of that term. As stated already, he may swear when excited, or get out more innocent interjections, simple or compound (acquired parts of emotional language).[13] (p. 320)

Jackson, therefore, linked the ability to sing in aphasia to the expression of emotional language, such as recurring utterances (automatic speech). Thus, Jackson contributed several early ideas about music in neurologic patients. He used examples of the preserved ability to sing to support his ideas about speechlessness and the differences between emotional and intellectual language. He drew from Herbert Spencer's ideas about the origin of music. He also speculated about the preservation of music reading in the presence of alexia and agraphia and that an impairment in singing familiar songs would parallel aphasia. It is curious that Jackson used the music analogy so consistently in his writings on aphasia because Jackson apparently had '…no particular taste for music or art in any form…', and he '…often admitted he could not distinguish the National Anthem from *Rule Britannia*…'[15] (p. 910).

An interest in studying automatic or nonpropositional speech has continued to the present day. In current literature, automatic speech is considered to include highly over learned sequences (e.g., nursery rhymes, expletives, memorised text, counting) that are not generated to convey ideas. Several authors consider automatic speech as a specific category of language use that is distinct from propositional language, which is used to convey ideas and requires phonological, lexical, and semantic access.[16] Several modern studies have investigated the apparent dissociation between singing and speaking words in severe expressive aphasia, arguing that patients do not produce more words during singing than in repetition tasks.[17,18] However, Straube *et al.*[19] found an increase in the number of words produced during singing when compared with speaking excerpts of familiar song lyrics under certain conditions in their single case study. Future studies should evaluate larger cohorts of patients rather than focusing on single case studies to determine the prevalence of this phenomenon.

Other Early Reports of Music in Neurology

In addition to the articles discussed above, there were other reports about music abilities and aphasia during the nineteenth century. In 1836, Behiér, who was most likely a physician in Paris at the Hôtel Dieu, described an aphasic who could sing (cited in Falret, 1866).[11] In 1864, M. Jules Falret, a French psychiatrist, noted in his *Dictionnaire Enclyclopédique des Sciences Médicales*[11] that some aphasics may retain the ability to sing with a text, while other aphasics can sing without a text The French Academy of Medicine published a series of papers about language that appeared in the *Bulletin de l'Académie Imperiale de Médicine (Paris)* (1864–1865). In this series, Armand Trousseau (1865),[12] a physician in Paris at the Hôtel-Dieu, described a musician who could not read or write language but could write music notes to dictation. In the same series, Jean-Baptiste Bouillaud (1865),[20] another physician in Paris, described an aphasic musician who could compose music and accompany vocalists on the piano. Bouillaud followed the work of Franz Joseph Gall who also contributed to ideas about music and brain by identifying a brain centre for the sense of sounds and music talent (organ 17) in the early nineteenth century.[21] However, Gall's work was not cited as a source for knowledge about music and the brain in any of the aphasia literature. In Berlin, Karl Maria Finkelnburg (1870)[22] studied a violinist with music agraphia and music alexia in his paper about 'asymbolia'. Adrien Proust (1872)[23] described another musician who could not hum a tune, but could read music notation, play scales and recognise tunes. Carl Wernicke (1874)[24] in his *Der aphasische Symptomencomplex* commented about a woman with aphasia (Susanne A) who could sing, but without the song text. Wernicke (1874) stated:

> Das Tyrolerlied (Wenn ich zu meinem Kinde geh'), das zufällig von einer anderen Kranken gesungen wurde, singt sie richtig nach, aber ohne Text. (Without text, she correctly sings the Tyrol song, *When I go to my child*, after it happened to be sung by another patient.)[24] (p. 40)

In summary, the early studies about music and aphasia revealed both preserved abilities but also impaired abilities involving music. The authors of these papers did not go into great detail describing the patterns of music. However, it is important to note that many of the early researchers on aphasia commented on music, in particular, singing abilities. These authors used observations of music to help evaluate the degree to which language was impaired in severe motor aphasia. Most were fascinated by the ability to sing words that the patients were not able to repeat or verbalise in conversational speech. These early studies primarily involved individuals without music training, but several neurologists studied musicians with aphasia. The early studies of musicians with aphasia provided hints that different aspects of music thinking could be preserved or impaired and set the stage for even more sophisticated analyses of music abilities that followed in subsequent years.

Knoblauch and Amusia

By the mid 1880s, a number of published case reports that described music abilities in neurologic patients had accumulated. In 1888, August Knoblauch, a German physician and anatomist, published a paper describing another patient and coined the term 'amusia' to classify impairments in music abilities after brain damage. The paper was first published in German and then in English two years later in *Brain*.[25,26] Knoblauch worked at the Asylum for the Insane at the University of Heidelberg under Wilhelm Erb, a well-known German neurologist. In his article, Knoblauch also proposed a diagrammatic model of music and the brain that was based on Ludwig Lichtheim's model of language. The sections below further describe Knoblauch's contributions to the neurology of music.

Knoblauch and his patient L.S.

When he was working in Erb's clinic in Heidelberg, Knoblauch observed a six-year-old girl (L.S.) who had severe motor aphasia and could only verbalise the word 'mama' and repeat a few other words. The girl suffered from 'scarlatina' (scarlet fever) that was associated with acute brain inflammation, seizures, loss of consciousness, and right-sided hemiplegia, and a severe aphasia. Knoblauch noted that the girl could sing *Weisst Du wie viel Sternlein stehen* with the song text but in an 'automatic way'. He also noted that she was unable to continue or start singing again if stopped. At this tie, she was also not able to recite the song text or voluntarily verbalise words from the song text. L.S. was later admitted to the clinical hospital at Heidelberg to undergo treatment (i.e., galvanisation of the head) and speech therapy. After some treatment, L.S. was able to repeat single words. In addition, she was able to sing with, 'purer intonation than at the beginning of the treatment',[25] (p. 320) and recite the text of the song without singing. After two months, L.S. was able to form complete sentences and was discharged from the hospital. Knoblauch's case is also interesting because he describes improvement in the patient's singing ability after treatment.

Diagrammatic model of music

Knoblauch used the example of his patient and also other published reports in the literature to propose a diagrammatic model of music (Fig. 1). He based the structure of his music model on Ludwig Lichtheim's (1885)[27] diagrammatic model of language.

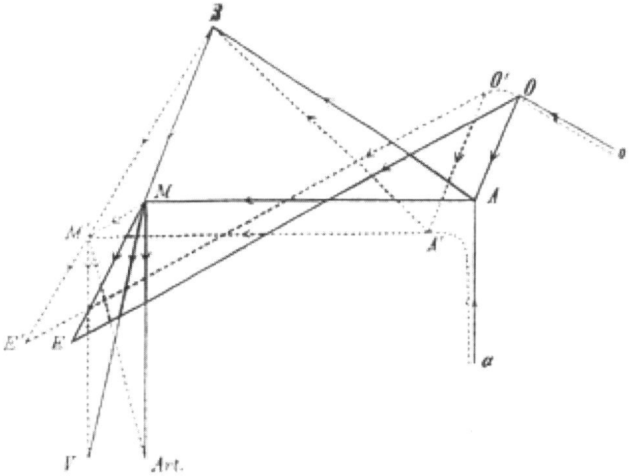

Figure 1. Knoblauch's (1888/1890) diagrammatic model of music (dashed lines) overlaid on Lichtheim's (1885) diagrammatic model of language (solid lines).

Knoblauch published his music model in conjunction with Lichtheim's language model where ***B*** is the centre for ideas, ***M'*** is the motor centre for tones, ***A'*** is the auditory centre for tones, ***E'*** is the motor centre for writing music notation, and ***O'*** is the visual centre for reading music notation. In addition, ***a*** refers to the auditory nerve, and ***V*** and ***Art.*** are involved in articulation. These centres closely parallel Lichtheim's centres for language. The centre for ideas is the only centre that serves both music and language functions. Knoblauch's model is the first diagrammatic model of music and represents how music and language were considered parallel functions in the brain.

Coining of the term 'amusia' and additional classifications

Knoblauch classified music impairments into motor (production) and sensory (perception) disorders, which was similar to how aphasia was classified. Knoblauch coined the term 'amusia' (*Amusie*) to refer to a motor impairment in the production of music after brain damage. He distinguished disorders of music production and comprehension and proposed nine types of amusia. For example, Knoblauch used the terms 'note deafness' (*Tonetaubheit*) to refer to a deficit in comprehending music and 'note blindness' (*Notenblindheit*) to refer to an impairment in reading music notation. The majority of these distinctions were hypothetical, as some had not yet been described in the literature.

Shortly after Knoblauch's paper, other authors drafted additional classification schemes for amusia, including Richard Wallaschek (1891),[28] M. Brazier (1893),[29] and Paul Blocq (1893).[30] In addition, there was a sharp increase in studying music in neurologic patients. By 1898, J.G. Edgren reviewed 51 cases involving music in neurologic patients,[31] and by 1920, Solomon Henschen summarised over 300 cases.[32] Over time, amusia became used as a more generalised term, referring to any impairment in music perception or production rather than only to motor impairments for music. Amusia is a term that continues to be used in current neurology and neuropsychology literature.

Conclusions

In summary, early interest in studying music in aphasia stemmed from the striking observation that some severely aphasic patients could sing text in the context of songs and the desire to better understand automatic speech. A more careful study of music abilities in aphasics revealed a complex pattern of music preservations and deficits. Music started to be considered a complex behaviour, like language. The early studies laid the groundwork for the first cognitive model of music and the coining of the term 'amusia' (Knoblauch), while stimulating a field of study that continues to present day.

Acknowledgements

This research was supported by an Arnold L. and Lois S. Graves Award in the Humanities (A.G.). The authors would like to thank the University of California libraries for helping obtain original sources (Valerie Wheat). We kindly thank Arne Brun, MD (Lund, Sweden) for locating and translating Dahlin (1745).

References

1. Dahlin, O. Berättelse om en Dumbe, som kan siunga [On a Mute who Can Sing]. *Swenska Wetenskaps Academiens Handlingar* **1745**: 114–115.
2. Benton, A. L. and Joynt, R. J. (1963). Three pioneers in the study of aphasia (Johann Schmidt, Peter Rommel, Johann A.P. Gesner). *J Hist Med Allied Sci* **18**: 381–384.
3. Gesner, J. (1770). Die Sprachamnesie [Speech Amnesia]. *Sammlung von Beobachtungen aus der Arzneygelahrtheit und Naturkunde [Collection of Observations from the Medical and Natural Sciences]*, Vol. 2, pp. 107–182. Nördlingen: CC Beck.
4. Luzzatti, C. (2002). Johann August Philipp Gesner (1738–1801). A review of his essay "The language amnesia" in the bicentennial anniversary of his death. *J Hist Neurosci* **11**: 29–34.
5. Benton, A. L. and Johann, A. P. (1965). Gesner on aphasia. *Med Hist* **9**: 54–60.
6. Jackson, J. H. (1866). Clinical remarks on emotional and intellectual language in some cases of disease of the nervous system. *A Mirror.*
7. Spencer, H. (1857). The origin and function of music. *Frazer's Magazine* **56**: 396–408.
8. Critchley, M. (1986). Hughlings Jackson. The man and his time. *Arch Neurol* **43**: 435–437.
9. Swash, M. (1986). John Hughlings-Jackson: A sesquicentennial tribute. *J Neurol Neurosurg Psychiatry* **49**: 981–985.
10. Jackson, J. H. (1871). Singing by speechless (aphasic) children. *The Lancet* **98**: 430–431.
11. Falret, J. (1866). Aphasie, aphémie, alalie. In Dechambre, A. (Ed.), *Dictionnaire Encyclopédique des Sciences Médicales*, Vol. 5, pp. 506–644. Paris: Victor Masson et Fils.
12. Trousseau, A. (1865). Discussion sur la faculté du langage articulé [Discussion on the faculty of spoken language]. *Bulletin de l'Académie Impériale de Médecine* **30**: 647–657; 659–675.
13. Jackson, J. H. (1878). On affections of speech from disease of the brain. *Brain* **1**: 304–330.
14. Jackson, J. H. (1879). On affections of speech from disease of the brain. *Brain* **2**: 203–222; 323–356.
15. Buzzard, E. F. (1934). Hughlings Jackson and his influence on neurology. *The Lancet* **2**: 909–913.
16. Van Lancker, D. (1987). Nonpropositional speech: Neurolinguistic studies. In Ellis, A. W. (Ed.), *Progress in the Psychology of Language*, Vol. 3. London: Erlbaum.
17. Racette, A., Bard, C. and Peretz, I. (2006). Making non-fluent aphasics speak: Sing along! *Brain* **129**: 2571–2584.

18. Hebert, S., Racette, A., Gagnon, L. and Peretz, I. (2003). Revisiting the dissociation between singing and speaking in expressive aphasia. *Brain* **126**: 1838–1850.
19. Straube, T., Schulz, A., Geipel, K., Mentzel, H. J. and Miltner, W. H. (2008). Dissociation between singing and speaking in expressive aphasia: The role of song familiarity. *Neuropsychologia* **46**: 1505–1512.
20. Bouillaud, J.-B. (1865). Discussion sur la faculté du langage articulé [Discussion on the faculty of spoken language]. *Bulletin de l'Académie Impériale de Médecine* **30**: 724–781.
21. Zola-Morgan, S. (1995). Localization of brain function: The legacy of Franz Joseph Gall (1758–1828). *Annu Rev Neurosci* **18**: 359–383.
22. Finkelnburg, D. C. (1870). Niederrheinische Gesellschaft in Bonn. *Berliner klinische Wochenschrift* 7: 449–450; 460–462.
23. Proust, A. (1872). De l'aphasie. *Archives Générale de Médicine* **9**: 147–166; 303–318; 653–685.
24. Wernicke, C. (1874). *Der aphasische Symptomencomplex. Eine psychologische Studie auf anatomischer Basis.* Breslau.
25. Knoblauch, A. (1888). Ueber Störungen der musikalischen Leistungsfähigkeit infolge von Gehirnläsionen. *Deutsches Archiv für Klinische Medecin* **43**: 331–352.
26. Knoblauch, A. (1890). On disorders of the musical capacity from cerebral disease. *Brain* **13**: 317–340.
27. Lichtheim, L. (1885). Ueber aphasie. *Deutsches Archiv für Klinische Medicin.* **36**: 204–268.
28. Wallaschek, R. (1891). Ueber die Bedeutung der aphasie für den musikalischen Ausdrucks-Vermögens. *Deutsche Zeitschrift für Musikwissenschaft* 7: 53–73.
29. Brazier, M. (1892). Du trouble des facultés musicales dans l'aphasie. *Revue Philosophique* **34**: 337–368.
30. Blocq, P. L'Amusie (1893). *Gazette Hebdomadaire de Médecine et de Chirurgie* **8**: 86–90.
31. Edgren, J. G. (1895). Amusie (musikalische Aphasie). *Deutsche Zeitschrift für Nervenheilkunde* **6**: 1–64.
32. Henschen, S. E. (1920). Kausuistik der amusie und akalkulie. *Klinische und anatomische beitrage zur pathologie des gehirns*, Vol. 5, pp. 42–136. Stockholm: Nordische Bokhandeln.

Chapter 3

The Creative Brain: Fundamental Features, Associated Conditions and Unifying Neural Mechanisms

Stavia Blunt

High intelligence, domain-specific knowledge and a conducive environment are essential — but not sufficient — for creativity. Other attributes are required. Certain personality traits, cognitive features and neuropsychiatric conditions are associated with creativity. Our understanding of these traits and conditions and their relationship to creativity may shed light on neural mechanisms which underlie the creative brain. A characteristic feature of creativity, and of certain attributes of the creative individual, is variability over time, in both the short and longer term. This fluctuation is mirrored by temporal variability in some of the associated neuropsychiatric conditions. These attributes, and the associated neuropsychiatric conditions, are related to activity in key transmitter systems: DA, NA, 5HT and GABA. The observed temporal variabilities and correlations suggest that a fundamental property of the creative brain is one of transmitter dysregulation, especially involving DA pathways. This dysregulation can be detrimental to creativity if too wildly fluctuant, but is the essential ingredient when it occurs in more limited degree. Transmitter dysregulation may arise from heterogeneous genetic abnormalities and environmental influences operating during development and adult life. Studies in the associated neuropsychiatric conditions — and in creativity itself — have focussed on genes affecting synaptic function and neuronal development. Many of these genes converge on final common pathways involving growth factor and stress activated kinases. These same routes may be fundamental to the creative brain. Dysfunction of neurotrophic factor pathways, whether primary or secondary to converging genetic defects, would account for many characteristics of the creative brain, including the association with certain neuropsychiatric conditions, dysregulation of specific transmitter systems, possible abnormal dendritic development, and the influences of sex hormones and environmental factors. Neurotrophic factor mechanisms would account for the observation that creativity is quintessentially a neurodevelopmental phenomenon.

Introduction

The nature of creativity has perplexed thinkers for centuries, but only in the last 60 years have neuroscientists begun to investigate its neural basis. The study of creativity has been fraught with difficulties[1]: of definition resulting from its subjective, involuntary and ephemeral nature, about its distribution and relationship to intelligence, and whether it can be measured. It has been approached using a range of methods. What has emerged, not so controversially, is that creativity results from complex genetic and environmental interactions operating throughout development and adult life. The exact neural mechanisms underlying creativity may vary across disciplines, but studies of creativity across all domains have revealed many parallels. For this reason, I discuss creativity and the creative brain in general.

First, some definitions: a creative person is one who readily comes up with new ideas; but novelty alone is not enough for creativity. *Creativity* is the ability to produce work that is both novel and appropriate.[1] *The creative process* is the series of mental events that leads to a creative product; it consists of several stages: preparation, incubation, inspiration and verification.[2]

Nature versus Nurture

Exactly what makes a person extraordinarily creative — a 'genius' — has puzzled people for centuries. Some view it as a heritable attribute, while others believe anyone can be a genius. Take these two views separated by over 300 years:

> *'Genius must be born and can never be taught'* (John Dryden — 1631–1700)
> *'Creativity is 100% nurture and 0% nature'* (Michael Howe — 1999)

In fact, both nature (genes) and nurture (environment) are important.

Heredity

Creative pedigrees

In a highly selected study of pedigrees of eminently creative people across various domains, Francis Galton found that the frequency of genius in kinships far exceeded reasonable explanation; in his view it was a male attribute.[3] Pedigrees that support his thesis include Galton's own family (Charles and Erasmus Darwin); the Bachs (20 eminent musicians over eight generations from 1550–1800); the Van Eycks, Bellinis, Monets and Titians (painters); the Coleridges, Tennysons and Wordsworths (poets); and more recently the Huxleys (scientists and novelists)[4] and the Jacksons (pop music). But not, of course the Brontes – who were all women! However, pedigree studies do not show how often no evidence for heritability exists nor do they remove the effects of environment. Galton himself later accepted the importance of environment, introducing the concept of 'nature versus nurture'.[5]

Case-control and twin studies

The evidence for the heritability of creativity has been controversial. Andreasen found that creativity is more common in relatives of creative writers compared with controls,[6] and the co-existence of *multiple forms* of creativity suggested a heritable 'general creativity factor'.[4] In some twin studies, the correlation of creativity traits is higher in identical than fraternal twins[7] but not in others.[8,9] These contradictions might be because several traits — which are individually highly heritable — are required *simultaneously* for creativity to occur ('emergenic' traits[10]). Other studies linking creativity with bipolar disorder (BPD) support a genetic element (see below).

Gender — Or 'degree of maleness'

Whether there are sex differences in creativity is not known. Galton's view of creativity as being a male attribute is superficially supported by the male preponderance of eminent people across all domains. But reaching eminence is a measure of creative *productivity* not *potential*, and many factors both intrinsic (e.g. personality) and extrinsic (e.g. opportunity, role in society, cultural attitudes) influence the realisation of this potential.[11] Furthermore, the interruption (and, possibly, biological changes) brought about by childbearing makes the immersion, commitment, isolation and motivation required for exceptional creative achievement difficult. Even in Western societies today, women are more likely to 'drop out' regardless of initial creative promise.[12]

Putting aside factors that influence *realisation* of creative potential, could gender — or, more specifically, sex hormones — affect neural mechanisms underlying creativity? Baron Cohen's 'extreme male brain' theory of autism[13] implicates high testosterone levels during development, which might be relevant to the pockets of *ability* rarely found in autistic *idiot savants*. For true creativity, however, including intellect, a more subtle effect may be required: it may not be gender *per se* that is important, but the 'degree of maleness'. Thus, musical creative ability correlated with testosterone levels — low normal for males, and high normal for females.[14] A 'maleness factor' also resonates with the male preponderance of some of the neuropsychiatric disorders associated with creativity, such as BPD-1, ADHD, schizophrenia, and Tourette's syndrome, which show a divergence in gender prevalence (M > F) at adolescence. Furthermore, given the influence of sex hormones (especially oestrogen) on monoamine systems, psychiatric and cognitive function,[15,16] changes in relative ratios of sex hormones across reproductive life might affect creative output, at least in women. This could be readily studied.

Intelligence and domain-specific intellectual precocity

In trying to understand the neural basis for creativity, its relationship to intelligence has been studied. Creativity requires high intelligence[4,17] and a correlation between the two exists up to an IQ of about 120. Thereafter the correlation tails off.[4,18] What instead emerges is that creativity is related to a very high ability in one or more *particular domains* i.e. 'domain specific intellectual precocity' or 'talent' — which becomes apparent early on in life. Creative individuals have an affinity for their subject, easily building up intense knowledge,[11,19,20] which, once acquired, is added to, tapped, sifted and re-associated in future creativity.

However, whilst high intelligence is a pre-requisite for exceptional creativity, the reverse is not the case: i.e. not all intelligent people are creative. Likewise across the span of creative people, intellectual precocity may be restricted to a single domain, or may exist in many (the 'polymath'). Thus, creativity and intelligence are related — but different — often considered as overlapping sets.[17]

Environment

Family background

Regardless of degree of final eminence or discipline, certain family and education factors persistently emerge as important in realisation of early talent.[11,12,19,21] Generally, creative people have comfortable family backgrounds which provide opportunity and nurture talent as it appears. The family has strong moral values, they live away from major cities, but retain contact with the field of interest; parents tend to be professional or entrepreneurial, placing high value on learning, and often have a history in the child's domain of interest. Diverse opportunities are presented, and omnivorous reading is encouraged.[5,19,21,22]

'Crystallising experiences' — e.g. a book of poetry or a mathematical text — often set the final path.[11] Religious and ethnic minorities produce a disproportionate number of eminent creators, though only when basic rights are enjoyed.[19] Certain 'stressors' are conducive to creativity: early parental loss,[23] other personal childhood trauma or being an 'outsider'.[11,12,19,21] However, too much hardship is detrimental.

Education, knowledge, training and the 'ten-year rule'

Across all creative disciplines, even the most talented individuals required years of preparation and immersion in their domain before they began to produce their major works.[20,24] This time of 'silence', in which the creative individual is building knowledge and skill has been called the 'Ten-year rule'. It applies to composers (Mozart, The Beatles), painters, poets, chess masters and even jazz musicians whose improvisations in fact rely on the use of memories internalised over years of practice, which are later 'automatically' sifted and re-assembled as they create new music.[20,24]

Timing and timeliness — The creative context and receptiveness of the field

To be useful and appreciated, creative outputs must be relevant to current practice, societal and cultural standards, and are usually edged with a dissatisfaction of and motivation to change the field. However, many creative individuals are 'ahead of their time', their material not appreciated until after their death.[12]

Talent, a conducive environment and training are necessary, but not sufficient for adult creativity. Thus, being a child prodigy does not guarantee adult creativity, though an intense early start certainly helps.[12,25] Something else is required for that special creative sparkle. What these neurobiological mechanisms might be can be gleaned by studying patterns of creativity, personality and cognitive features, and the neuropsychiatric conditions with which creativity is commonly associated.

Variability of Creativity and Associated Personality and Cognitive Attributes

Variability and periodicity of creativity

Unlike intelligence — which is constant — the ability to create waxes and wanes involuntarily — even in highly creative people. The literature is full of examples of eminent people whose creativity has fluctuated, often going for years without creative output, only to be replaced by a flurry of creativity. This long-term variability correlates with certain mood, cognitive and behavioural changes, demonstrated most vividly in patients with BPD. But creativity also fluctuates in the short term as demonstrated by the sudden illumination characteristic of the insight phase of the creative process. This process correlates with changes in cognitive function and level of arousal. This variability over the short and long term may be a fundamental property of the creative brain and is discussed in detail below.

Cognitive styles and the creative personality

Several studies confirm that the 'creative personality' exists (Table 1).[26–29]

Compared with creative scientists, artists are more anxious, emotionally labile, impulsive, asocial and show higher psychoticism; scientists are more conscientious and arrogant.[27] While engaged in their activity, creative people are absorbed, goal directed, tenacious, obsessive, and obtain pleasure and reward from the activity itself.[27–30]

Table 1. Personality traits characteristic of creative people.

• Open to new experiences	• Meticulous
• Less conventional	• Dominant
• Self-confident & accepting	• Autonomous
• Resourceful	• Hostile
• High drive	• Impulsive
• Perseverance	• Asocial
• Ambitious	• Higher psychoticism[29]

Creative people also show paradoxical features, being at once emotionally labile, yet controlled and stable; primitive, yet more cultured; naïve and childlike, yet intensely knowledgeable; more destructive, but also more constructive; crazier yet saner than the average person.[26,31]

Cognitive styles — Day- and night-dreams, arousal and the creative process

Numerous introspective accounts describe the importance of day — or night-dreaming in the creative process — notably the stage of creative inspiration[4,32] (Appendix 1). Take this account of Wolfgang Amadeus Mozart's[33]:

> When I am completely in myself, entirely alone, and of good cheer — say travelling in a carriage, or walking after a good meal, or during the night when I cannot sleep; it is on such occasions that my ideas flow best and most abundantly. Whence and how they come, I know not; nor can I force them. Those pleasures that please me I retain in memory, and am accustomed as I have been told, to hum them to myself. If I continue in this way, it soon occurs to me how I may turn this or that morsel to account, so as to make a good dish of it, that is to say, agreeably to the rules of counterpoint, to the peculiarities of the various instruments etc. All this fires my soul, and, provided I am not disturbed, my subject enlarges itself, becomes methodized and defined, and in the whole, though it be long, stands almost complete and finished in my mind, so that I can survey it, like a fine picture or a beautiful statue, at a glance. Nor do I hear in my imagination the parts successively, but I hear them, as it were, all at once. What a delight this is I cannot tell! All this inventing, this producing, takes place in a pleasing lively dream. When I proceed to write down my ideas, I take out of the bag of my memory what has been previously collected... the committing to paper is quickly done — for everything is already finished.

Famously, Samuel Taylor Coleridge obtained his inspiration for the *Kubla Khan* during a (night) dream[34]:

> The author continued for about three hours in a profound sleep, at least of the external senses, during which time he has the most vivid confidence that he could not have composed less than 200–300 lines... all the images rose up before him as things, without any sensation or consciousness of effort.

> On awakening, he appeared to himself to have a distinct recollection of the whole, and taking his pen, ink and paper, instantly and eagerly wrote down the lines that are here preserved. At this moment he was unfortunately called out by a person on business and detained by him above an hour, and on his return to his room, found, to his no small surprise and mortification, that though he still retained some vague and dim recollection of the general purport of the vision, yet, with the exception of some eight or ten scattered lines and images, all the rest had passed away like the images on the surface of a stream to which a stone had been case.

Mozart and Coleridge were describing the inspiration and verification phases of the creative process. Both had been immersed in their work (preparation, incubation) leading up to this inspiration, which came as an unbidden illumination during day- and night-dreaming respectively. During this stage of 'free associations', they were using high intellect and tapping into knowledge built up over years; both obtained intense pleasure and reward from the process. Inspiration was followed, with urgency and compulsion, by verification, where the ideas were committed to paper — or, in the case of Coleridge, only partly committed — before they were lost forever. This verification phase required focused, uninterrupted attention.

What kind of thought processes characterise day- and night-dreaming? Do creative people day-/night-dream more often or more effectively than others, or do they simply remember their dreams better because of the creative ideas they endow? A closer look at relevant aspects of the different types of thought is required.

Brief overview of cognitive processes

Psychology and physiology: The main dimension along which thinking varies is the primary process-secondary process continuum[35,36]; this correlates physiologically with level of arousal, where arousal is a continuum from sleep through alert wakefulness to a state of emotional tension.[26] Primary process thought (aka free associative[4] or defocused episodic memory[6,26,37]) is autistic, involuntary, time-related autobiographical memory that enables past experiences to be mulled over and recombined; novel associations and future plans to be made — i.e. the sort of thought that results in creative insight. It occurs during low arousal states, such as day-/night-dreaming, but also during psychotic thought.[26] Secondary process is voluntary, logical, reality- and goal-orientated thought that occurs when concentrating on a task at hand[4,26]; it occurs during higher states of arousal.

Functional anatomy: Activity in the frontal lobes is important in the development of alternative solutions to problems (divergent thinking) and in directing primary process thought. The frontal lobes have strong associative connections with polymodal regions of the temporal and parietal lobes where concepts and knowledge are stored; the scope of activation of these networks determines the extent of novel associations. Functional imaging shows that primary process thought ('defocussed episodic memory') occurs with activity in associative cortex of prefrontal, temporal and parietal lobes, including the posterior cingulate and retrosplenial cortexes.[38] Andreasen's PET findings foreshadowed fMRI studies (using paradigms such as 'conscious resting state' or 'stimulus independent thought'[38–40]) which demonstrated connections between these regions (the default mode network, DMN), and showed that people differ in their propensity to 'day-dream', and to recall these thoughts.[39–41] These areas are still maturing in the second decade of life[38,42] — a factor which might increase susceptibility to environmental and genetic interactions. Significantly, abnormal activity in the DMN has been found in schizophrenia,[43] depression[44] and ADHD.[45]

Neurochemistry: The degree of co-activation across associative networks depends on level of cortical arousal, which is modulated by mesocortical DA and coerulo-cortical NA innervation of prefrontal cortex and hippocampus.[46] High DA and NA transmission results in reduced cortical inhibition and high arousal; low transmission results in high cortical inhibition and low arousal. During day- and night-dreaming, low arousal is associated with reduced mesocortical DA transmission. DA and NA innervation of prefrontal cortex is also important in attentional focus: DA in attentional set selection, and NA in switching between sets.[46] Reduced 5HT transmission in the prefrontal cortex results in perseveration, repetitive responding to a reward stimulus and deficient inhibitory control.[46] Differential function within these systems may be critical to cognitive styles seen in creative people, including the ability to switch between arousal levels, vary attentional focus (wide vs narrow), and stick to a task.

Cognitive styles characteristic of creative people — A lower default DA tone?

The stereotype of the creative person as day-dreamer may have some foundation, since they can access primary process thought more readily than non-creative people (and possibly more effectively[47]), they remember those thoughts better, and use them more often in creative work.[26] This ability to switch more easily between primary and secondary process thought is linked to a *marked and involuntary variability* in cortical arousal (as measured with EEG).[26,29,36] Thus, in creative people, while baseline levels of arousal are higher than non-creative people (i.e. reduced cortical inhibition = higher DA/NA transmission), during the inspiration phase of creative thinking (e.g. day-dreaming), *their arousal level is lower than non-creative people* (i.e. high cortical inhibition = lower DA/NA transmission); arousal during verification is similar to that of non-creative people. These changes in DA/NA transmission have implications for transmitter dysregulation discussed later. Figure 1 shows a schematic diagram of how these changes relate to each other.

Neuropsychiatric Conditions Associated with Creativity

Affective disorders

Behind the proverbial image of the 'mad genius' lies some truth: a disproportionate number of eminent people across history are known or believed to have suffered from mental illness[32,48–50] and studies confirm that all forms of psychopathology are increased across all fields of creativity.[27] Poets are especially severely affected and have a high rate of suicide.[32,48–50] A particularly strong link exists between creativity and affective disorders, especially BPD (Table 2).

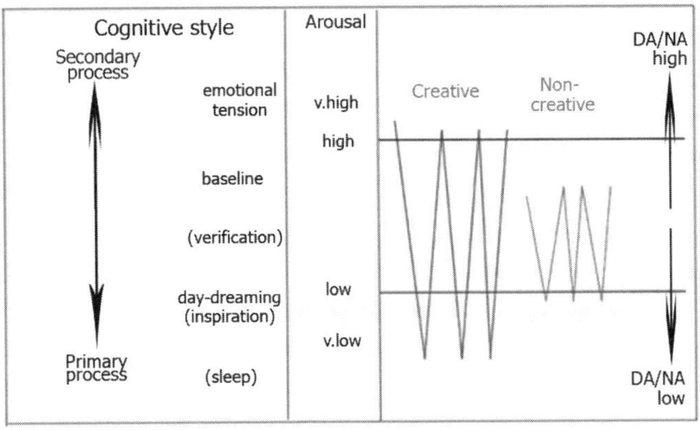

Figure 1. Schematic representation of cognitive styles, level of arousal and cortical DA/NA transmission.

Table 2. Evidence for association between creativity and affective disorders.[4,6,28,31,48–54]

- Much higher prevalence of affective disorders in creative people — especially BPD.*
- Creativity & affective disorders more common in *relatives* of creative people.
- Creativity in adopted children correlates with mental illness in biological parents.
- Strong co-heritability between BPD and creativity in adults and children.
- Correlation between affective personality traits and creativity in mentally healthy people.

* (e.g. of 80 writers with affective disorder, 43% had suffered manic/hypomanic episodes[6]).

Temporal correlation between mood and creative output

The ability to create fluctuates over time. Similarly, a characteristic feature of all mood disorders is an involuntary waxing and waning which has constitutional forces but is also modulated by environmental factors. Temporal variation in mood correlates strongly with creative activity: major works and creative productivity often occur during *recovery* from depression, or hypomania, and there is seasonal and possibly diurnal variation.[48–50,55–57]

Certain mood, cognitive and behavioural changes seen in hypomania are characteristic of intensely creative episodes: enthusiasm, energy, self-confidence, speed of mental association, fluency of thoughts, rising mood, ability to concentrate, emotional intensity, sense of wellbeing, rapid thinking, expansiveness, (reduced) need for sleep, sensory awareness, restlessness, impulsivity, irritability, sexuality, talkativeness, drinking, religious thoughts/feelings, spending of money, anxiety, suspiciousness, argumentativeness, sociability.[32,50] On the other hand, while productivity declines during depressive phases, the depressive experience is invaluable to the creative process.[32]

The above data suggests (and creative artists vouch for it) that moods — or some process associated with them — are an integral part of the creative process. On this basis, treatment of the mental disorder might affect creativity, positively or negatively. However, there are few studies addressing this. It is likely that the effect of treatment may depend on how severe the illness is to start with.[4,32]

Not too mad or too sad — and other essential ingredients!

While creative people frequently suffer from affective illness, the reverse does not necessarily follow: i.e. not all people with affective illness are creative. As noted by Myerson and Boyle in their study of eminent people[58]:

> It does not necessarily follow that the individuals who appear in these records were great because they had mental disease: The manic drive in its controlled form and phase is of value only if joined to ability. A feeble minded person of hypomanic temperament would simply be one who carried on more activity at a feeble-minded level, and this is true also of mediocrity; so the bulk of manic-depressive temperaments are of no special value to the world… If however, the hypomanic temperament is joined to high ability — an independent characteristic — then the combination may well be more effective than the union of high ability with normal temperament and drive might be.

There may be two reasons for this: first, there is a trade-off between severity of mental disorder and creativity, and second, other ‘essential ingredients' (of which domain-specific intellect (talent) and knowledge are two) are required.

Schizophrenia and creativity

A possible association between schizophrenia and creativity has been proposed as one evolutionary explanation for the persistence of schizophrenia in the population.[59] But is there an association? While several eminent people are known to have had schizophrenia, and some have linked their schizophrenic thought processes to their creativity,[63] full-blown, untreated schizophrenia is probably too disruptive for useful creativity. However, an association with creativity *is* suggested by several findings: Compared with controls, schizophrenia is more common in the families of creative people; creative people show higher scores of psychoticism; and healthy individuals show correlations between schizotypy and creative thinking scores.[26,29,60] Furthermore, in schizophrenic patients, older neuroleptics were notorious for destroying creative thought, drive and novel exploration, whereas newer neuroleptics may actually facilitate the re-emergence of creativity.[61]

Tourette's syndrome (TS), obsessive compulsive disorder (OCD) and attention deficit hyperactivity disorder (ADHD)

TS, OCD and ADHD are neurobehavioural developmental conditions with some overlap and comorbidity, including with depression and BPD.[32,62,63] Though less strong than for affective disorders, an association between TS, OCD, ADHD and creativity is suggested[64–68]: notable examples of eminent people known or believed to have suffered from TS/OCD include Dr Samuel Johnson,[67] and Mozart — whose tics and coprolalia correlated with intense performance schedules and other stressors.[68] A modern example is Michael Jackson, one of the most successful and creative pop-singers and dancers ever. He had features of OCD and possible TS, with vocal tics and 'noises' in his songs, and tic-like movements in his dance routines. Some affected musicians believe the disorder is causally related to their creativity, and report an amelioration of their tics by music[66] (plus personal cases in two professional drummers). As with affective disorders and creativity, tics occur in bouts over time.[69]

Migraine

The association between migraine and mood disturbances is well established. However, migraine is also linked with creativity, as noted in Friedman's excellent review[70]: 'Migraine, as we know, seems especially to afflict outstanding persons: the innovators, the meticulous planners, the achievers, the self-driven men and women who seem to strive constantly for perfection'. Certainly, many eminent people have used their migraines in their creative work, including Charles Dickens, Shakespeare, Lewis Carroll, and numerous artists.[71] But whether there is a causative connection between migraine and creativity is unclear. That some neural processes or neurochemical changes occurring in migraine might be conducive to creativity is suggested by introspective accounts of enhanced creative elements during migraine prodrome (Podoll, this volume) or in the immediate recovery from migraine (several personal cases).

Synaesthesia and creativity

Synaesthesia is the experience of sense other than the one that is being stimulated. It is automatic and present from childhood but may decline in later life. There is a genetic element.[72,73] Although its prevalence in creative people is debated, several artists, writers and musicians who are synaesthetes feel it adds to their creativity; similarly, synaesthetes more frequently take on artistic occupations and show higher scores on some psychometric creativity tests.[74,75] The sensory pairings can range in complexity, and functional imaging demonstrates activation of brain regions that are consistent with abnormal connectivity (e.g. colour-word associations activate language and visual associative cortex).[76] Such connections may be vestigial remnants — e.g. resulting from abnormal dendritic pruning during development; and/or they may exist in all people, but normally in a functionally inhibited state.[72] The possibility that synaesthesia might result from abnormal dendritic pruning resonates with previous theories of creativity[77] and is consistent with proposed synaptic and neurotrophic factor mechanisms discussed below.

Implications for the association between creativity and the above disorders

The association between creativity and the above neuropsychiatric conditions suggests common genetic roots and neurobiological mechanisms. As discussed later, these include neurodevelopmental processes and involvement of DA, NA, 5HT and GABA. The temporal variation in creativity and these associated disorders may implicate a disturbance in transmitter regulation.

Fronto-temporal circuits, visual and musical creativity

Some patients with fronto-temporal dementia (FTD) acquire visual, musical or transmodal creativity as the disease evolves.[79] This unveiling of latent creativity may result from degeneration of dominant fronto-temporal regions which normally inhibit non-dominant posterior parietal and temporal neocortex — areas which subserve heteromodal and polysensory integration. These are the same regions implicated in musical skill, supporting a role in innate musical talent or long-term practice.[80] The unleashed creativity can become obsessive, a feature which might be related to reduced 5HT function. Although FTD is an acquired disease, these cases may tell us something about visual and musical creativity in general, including that which is 'in-born'. Thus if the dominant temporal lobe — or its trans-hemispheric callosal connections — fail to develop normally (including at the dendritic level), this may allow the latent creative potential seated in the non-dominant parieto-temporal neocortex to over-develop, enlarge or be unleashed. The role of the non-dominant hemisphere in visual and musical creativity is supported by other observations.[26]

Temporal lobe epilepsy (TLE)

Many great people have suffered from epilepsy. Given the stigma attached to epilepsy, as well as the frequent misdiagnoses, it is likely that many more suffer(ed) without the world knowing. Several undoubtedly used epileptic experiences in their work (e.g. Charles Dickens, Tolstoy, Dostoevsky, Van Gogh), but whether epilepsy can enhance creativity is debated.[81,82] Extrapolating from studies in FTD, it seems plausible that epileptic disturbance — at least if it disturbs dominant temporal lobe function — might be capable of enhancing musical and visual creativity.

Neurobiological Mechanisms Underlying Creativity

This discussion focuses on key creative personality and cognitive attributes and the neuropsychiatric conditions with which creativity is especially associated — affective disorders, BPD, schizophrenia, TS/OCD and ADHD. When considering possible neural mechanisms, several common threads emerge between these conditions which may be of fundamental importance to the creative brain:

1. All are neurodevelopmental syndromes involving heterogeneous genetic and environmental influences, with areas of genetic overlap.
2. The symptoms fluctuate over the short and longer term.
3. The same major neurotransmitter systems are implicated — DA, NA, 5HT and GABA/glutamate.[83–88]

Variability and periodicity of creativity

Eysenck suggested that personality traits are central to creativity[29]; and that creativity is not an ability akin to intelligence, but a cognitive style related to psychoticism. Central to his theory is hippocampal function, DA and 5HT — the former heightening, the latter lowering creativity. However, a key feature of creativity is its *variability*. Thus, creativity is not continuously present. Creative people may spend months without a creative thought, when suddenly, as if from nowhere, this dearth of activity is replaced by a hive of fruitful creativity. There is also variability over a shorter time-frame, as evidenced by the marked and ready changes in level of cortical arousal.[26] This variability of creativity runs hand-in-hand with mood and behavioural changes and symptom fluctuations in conditions with which it is associated (e.g. BPD) and may be explained by abnormal regulation of transmitter release in key systems and regions of the brain (cortex, limbic system, striatum).

Is transmitter dysregulation a fundamental property of the creative brain?

Transmitter dysregulation in conditions associated with creativity

The notion of DA 'dysregulation' was highlighted in Parkinsonian patients (PD) who developed compulsive medication use with disabling motor and behavioural features.[89] These excessive DA-mediated features result from sudden, abnormally high DA transmission due to a disturbance in the relationship between pre- and post-synaptic receptor sites. Normal synaptic transmission relies on a homeostatic relationship between these sites, where post-synaptic receptor supersensitivity is modified by the amount of transmitter exposure; in turn pre-synaptic transmitter release is regulated by sensitivity of post-synaptic receptors. Normally, these changes operate within a small range, reflected in a narrow range of behaviours. However, a disturbance in this equilibirum — due, for example, to inappropriate production of transmitter, or to abnormal function/sensitivity of post-synaptic receptors — results in extremes of transmission, or dysregulation, with correspondingly wide clinical changes. The speed and magnitude of the fluctuation in transmission, and the duration of normal equilibrium between the extremes, may depend on the underlying cause. Reports of poetic and artistic talent (and high drive and productivity) being unmasked[90] and enhanced[91] by DA-rgic treatment in patients with PD, who showed other behavioural features of DA dysregulation, is direct evidence for a role of DA dysregulation in the workings of the creative brain.

DA (and 5HT/GABA) dysregulation in different brain regions may also explain the paradoxical features seen in BPD[92] and schizophrenia,[93] which suggest that, at times, DA transmission is too high (e.g. mania, high energy, psychosis) or too low (e.g. depression, fatigue, anhedonia, negative affect, social withdrawal).

Similarly, in TS, the controversy about whether the primary DA abnormality is one of deficiency,[94] or excess,[62] could be reconciled by the notion of dysregulation. Nomura and Segawa suggest that the motor and behavioural symptoms of increased DA transmission result from a supersensitivity of DA receptors which is *secondary* to a chronic developmental DA deficiency[94] (cf. PD); the release, then, of even small amounts of DA into the synaptic space results in an exaggerated response. A cyclical dysregulation could explain the short- and long-term time-course of TS symptoms. Transmitter dysregulation (DA, NA, 5HT) might underlie some of the conflicting features seen in ADHD, and perhaps may be relevant to migraine also. As discussed below, a cyclical dysregulation of DA (and possibly 5HT) might explain many characteristic features of the creative brain.

Creative personality traits

Several creative personality traits suggest increased DA transmission,[29] at least transiently: confidence, drive, impulsiveness, psychoticism, dominance and hostility; whilst reduced 5HT[95,96] and increased glutamate[96] transmission is suggested by neuroticism, anxiety and obsessiveness (the latter have also been linked with abnormal expression of BDNF[97]). However, creative individuals also display paradoxical traits[31] which could conceivably be explained by fluctuations of DA/5HT/GABA transmission. Further support for DA dysregulation, in the short and long term respectively, comes from studies of cognitive changes during the creative process, and mood and behaviour changes during creative productivity (see below).

Creative inspiration: Low default dopamine tone?

Creative insight involves the discovery of a connection between two or more ideas/images not previously associated. It does not arise from logical voluntary reasoning, but as *an involuntary, sudden revelation* (whence, no doubt, came ideas of divine intervention and muses as sources of such inspiration). Creative insight occurs during primary process thought (typical of day- or night-dreaming), when cortical arousal is low and DA transmission is low.

Creative people show a marked variability in level of arousal ranging from a higher basal arousal at rest, to a lower arousal during the inspiration phase of creative process.[26] The cognitive correlates of this are that they can more readily switch between secondary and primary process thought; the extent of associative network activation may be greater; and they have greater recall of primary process thought. Given the role of DA/NA transmission on cortical arousal, it follows that the neurochemical correlate of this variability in arousal is an ability to switch from a higher basal cortical DA/NA transmission to a lower DA/NA transmission during the inspiration ('default mode') phase. This may suggest that creative people naturally possess a lower 'default' DA transmission tone. Conversely, their higher basal level of arousal might result from exaggerated DA transmission due to stimulation of receptors, which are supersensitive secondary to low default DA tone — analogous to the dysregulation proposed in BPD and TS. The *involuntary* nature of creative inspiration (and the accompanying changes in arousal and DA transmission) further raises the possibility of basal ganglia or cerebellar involvement in its direction.

Creative verification: Increased mesolimbic DA transmission?

The verification phase of the creative process is characterised by a different style of thinking and behaviour from that of inspiration.[26] The level of cortical arousal is higher than during inspiration, but does not differ between creative and non-creative subjects[26]; however, increased transmission in mesolimbic DA systems is suggested by the driven, goal-directed, internally-rewarding behaviour,[26,98,99] and might also underlie novelty seeking behaviour and — by virtue of mesolimbic DA-mediated opiate release[99] — the intrinsic pleasure derived from the creative activity itself. Differential increases in cortical DA and NA transmission may underlie the ability of creative people to direct attention and focus intently on the task, whilst relatively low 5HT transmission might explain their perseverance and tenacity.[46]

Creative productivity: High, or increasing DA transmission?

Intense creative productivity is accompanied by mood, behavioural and cognitive changes that suggest high or *increasing* DA transmission (see earlier in this chapter); anxiety, obsessiveness and perseverance also increase — suggestive of reduced 5HT transmission. These changes may be florid — as seen in BPD. Dysregulation of DA and 5HT may be relevant not only to the mood and behavioural changes associated with BPD, but also to creative productivity. DA dysregulation could also underlie the link between psychoticism and creativity, as demonstrated graphically by the mathematician John Nash who suffered from schizophrenia, who felt that his mathematical inspirational genius and his psychotic symptoms were caused by the same thought processes. Thus, when asked, *'How could you, a man devoted to reason and logical proof, believe that extraterrestrials are sending you messages?'* — Nash answered, *'Because... the ideas I had about supernatural beings came to me the same way that my mathematical ideas did. So I took them seriously'*.[100]

What processes might cause transmitter dysregulation?

Transmitter dysregulation may be the 'end product' of a range of causes. Any process that affects the normal relation between production of transmitter pre-synaptically and the response of the post-synaptic neuron could result in dysregulation. The primary problem could be at the level of the synapse (e.g. impaired synthesis of DA; sensitivity of DA receptors) or involving other transmitter systems (such as GABA) further up- or down-stream. Alternatively, it could result from developmental problems affecting, for example, dendritic pruning. At a molecular level it could result from defects in genes controlling metabolic enzymes, receptors, neurotrophic factors or other aspects of neurodevelopment. Regulation of transmission throughout life could also be affected by environmental factors.

Which genes might be most relevant to creativity?

Many heterogeneous genes have been implicated in the conditions associated with creativity; furthermore, there are areas of genetic overlap — particularly between BPD and schizophrenia. Key candi date genes include those which affect DA, 5HT or GABA metabolism or receptor function. Comparable gene candidates have been proposed in creativity — including DADR2, and the 5HT transporter SLC6A4.[101,102] Abnormalities in genes controlling neural development have also been implicated in the conditions associated with creativity, while aberrations at the stage of dendritic pruning have been proposed in creativity itself[77] — which might also be relevant to associative ('day-dreaming') thought and synaesthesia. Theoretically, any of these heterogeneous genetic defects could affect synaptic function in such a way that it results in transmitter dysregulation, and thus be of relevance to the workings of the creative brain. But it is also possible that these heterogeneous defects could be working via common final pathways.

Convergence of multiple genes on neurotrophic factors: Implications for creativity

In a masterful series of reviews, Carter has described how genes and factors associated with BPD and schizophrenia converge on pathways involving growth factors and stress activated kinases, which are also influenced by pertinent environmental factors — stress, viruses, malnutrition and circadian factors.[103–105] Directly or indirectly, growth factors may underlie the neurochemical features of these conditions, and — I suggest — of creativity itself.

The notion that neurotrophic factors might be the final common pathway in the creative brain is consistent with our knowledge of their function in the nervous system.[105] Thus, abnormalities in BDNF have been described in disorders with which creativity is associated: depression, BPD, schizophrenia, TS and OCD, and ADHD, as well as personal traits associated with creativity.[97,107–110] BDNF regulates development and adult maintenance of specific cell groups — notably the monoamines DA, NA and 5HT, and GABA, which, as we have seen, are of special relevance to the creative brain.[110] BDNF also regulates dendritic pruning[111] where its action is modulated by testosterone and oestradiol.[112]

Summary

The creative brain is a neurodevelopmental phenomenon resulting from genetic and environmental interactions. In addition to high intelligence, 'talent' and a conducive environment, further requisite neural attributes involve activity in key transmitter systems (DA, NA, 5HT and GABA). I have presented a hypothesis in which transmitter dysregulation, notably involving DA pathways, is a fundamental property of the creative brain. DA dysregulation is consistent with the short and long term variability in creativity and associated personality, cognitive, mood and behavioural properties, and with the temporal variations seen in associated neuropsychiatric conditions. A low 'default DA tone' might explain the propensity towards 'day-dreaming' during creative insight, while enhanced drive and focussed attention during creative verification might result from activation of supersensitive DA receptors in the mesolimbic system and prefrontal cortex. Variations in transmission in the longer term could explain the cognitive, behavioural and mood changes behind creative productivity.

Genes implicated in transmitter dysregulation include those that affect synaptic function and neural development. These genes converge on final common pathways involving neurotrophic factors and stress-activated kinases. The involvement of a final common pathway converging on growth factors, especially BDNF, would explain many key features of the creative brain, including the association with a particular group of neuropsychiatric disorders, dysregulation of specific transmitter systems, possible dendritic and sex hormone influences, and the fact that creativity is quintessentially a neurodevelopmental phenomenon.

Acknowledgements

I am grateful to Dr Larry Rifkin and George Leggatt for helpful comments during the preparation of this paper.

Appendix 1

Famous examples of eminent people who have derived inspiration from dreams or day dreams.[A1–A11]

Writers/musicians/artists/sportsmen	Scientists/mathematicians
• Bronte sisters — especially Charlotte	• Archimedes — 'eureka' — buoyancy
• Leo Tolstoy — chapters on 'Daydreams'	• Henri Poincaré — work on Fuchsian functions at night during insomnia; another stepping onto a bus
• Mary Shelley — Frankenstein	• Albert Einstein
• Robert Louis Stevenson — 'A chapter on dreams'	• Srinivasa Ramanujan — elliptical integrals, many in dreams
• John Keats — Mansion of many apartments	• Friedrich Kekule — discovered the structure of benzene ring in a dream
• William Wordsworth — Tintern Abbey	• *Otto Loewi — chemical neurotransmission
• Anthony Trollope — day-dreams into fiction	• Elias Howe — sewing machine needle with hole
• Samuel Taylor Coleridge — Kubla Khan	
• Stephen King	
• Alois Vogel	
• Jean Cocteau	
• August Wilson	
• Neil Simon	
• Jack Niklaus' new golf swing	
• Tchaikovsky	
• Mozart	
• Gretchen Lanes (artist)	
• Paul McCartney — 'Yesterday'	

*Dr. Loewi noted: "Most so-called 'intuitive' discoveries are such associations made in the subconscious."

A1. Klinger, E. *Daydreaming*. LA: Jeremy Tarcher (1990).
A2. Mary Wollstonecraft Shelley, from her introduction to *Frankenstein*.
A3. Robert Louis Stevenson. A Chapter on Dreams, *Across the Plains*. Chattus & Windus (1892).
A4. John Keats. *The Fall of Hyperion: A Dream* (1819).
A5. Epel, N. *Writers Dreaming: 26 Writers Talk About Their Dreams and the Creative Process*. Vintage (1994).
A6. *A Popular History of American Invention* (Vol. 2). New York: Scribner's Sons (1924).
A7. Jack Nicklaus, as told to a *San Francisco Chronicle* reporter, 27 June 1964.
A8. Ranganathan, S. R. *Ramanujan, the Man and the Mathematician* (1967).
A9. Roberts, R. M. *Serendipity: Accidental Discoveries in Science*. Wiley & Sons (1989).
A10. Otto Loewi. An Autobiographical Sketch, *Perspectives in Biology and Medicine, Autumn* (1960).
A11. Miles, B. *Paul McCartney: Many Years From Now*. NY: Henry Holt (1997).

References

1. Sternberg, R. J. and Lubart, T. I. (1999). The concept of creativity: Prospects and paradigms. In Stemberg, R. J. (Ed.), *Handbook of Creativity*, pp. 3–15. Cambridge: CUP.
2. Wallas, G. (1926). *The Art of Thought*. New York: Harcourt, Brace and World.
3. Galton, F. (1962). *Hereditary Genius: An Inquiry into Its Laws and Consequences*. Cleveland: World Publishing — Original work published 1869.
4. Andreasen, N. C. (2005). *The Creating Brain — The Neuroscience of Genius*. Dana.
5. Galton, F. (1874). *English Men of Science — Their Nature and Nurture*. Macmillan.
6. Andreasen, N. C. (1987). Creativity and mental illness — Prevalence rates in writers and their first-degree relatives. *Am J Psych* **144**: 1288–1292.
7. Canter, S. (1973). Personality traits in twins. In Claridge, G. *et al.* (Eds.), *Personality Differences and Biological Variations*, pp. 21–51. New York: Pergamon Press.
8. Waller, N. G. *et al.* (1993). Creativity, heritability, familiality: Which word does not belong? Psychol Inquiry **4**: 235–237.
9. Bullough, V., Bullough, B. and Mauro, M. (1981). History and creativity: Research problems and some possible solutions. *J Creative Behaviour* **15**: 102–116.
10. Lykken, D. T. (1981). Research with twins: the concept of emergenesis. *Soc Psych Res* **19**: 361–372.

11. Feldman, D. H. The development of creativity. *Ibid.* (1), pp. 169–186.
12. Csikszentmihalyi, M. Implications of a systems perspective for the study of creativity. *Ibid* (1), pp. 313–335.
13. Baron-Cohen, S. (1999). The extreme-male-brain theory of autism. In Flusberg, H. (Ed.), *Neurodevelopmental Disorders.* MIT Press.
14. Hassler, M. (1992). Creative musical behaviour and sex hormones: Musical talent and spatial ability in the two sexes. *Psychoneuroendocrinology* **17**: 55–70.
15. Cyr, *M et al.* (2002). Estrogenic modulation of brain activity: Implications for schizophrenia and Parkinson's disease. *J Psychiatr Neurosc* **27**: 12–28.
16. Craig, M. C. and Murphy, D. G. (2007). Estrogen: Effects on normal brain function and neuropsychiatric disorders. *Climacteric Suppl* **2**: 97–104.
17. Sternberg, R. J. and O'Hara, L. A. Creativity and intelligence. *Ibid.* (1), pp. 251–272.
18. Terman, L. W. (1925–1959). *Genetic studies of Genius*, Vols. 4 and 5. Stanford: SUP.
19. Simonton, D. K. Creativity from a historiometric perspective. *Ibid.* (9): 116–133.
20. Weisberg, R. W. Creativity and knowledge. *Ibid.* (1), pp. 226–250.
21. Gardner, H. (1993). *Creating Minds.* New York: Basic.
22. Goertzel, M. G, Goertzel, V. and Goertzel, T. G. (1978). *Three Hundred Eminent Personalities: A Psychosocial Analysis of the Famous*, San Francisco: Jossey-Bass.
23. Eisenstadt, J. M. (1978). Parental loss and genius. *American Psycho* **33**: 211–223.
24. Hayes, J. R. (1989). Cognitive processes in creativity. In Glover, J. A. *et al.* (Eds.), *Handbook of Creativity*, pp. 135–145. New York: Henry Holt.
25. Howe, M. J. A. Prodigies and creativity. *Ibid.* (1), pp. 431–446.
26. Martindale, C. Biological bases of creativity. *Ibid.* (1), pp. 137–152.
27. Feist, G. J. Influence of personality on artistic and scientific creativity. *Ibid.* (1), pp. 273–296.
28. Post, F. (1994). Creativity and psychopathology: A study of 291 world-famous men. *Br J Psych* **165**: 22–34.
29. Eysenck, H. (1995). *Genius: The Natural History of Creativity.* Cambridge: CUP.
30. Collins, M. A. and Amabile T. M. Motivation and creativity. *Ibid.* (1).
31. Barron, F. (1963). *Creativity and Psychological Health.* NY: Van Nostrand.
32. Jamison, K. R. (1993). *Touched with Fire: Manic Depressive Illness and the Artistic Temperament.* Simon and Schuster.
33. Holmes, E. (1878). *The Life of Mozart.* London: Chapman and Hall.
34. Coleridge, S. T. (1816). ESQ. In *Christabel &c 3rd Edition.* London: John Murray.
35. Fromm, E. (1978). Primary and secondary process in waking and in altered states of consciousness. *J Altered States of Consciousness* **4**: 115–128.
36. Kris, E. (1953). *Psychoanalytical Explorations in Art.* NY: IUP.

37. Tulving, E. (2002). Episodic memory: From mind to brain. *Ann Rev Psyc* **53**: 1–25.
38. Andreasen, N. C. *et al.* (1995). Remembering the past: Two facets of episodic memory explored with positron emission tomography. *Am J Psych* **152**: 1576–1585.
39. Greicius, M. D. *et al.* (2003). Functional connectivity in the resting brain: A network analysis of the default mode hypothesis. *Proc Natl Acad Sci* **100**: 253–258.
40. Mason, M. F. *et al.* (2007). Wandering minds: The default network and stimulus-independent thought. *Science* **315**: 393–395.
41. Buckner, R. L. *et al.* (2008). The brain's default network: Anatomy, function, and relevance to disease *Ann NY Acad Sci* **124**: 1–38.
42. Fair, D. A. *et al.* (2008). The maturing architecture of the brain's default network. *Proc Natl Acad Sci* **105**: 4028–4032.
43. Garrity, A. G. *et al.* (2007). Aberrant "default mode" functional connectivity in schizophrenia. *Am J Psych* **164**: 450–457.
44. Greicius, M. D., Flores, B.H. and Menon, V. (2007). Resting-state functional connectivity in major depression: Abnormally increased contributions from subgenual cingulate cortex and thalamus. *Biol Psychiatry* **62**: 429–437.
45. Castellanos, F. X., Margulies, D. S. and Kelly, C. (2008). Cingulate-precuneus interactions: A new locus of dysfunction in adult attention-deficit/hyperactivity disorder. *Biol Psychiatry* **63**: 332–337.
46. Robbins, T. W. and Roberts, A. C. (2007). Differential regulation of fronto-executive function by the monoamines and acetylcholine. *Cerebral Cortex* **17**: 1151–1160.
47. Carlsson, I., Wendt, P. and Risberg, J. (2004). On the neurobiology of creativity. Differences in frontal activity between high and low creative subjects. *Neuropsychologia* **38**: 873–885.
48. Juda, A. (1949). The relationship between highest mental capacity and psychic abnormalities. *Am J Psych* **106**: 296–307.
49. Ludwig, A. M. (1994). Mental illness and creative activity in female writers. *Am J Psych* **151**: 1650–1656; and Ludwig, A. M. (1995). *The Price of Greatness*. NY: Guilford.
50. Jamison, K. R. (1989). Mood disorders and patterns of creativity in British writers and artists. *Psych* **52**: 125–134.
51. Richards, R., Kinney, D. K. and Lunde, I. (1988). Creativity in manic-depressives, cyclothymes, their normal relatives and controls. *J Abnorm Psychol* **97**: 281–288.
52. Santosa, C. M. *et al.* (2007). Enhanced creativity in bipolar disorder patients: A controlled study. *J Affect Disord* **100**: 31–39.
53. Strong, C. M. *et al.* (2007). Temperament-creativity relationships in mood disorder patients, healthy controls and highly creative individuals. *J Affec Disord* **100**: 41–48.

54. Simeonova, D. I., Chang, K. D., Strong, C. and Ketter, T. A. (2005). Creativity in familial bipolar disorder. *J Psychiatry Research* **39**: 623–631.
55. McDermott, J. F. (2001). Emily Dickinson revisited: A study of periodicity in her work. *Am J Psych* **158**: 686–690.
56. Wintersgill, P. (1994). Music and melancholia. *J Roy Soc Med* **87**: 764–766.
57. Kraepelin, E. (1989). Manic-depressive insanity and paranoia. In Robertson, G. M. (Ed.), *The Classics of Medicine Library.*
58. Myerson, A. and Boyle, R. D. (1941). The incidence of manic-depressive psychosis in certain socially important families. *Am J Psych* **98**: 11–21.
59. Nettle, D. and Clegg, H. (2006). Schizotypy, creativity and mating success in humans. *Proc R Soc* **273**: 611–615.
60. Bentall, R. P. (2004). *Madness Explained. Psychosis and Human Nature.* Penguin.
61. Murry, P. and Torrecuadrada, J. L. (1997). Creativity and antipsychotic drugs. *Encephale* **4**: 17–19.
62. Leckman, J. F. and Cohen, D. J. (1999). Evolving models of pathogenesis. In Leckman and Cohen (Eds.), *Tourette's Syndrome: Tics, Obsessions, Compulsions.* John Wiley & Sons.
63. Tamam, L., Karakus, G. and Ozpoyraz, N. (2008). Comorbidity of adult attention-deficit hyperactivity disorder and bipolar disorder: Prevalence and clinical correlates. *Eur Arch Psych Clin Neurosc* **258**: 385–393.
64. Underwood, A. (2005). The gift of ADHD? *Newsweek* **145**: 48.
65. Rucklidge, J. J. *et al.* (2006). An investigation into the relationship among ADHD symptomatology, creativity, and neuropsychological functioning in children. *Child Neuropsychol* **12**: 421–438.
66. Sacks, O. (1992). Tourette's syndrome and creativity. *BMJ* **305**: 1515–1516.
67. Murray, T. J. (1979). Dr Samuel Johnson's movement disorder. *BMJ* **1**: 1610–1614.
68. Simkin, B. (1992). Mozart's scatological disorder. *BMJ* **305**: 1563–1567.
69. Leckman, J. F., King, R. A. and Cohen, D. J. Tics and tic disorders. *Ibid.* (62).
70. Friedman, A. P. (1972). The headache in history, literature, and legend. *Bull NY Acad Med* **48**: 661–681.
71. Podoll, K. (2006). Migraine art in the internet: A study of 450 contemporary artists. *Int Rev Neurobiol* **74**: 89–107.
72. Ramachandran, V. S. and Hubbard, E. M. (2003). The phenomenology of synaesthesia. *J Conscious Stud* **10**: 49–57.
73. Harrison, J. (2007). *Synaesthesia — The Strangest Thing.* Oxford: OUP.
74. Robertson, L. C. and Sagiv, N. (2004). *Synaesthesia: Perspectives from Cognitive Neuroscience.* Oxford: OUP.
75. Ward, J. *et al.* (in press). Synaesthesia, creativity and art. *Br J Psychol.*
76. Paulesu, E., Harrison, J., Baron Cohen, S. *et al.* (1995). The physiology of coloured hearing. A PET activation study of colour-word synaesthesia. *Brain* **118**: 661–676.

77. Haier, R. J. (1993). Cerebral glucose metabolism and intelligence In Vernon, P. A. (Ed.), *Biological Approaches to the Study of Human Intelligence.* Norwood.
78. Seeley, W. W., Matthews, B. R., Crawford, R. K. *et al.* (2008). Unravelling *Bolero*: Progressive aphasia, transmodal creativity and the right posterior neocortex. *Brain* **131**: 39–49.
79. Miller, B. L. *et al.* (1996). Enhanced artistic creativity with temporal lobe degeneration. *Lancet* **348**: 1744–1745.
80. Gaser, C. and Schlaug, G. (2003). Brain structures differ between musicians and non-musicians. *J Neurosc* **23**: 9240–9245.
81. LaPlante, E. (1993). *Seized.* New York: HarperCollins Publishers.
82. Engel, J. Jr. (1989). *Seizures and Epilepsy.* Philadelphia: F. A. Davis Co.
83. Leonard, B. E. (2007). Psychopathology of depression. *Drugs Today* **43**: 705–716.
84. Toda, M. and Abi-Dargham, A. (2007). Dopamine hypothesis of schizophrenia: Making sense of it all. *Curr Psychiatry Rep* **9**: 329–336.
85. Anderson, G. M., Leckman, J. F. and Cohen, D. J. Neurochemical and neuropeptide systems. *Ibid.* (62).
86. Muller-Oerlinghausen, B. *et al.* (2002). Bipolar disorder. *Lancet* **359**: 241–247.
87. Pliszka, S. R. (2005). The neuropsychopharmacology of ADHD. *Biol Psych* **57**: 1385–1390.
88. Torrey, E. *et al.* (2005). Neurochemical markers for schizophrenia, bipolar disorder, and major depression in postmortem brains. *Biol Psychiatr* **57**: 252–260.
89. Evans, A. H. and Lees, A. J. (2004). Dopamine dysregulation syndrome in Parkinson's disease. *Curr Opin Neurol* **17**: 393–398.
90. Achrag, A. and Trimble, M. (2001). Poetic talent unmasked by treatment of Parkinson's disease. *Mov Dis* **16**: 1175–1176.
91. Walker, R., Warwick, R. and Cercy, S. P. (2006). Augmentation of artistic productivity in Parkinson's disease. *Mov Dis* **21**: 285–286.
92. Berk, M. *et al.* (2007). Dopamine dysregulation syndrome: Implications for a dopamine hypothesis of bipolar disorder. *Acta Psychiatr Scand* **116**: 41–49.
93. Meisenzahl, M. *et al.* (2007). The role of dopamine for the pathophysiology of schizophrenia. *Int Rev of Psych* **19**: 337–345.
94. Nomura, Y. and Segawa, M. (2003). Neurology of Tourette's syndrome as a developmental dopamine disorder: A hypothesis. *Brain Dev* (Suppl. 1): S37–S42.
95. Reimold, M. Batra, A. and Knobel, A. (2008). Anxiety is associated with reduced central serotonin transporter availability in unmedicated patients with unipolar major depression: A [(11)C]DASB PET study. *Mol Psych* **13**: 606–613.
96. Goddard, A. W. *et al.* (2008). Serotoninergic mechanisms in the treatment of obsessive-compulsive disorder. *Drug Discovery Today* **13**: 325–332.
97. Foster, J. A. and MacQueen, G. (2008). Neurobiological factors linking personality traits and major depression. *Can J Psych* **53**: 6–13.

98. Flaherty, A. W. (2005). Frontotemporal and dopaminergic control of idea generation and creative drive. *Comp Neurol* **493**: 147–153.
99. Ikemoto, S. (2007). Dopamine reward circuitry: Two projection systems from the ventral midbrain to the nucleus accumbens-olfactory tubercle complex. *Brain Res Rev* **5**: 627–678.
100. Nasar, S. (1998). *A Beautiful Mind*. London: Faber & Faber London.
101. Reuter, M., Roth, S., Holve, K. and Hennig, J. (2006). Identification of first candidate genes for creativity: A pilot study. *Brain Research* **1069**: 190–197.
102. Bachner-Melman, R., *et al.* (2005). AVPR1a and SLC6A4 gene polymorphisms are associated with creative dance performance. *PLoS Genetics* **1**: 394–403.
103. Carter, C. J. (2007). Multiple genes and factors associated with bipolar disorder converge on growth factor and stress activated kinase pathways controlling translation initiation: Implications for oligodendrocyte viability. *Neurochem Int* **50**: 461–490.
104. Carter, C. J. (2007). eIF2B and oligodendrocyte survival: Where nature and nurture meet in bipolar disorder and schizophrenia? *Schizophr Bull* **33**: 1343–1353.
105. Carter, C. J. (2006). Schizophrenia susceptibility genes converge on interlinked pathways related to glutamatergic transmission and long-term potentiation, oxidative stress and oligodendrocyte viability. *Schizophr Res* **86**: 1–14.
106. Klaffke, S. *et al.* (2006). BDNF: A genetic risk factor for OCD and Tourette syndrome? *Mov Dis* **21**: 881–883.
107. Hashimoto, K., Shimizu, E. and Iyo, M. (2004). Critical role of BDNF in mood disorders. *Brain Res Reviews* **45**: 104–114.
108. Grados, M. A. and Walkup, J. T. (2006). A new gene for Tourette's syndrome: A window into causal mechanisms? *Trends Genetics* **22**: 291–293.
109. Conner, A. C. *et al.* (2008). Neurotrophic factor-related gene polymorphisms and adult attention deficit hyperactivity disorder (ADHD) score in a high-risk male population. *Am J Med Genet B* **147**: 1476–1480.
110. Huang, E. J. and Reichardt, L. F. (2001). Neurotrophins: Roles in neuronal development and function. *Annu Rev Neurosci* **24**: 677–736.
111. Singh, K. K. *et al.* (2008). Developmental axon pruning mediated by BDNF-p75NTR-dependent axon degeneration. *Nat Neurosci* **11**: 649–658.
112. De Vries, G. and Simerley, R. B. (2002). In Pfaff, D. W. *et al.* (Eds.), *Hormones, Brain and Behaviour: Development of Hormone-Dependent Neuronal Systems*. San Diego: Academic Press.

Chapter 4

The Neurologist in the Concert Hall and the Musician at the Bedside

George K. York III

Consideration of the anatomy, physiology and pathology of music from the point of view of the neurologist in the concert hall and the musician at the bedside illuminates the neurology of music. The cerebral localisation of musical composition, performance and perception shows that music occupies a discrete niche in the hierarchy of higher cortical function. Our emotional response to music recapitulates the neurological reaction to other intensely pleasurable experience. Transient or permanent symptoms of neurological disease may take the form of increased, decreased or absent appreciation of musical forms. Seizures or hallucinations may have musical components. A sound understanding of the neurology of music enlivens the concert hall and bedside.

Introduction

The audiences in the Festival Hall, Covent Garden or the Royal Albert Hall have their spirits lifted and their intellects tickled when they hear a well-performed *Rigoletto* or the strains of the *Sultans of Swing*. If the listener happens to be a neurologist, he or she brings a unique scientific perspective to the musical experience. In the interludes between flights of fancy, the cultivated neurologist might well muse on the ways in which the nervous system makes music possible. Astute neurologists will ask: 'Where in the composer's nervous system did the music arise in the first place? Which parts of the nervous system does the performer enlist when he or she makes those sublime sounds? Which parts do we activate when we hear music that sends chills up our spine? How does all of this happen?' The answers to these questions, to the extent that they are known, enrich our lives and stimulate our minds. In a world in which aesthetics is subordinated to commerce, physicians and others seek insight wherever it appears, whether in the art gallery, the stadium or the concert hall.

Neurologists are privileged to examine patients with curious signs and symptoms, and it can be hard to separate the mental from the physical, the scientific from the speculative. Sometimes these signs and symptoms involve music, at which time an understanding of the subtleties of musical production and perception aid the clever diagnostician. Many apparently normal people have spontaneous musical experiences which may be a cause of concern to themselves and others. Most people love music and seek it out as a form of entertainment and illumination, but there may be times in which we can get too much of a good thing. For instance, most of us have had the experience of a tune running though our head despite all efforts to suppress it. For a few people, an insatiable craving for music is a sequel of serious brain illness.

This essay explores music from the point of view of the informed neurologist. Though this perspective is not necessarily privileged, it is one that can be simultaneously scientific and artistic. The practice of medicine involves both art and science, and it may be difficult to tell the difference. The interface between the two is a particularly productive source of insight because it can tell us things about both disciplines. This story is best told by the neurologist in the concert hall and the musician at the bedside.

The Neurologist in the Concert Hall

The quizzical neurologist, when faced with *La Traviata* or *Blowin' in the Wind*, immediately wonders how Verdi and Bob Dylan realised their musical ideas. He knows about the dissociation of language and composing, as the Russian composer Vissarion Shebalin showed when he composed stunning music after being rendered aphasic by a stroke. When jazz musicians or pianists improvise, they suppress their limbic systems, notably their amygdalae, and they activate their frontal cortices while deactivating their dorsolateral prefrontal cortices.[1,2]

Composers create their music while in a slightly altered mental state, similar to the reverie experienced by the dreamy neurologist. They rely on their **musical memory**, at least in part, in writing music. When musicians are given the task of remembering either familiar or unfamiliar melodies, they activate the superior aspects of both temporal lobes, most prominently the non-dominant one.[3] Of course, composers also rely on their **musical imagination** to pluck new melodies from the recesses of their minds. This imagination relies on the frontal and temporal regions bilaterally, perhaps more on the dominant side. For example, Maurice Ravel became unable to compose music in the course of a dementing illness characterised by aphasia and apraxia, possibly frontotemporal dementia or Alzheimer's disease.[4,5] In an earlier stage of his illness, he confided to his friends that he could hear new musical ideas but could not write them down, a sort of musical agraphia in which the mechanical part of composing was affected but the mental aspect remained intact.

A possibly apocryphal story tells of a tourist approaching the pianist Arthur Rubenstein on the streets of New York and asking, 'How do I get to Carnegie Hall?' Rubenstein replied, 'Practice, practice, practice.' This old joke will cross the neurologist's mind when he or she contemplates the biology of **musical performance**. The representation of the left fingers in the sensory cortex of violinists is enlarged compared to non-violinists. No similar effect is seen on the right.[6] This implies that repeated practicing enlarges the cortical representation of a violinist's fingering hand, but not of the bowing hand. Indeed, musicians who started practicing at an early age have enlarged cortical auditory representations for piano tones but not for pure tones, irrespective of their primary instrument.[7]

Musicians' brains are morphologically different from those of non-musicians, particularly in the auditory and motor cortical areas and the cerebellum.[8,9] This morphological difference might be partly inherited, since the children of musicians detect off-key sounds better than the children of non-musicians.[10] However, there can be no doubt that assiduous practice from an early age produces both functional and structural differences in musicians' brains. In this as in many other ways, musicians are different from the rest of us.

Neurologists at a piano recital might hear the performer play Bartok's short pieces entitled *Mikrokosmos*. The sophisticated neurologist would know that when pianists played one of the pieces, entitled *Triolak*, while encased in an MRI scanner, they activated a bilateral frontotemporal network comprised of the primary sensory and motor cortices, the cerebellum, the supplemental and dorsolateral premotor cortices, the intraparietal sulcus and the extrastriate visual cortex. Simply imagining playing the piece activated the same network, excluding the motor cortices.[11] Such findings are not surprising; playing the piano is a complex activity involving sensation, fine motor movement, hearing, vision and cognition. These observations might divert our attention during less arresting recitals.

The anxious neurologist, waiting in the audience for the Rolling Stones to appear, might well hear Sir Mick singing *Satisfaction* in his or her mind. When investigators scanned subjects listening to recorded songs into which two to five second gaps of silence had been inserted in the tunes, the silent gaps in familiar tunes like *Satisfaction* induced strong activation of the left auditory association cortex.[12] This suggests that musical imagery, music playing in the mind without the benefit of any external stimulation, involves the activation of the dominant auditory association cortex.

Part of the wonder of the musical experience is the amazement we feel at the creativity of the composers and performers. Neurologists in the concert hall apply their unique way of thinking by asking, 'Where in the nervous system does musical creativity arise?' If, as Semir Zeki says, painters are neurologists of a sort whose province is the visual brain, then musicians are neurologists who manipulate the auditory brain for the aesthetic pleasure of us all. It would be a mistake, however, to infer that **musical creativity** is localised in the auditory brain any more that artistic creativity is located in the visual brain.

Neuroscientists have had trouble designing a task that reveals human creativity unambiguously, leading them to difficulties in the scanning room.[13,14] Following the sensible admonitions of Jacksonian cerebral localisation, the musical neurologist accepts that creativity, like all mental functions, occurs concomitant to sensorimotor events and are thus not localisable.[15]

Neurologists naturally want to split complex stimuli into their component parts, and the nervous system dissociates music into component elements of pitch, timbre, melody, harmony and rhythm. This allows us to consider the neurophysiology of musical perception from a number of points of view. For instance, sound stimulates the cochlea in different places depending on its frequency. We perceive this as the **pitch** of the sound, and it is tonotopically represented in the primary auditory cortex.[16] Indeed, we perceive the fundamental pitch of a complex chord even when the fundamental tone is removed. Careful study has revealed that the perception of **timbre** is independent of the perception of pitch. Patients with right temporal lobectomies could not perceive timbre, yet their perception of pitch was unimpaired.[17] Those with left temporal lobectomies had only minimal troubles with timbre. Even imagining the same tune played on different instruments activated subtly different parts of the temporal lobes.[18]

When Eric Clapton plays the opening **melody** of his rock anthem *Layla*, the neurologist in the audience immediately recognises that his or her response occurs primarily in the right temporal lobe. Even imagining Clapton's melody evokes a similar response.[19] A familiar song playing through our minds activates the frontal supplementary motor cortex.[20] The same thing occurs in musically unsophisticated children.[21] Listening to vocal or instrumental **harmony** activates both primary auditory cortices with a modest predominance in the right superior temporal gyrus.[22] Hearing slightly impure chords evokes a discrete electrical response in musicians, but not in the listening public.[23] When researchers play a highly disharmonious chord, however, the musically uninformed neurologist's medial frontal cortex lights up immediately.[24] Even four-month-old infants turn toward harmonious melodies and turn away from dissonant ones.[25]

Other physicians sometimes accuse neurologists of marching to the tune of a different drummer, both medically and socially. However that may be, **rhythm** forms the background of Western music, and untrained listeners activate both primary auditory cortices when given rhythmic stimulation. By contrast, trained musicians activate left perisylvian areas during passive rhythm perception.[26] Six-month-old infants learn complex rhythmic patterns more readily than adults.[27]

Music appeals to the emotions as well as to the intellect. The **emotional response to music** has neurobiological concomitants, and the musical neurologist is attuned to his or her own limbic system. When subjects heard pleasurable music, blood flow increases in the auditory cortices in the superior temporal gyri and the right orbitofrontal gyrus, while blood flow decreases in the right amygdala. As the music becomes progressively dissonant, blood flow to the right amygdala increases, correlating with progressively negative emotion. At the same time, blood flow to the orbitofrontal cortex and the subcallosal cingulate decreases, associated with the recognition that the music is unpleasurable.[28] When subjects heard music so intensely pleasurable that it sent chills down their spines, blood flow in a network of cognitive, emotive and autonomic areas increase; blood flow to limbic structures, especially the amygdalae, decreases.[29]

Blood and Zatorre have commented that intensely pleasurable music activates the nucleus accumbens, similar to that seen with sexual activity and cocaine consumption.[30] This may explain the appeal of sex, drugs and rock and roll to the enthusiastic music fan. Coincidentally, eating chocolate stimulates the same area, which gives older and presumably wiser neurologists a surrogate for the pleasures of Soho. Others simply buy a couple of Mars bars before the concert.

The Musician at the Bedside

Sooner or later the neurologist must leave the friendly confines of the concert hall and return to the rigors of the bedside. In this setting he or she may be confronted with patients whose neurological diseases involve music. Acquired **amusia** can affect a number of different modalities of musical perception. Just as we can discriminate between expressive and receptive aphasia, so also we can distinguish between expressive amusia in people who cannot sing or produce music and receptive amusia in people who cannot perceive music.[31] For example, Oliver Sacks tells the story of his own loss of perception of pitch in the course of a migraine attack.[32] A patient with right hemispheric focal seizures developed expressive amusia which was cured with anticonvulsants.[33] A similar deficit has been reported in a left-handed musician with a left temporal hemorrhage.[34] 133 years ago William Gowers described a patient with a left middle cerebral artery occlusion who was severely aphasic but could still sing, indicating at least some dissociation between spoken and sung language.[35] A more recent study of this phenomenon showed that singing lyrics did not help the production of spoken lyrics in an aphasic patient.[36] With lesions in different locations producing different disorders of producing and perceiving music, Stewart et al have produced a network model of musical listening that helps to organise our thinking in this arena.[37]

A selective loss of musical emotion can be seen also. A radio announcer with a fondness for Rachmaninov preludes had a dominant hemisphere stroke characterised by global aphasia and right hemiparesis. He recovered well, but lost his emotional response to Rachmaninov. Brain imaging showed an infarction of the left insula and amygdala.[38] Patients with resections of the parahippocampal cortex rated dissonant music as pleasurable, while unoperated subjects rated it unpleasurable.[39]

The American general and president Ulysses S. Grant told a reporter, 'I only know two tunes. One is *Yankee Doodle*, and the other one isn't.'[40] Patients with various brain diseases become, in a matter of speaking, 'tone-deaf', but in some people (such as President Grant) it is present from birth. **Congenital amusia** is estimated to affect about 5% of the population, not all of whom are punk musicians; the revolutionary physician Che Guevara and the Nobel economist Milton Friedman are said to have been be so afflicted. These subjects show deficits in pitch perception, musical memory and rhythm, while perceiving the spoken voice normally.[41,42] Twin studies show that this congenital amusia is a highly heritable trait.[43]

A curious symptom is the appearance of an insatiable craving for music. For instance, a patient with right fronto-temporal epilepsy developed an intense need to listen to classical music after being given therapeutic doses of lamotrigine.[44] A patient with a dominant hemisphere stroke developed similar **musicophilia**.[45] Another patient with dementia and asymmetric atrophy of the anterior temporal lobes developed an insatiable appetite for polka music, which the authors argued was not in itself a pathological condition.[46]

Sometimes patients with reflex epilepsy can have their seizures elicited by music. This **musicogenic epilepsy** occurs only when listening to specific types of music. For example, in Critchley's classical case, recorded dance music evoked no response but Tchaikovsky's *Waltz of the Flowers* produced a seizure within 15 seconds of the beginning of the recording.[47] Fox-trots and Strauss's *Thousand and One Nights* evoked seizures in other individuals. Another patient had a seizure at precisely 8:59 pm. Further inquiry revealed that BBC Radio played the Bow Street bells at exactly this time. Similar church bells also evoked a seizure, but other music did not. When the patient listened to recording of church bells while undergoing an electroencephalographic recording, the patient had a left temporal focal seizure with secondary generalisation.[48] Charles Symonds used musicogenic epilepsy to illustrate Jacksonian principles of excitation, inhibition and reflex physiology in the brain.[49]

Another variation occurs in epileptic patients with musical ictal manifestations. Patients with temporal lobe epilepsy sometimes sing as a form of automatism. One patient with a left frontal tumor sang about her life to various familiar melodies. These spells ceased with anticonvulsant treatment.[50] A patient with right temporal lobe epilepsy sang the blues during his seizures, which again responded to anticonvulsants.[51] A third patient sang or chanted wordlessly while rhythmically slapping his thigh, a sort of ictal rap music. His ictal rapping ceased after right frontal lobe resection, but he was left with severe amusia postoperatively.[52] Sacks describes a man with **musical seizures** whose ictal manifestation was hearing a solo violin and another patient, a musician with a brain tumor, whose seizures consisted of hearing unrecognisable music.[53]

When older people lose their sight or their hearing, they may develop hallucinations involving the affected sense. Patients with poor eyesight who are mentally normal may develop the Charles Bonnet syndrome, with vivid, complicated visual hallucinations. In the same way, people with poor hearing may develop complex auditory hallucinations which may be musical. Sacks describes one woman who was tormented by repetitive carols and popular songs. Practicing specific songs replaced her unwanted music with more satisfying melodies, but her **musical hallucinations** persisted until she had a cochlear implant. After the implant her musical hallucinations disappeared, but she was left with severe amusia as result of the insensitivity of the implant to changes in pitch.[54]

When trained musicians develop neurological disease they may exhibit unusual signs and symptoms. For instance, Brust described a musician with a dominant hemisphere infarction who, though only minimally aphasic, became unable to read or write musical notation.[55] A professional baritone lost his ability to perceive rhythm after a dominant hemisphere stroke.[56] In another example, an accomplished musicologist with a right temporal stroke could not recognise melodies or individual instruments despite being able to recognise and reproduce single notes. He also had transient musical hallucinations.[57]

Commentary. Thinking Neurologically About Art and Artists

The diagnosis of neurological disease involves a mode of thinking not seen in the diagnosis of digestive or circulatory disease. Higher cortical function is quite unlike higher digestive function, and the neurology of cognition and emotion plays an important role in a neurologist's life. As a result, the committed neurologist is attuned to social, psychological and aesthetic issues that do not enter into the diagnostic thinking of endocrinologists or nephrologists. This is not to say that there are no highly cultured and artistically talented cardiologists or general surgeons, far from it. The practice of medicine routinely attracts men and women whose high standards of performance encompass their professional and personal lives. These physicians sing and play instruments, and also paint, act and write plays, while simultaneously taking excellent care of their patients. They are intimately familiar with the pleasures of the concert hall as well as the library, the theatre and the picture gallery, and they carry on a proud tradition that considers the practice of medicine an inherently aesthetic act.

In this tradition, however, neurologists hold a unique place because of their way of thinking. When confronted with a problem of neurological diagnosis, they routinely address three questions: What is it? Where is it? How does it work? The method of systematic analysis of pathology, anatomy and physiology is deeply engrained in neurologists, so that they apply the same method to their daily experiences. This leads neurologists, when confronted with music, to ask themselves the same three questions about the composer, the musician and the audience: What is it? Where is it? How does it happen? Fortunately, neuroscientists ask the same questions and, in trying to understand how the nervous system works, employ their increasingly sophisticated instruments to the study of music.

The dreamy neurologist in the concert hall, imagining an orchestra full of players with misshapen brains producing a profusion of sounds that stimulate the cortices and the limbic systems of the two thousand people in the audience, might be accused of over-intellectualising a process that needs no scientific justification. No knowledge is bad, however, especially when it can be used for the patient's benefit. The musician at the bedside, flush with knowledge about how the nervous system responds to music, can help to establish the presence of pathology in the nervous system. Current opinion to the contrary notwithstanding, scans and electrical tests do not always make the correct diagnosis. It is left to the neurologist with a keen understanding of neurophysiology to reveal what the scans do not, and a grasp of the neurophysiology of musical production and perception assists this effort.

Just as the musician in the concert hall does not need the neurologist to make music, so also the neurologist at the bedside does not need a musician to make a diagnosis. On the other hand, reversing roles helps both the musician and the neurologist understand their art better. Furthermore, understanding the anatomy, physiology and pathology of music make the concert hall and the ward more interesting places for enquiring minds. In the final analysis, this is all the justification we need to pursue the neurology of music.

References

1. Limb, C. J. and Braun, A. R. (2008). *PLoS ONE* **3**: e1679.
2. Bengtsson, S. L., Csikszentmihályi M. and Ullén, F. (2007). *J Cogn Neurosci* **19**: 830–842.
3. Platel, H. (2005). *Ann NY Acad Sci* **1060**: 136–147.
4. Dalessio, D. J. (1984). *J Amer Med Assoc* **252**: 3412–3413.
5. Amaducci, L., Grassi, E. and Boller, F. (2002). *Eur J Neurology* **8**: 75–82.
6. Elbert, T. *et al.* (1995). *Science* **270**: 305–307.
7. Pantev, C. *et al.* (1998). *Nature* **392**: 811–814.
8. Gaser, C. and Schlaug, G. (2003). *J Neurosci* **23**: 9240–9245.
9. Hutchinson, S. *et al.* (2003). *Cerebral Cortex* **13**: 943–949.
10. Magne, C., Schön, D. and Besson, M. (2006). *J Cogn Neurosci* **18**: 199–211.
11. Meister, I. G. *et al.* (2004). *Cogn Brain Res* **19**: 219–228.
12. Kraemer, D. J. M., Macrae, C. N., Green, A. E. and Kelley, W. M. (2005). *Nature* **434**: 158.
13. Dietrich, D. (2004). *Psychonomic Bull* **11**: 1011–1026.
14. Damasio, A. R. (2001). In Pfenninger, K. H. and Shubik, V. R. (Eds.), *The Origins of Creativity*, pp. 59–68. Oxford: Oxford University Press.
15. York, G. K. (2004). In Rose, F. C. (Ed.), *Neurology of the Arts*, pp. 1–11. London: Imperial College Press.
16. Pantev, C., Hoke, M., Lütkenhöner, B. and Lehnertz, K. (1989). *Science* **246**: 486–488.
17. Samson, S., Zatorre, R. J. and Ramsay, J. O. (2002). *Brain* **125**: 511–523.
18. Halpern, A. R., Zatorre, R. J., Bouffard, M. and Johnson, J. A. (2004). *Neuropsychologia* **42**: 1281–1292.
19. Zatorre, R. J., Evans, A. C. and Meyer, E. (1994). *J Neurosci* **14**: 1905–1918.
20. Halpern, A. S. and Zatorre, R. J. (1999). *Cerebral Cortex* **9**: 697–704.
21. Overy, K. *et al.* (2004). *NeuroReport* **15**: 1723–1726.
22. Beisteiner, R. *et al.* (1999). *Neurosci Letters* **277**: 37–40.
23. Koelsch, S., Schröger, E. and Tervaniemi, M. (1999). *NeuroReport* **10**: 1309–1313.
24. Janata, P. *et al.* (2002). *Science* **298**: 2167–2170.
25. Zentner, M. R. and Kagan, J. (1998). *Infant Behav Development* **21**: 483–492.
26. Limb, C. J. *et al.* (2006). *The Anatomical Record Part A* **288A**: 382–389.
27. Hannon, E. E. and Trehub, S. E. (2005). *Proc Natl Acad Sci* **102**: 12639–12643.

28. Blood, A. J., Zatorre, R. J., Bermudez, P. and Evans, A. C. (1998). *Nature Neurosci* **2**: 382–387.
29. Blood, A. J. and Zatorre, R. J. (2001). *Proc Natl Acad Sci* **98**: 11818–11823.
30. *Ibid.*
31. Benton, A. L. (1977). In Critchley, M. and Henson, R. A. (Eds.), *Music and the Brain*, pp. 178–197. London: William Heinemann.
32. Sacks, O. (2007). *Musicophilia,* pp. 101–103. New York: Knopf.
33. Bautista, R. E. D. and Ciampetti, M. Z. (2003). *Epilepsia* **44**: 466–467.
34. Piccirilli, M., Sciarma, T. and Luzzi, S. (2007). *J Neurol Neurosurg Psychiat* **69**: 541–545.
35. Gowers, W. (1875). *Lancet* **2**: 794.
36. Hébert, S., Racette, A., Gagnon, L. and Peretz, I. (2003). *Brain* **126**: 1838–1850.
37. Stewart, L., von Kreigstein, K., Warren, J. D. and Griffiths, T. D. (2006). *Brain* **129**: 2533–2553.
38. Griffiths, T. D. *et al.* (2004). *J Neurol Neurosurg Psychiat* **75**: 344–345.
39. Gosselin, N. *et al.* (2006). *Brain* **129**: 2585–2592.
40. Münte, T. F. (2002). *Nature* **415**: 589–590.
41. Ayotte, J., Peretz, I. and Hyde, K. (2002). *Brain* **125**: 238–251.
42. Foxton, J. M. *et al.* (2004). *Brain* **127**: 801–810.
43. Drayna, D. *et al.* (2001). *Science* **291**: 1969–1972.
44. Rohrer, J. D., Smith, S. J. and Warren, J. D. (2006). *Epilepsia* **47**: 939–940.
45. Jacome, D. E. (1984). *J Neurol Neurosurg Psychiatr* **47**: 308–310.
46. Boeve, B. F. and Geda, Y. E. (2001). *Neurology* **57**: 1485.
47. Critchley, M. (1937). *Brain* **60**: 13–27.
48. Poskanzer, D. C., Brown, A. E. and Miller, H. (1962). *Brain* **85**: 77–92.
49. Symonds, C. (1959). *Brain* **82**: 133–146.
50. Vidailhet, M., Serdaru, M. and Agid, Y. (1989). *J Neurol Neurosurg Psychiat* **52**: 1306.
51. Meierkord, H. and Shorvon, S. (1991). *J Neurol Neurosurg Psychiat* **54**: 1114–1116.
52. McChesney-Atkins, S. *et al.* (2003). *Epilepsy Behav* **4**: 343–347.
53. Sacks, *op. cit.*, 18–27.
54. *Ibid.*, 49–59.
55. Brust, J. C. M. (1980). *Brain* **103**: 367–392.
56. DiPietro, M., Laganaro, M., Leeman, B. and Schnider, A. (2004). *Neuropsychologia* **42**: 868–877.
57. Sparr, S. (2002). *Neurology* **59**: 1659–1660.

Chapter 5

The Human Nervous System — A Clavichord? On the Use of Metaphors in the History of Modern Neurology

Frank Stahnisch

Metaphors derived from everyday-life and existing technological applications have always played a significant role in the clinical and research practice of neurology. A prominent place has thereby been given to metaphors originating from musicology and in particular the design of musical instruments. The musical notions which are used in neurological handbooks and articles are legion: the 'string chord', the 'harp', the 'clavichord', and the 'well-tempered clavier' are but a few. Drawing on sources from neurological research publications, personal and biographical narratives, and high literature — spanning the seventeenth, eighteenth, and nineteenth centuries — this chapter outlines various social and epistemological implications that are crucial to the historical use and understanding of musicological metaphors in neurological research and practice.

Introduction

Historiographic interest in this subject stems from two distinct but interrelated observations: First, throughout the history of modern neurology, the sheer quantity of musical instruments which have been used as metaphors to explain the structure and function of the human nervous system is remarkable (see also Smith, 2004, 191–192). In fact, musical metaphors used to explain the brain seem to outnumber all other metaphorical descriptions by far, such as those employed to illustrate other bodily systems — e.g. the physiological functions of the liver, the heart, or the lungs which have frequently been explained by the action of sponges, pumps, or air bellows. And second, it is of prime epistemological interest (for the general approach compare, for example, Rheinberger, 2007) why technological instruments — such as computer analogues, the telephone (see Fig. 1), or the telegraph — are used so widely as what philosophers and sociologists of modern media and science call 'metaphorical proxies' (Searle, 1981, 254; Pickering, 1995, 19f; Kittler, 2006, 40–45) in the process of neuroscientific discovery and explanation. Therefore, the question that begs an answer here is: what are the specific properties of musical and particularly string instruments that account for their repeated use in describing the actions of the pivotal organ of the human body — the brain?

The Research Question

In contemporary philosophy of technology (such as in the scholarly works of Dierkes, Marz and Hoffmann, 1992; Ott, 1994, or Tresch, 2007), the theoretical views of the German cultural philosopher Martin Heidegger (1889–1976) have undergone an important renaissance. This is due in particular to Heidegger's view that technological devices are intrinsically metaphoric when they are used as pictorial analogues ('*Bild von etwas*'), to represent structure-function relationships (Schirmacher, 1983, 251–259); in other words, they are a characteristic feature of explanations in the empirical sciences (Heidegger, 1962, 37–47). In his collection of academic papers on the philosophy of technology, *Technique and the Turn* (*Die Technik und die Kehre*) of the year 1962, Heidegger did not only describe modern cultural life as heavily dependent on technology *per se*, but even went so far as to reduce numerous research advances to instances of technological innovation: according to him, the characteristic

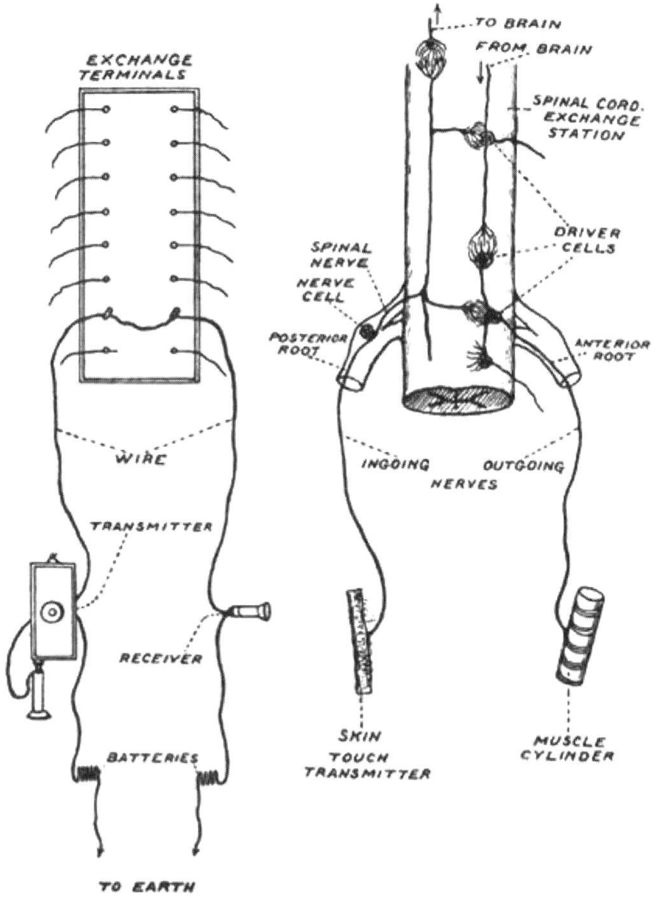

Figure 1. A comparison of the arrangement of two communication units of a telephone system (left) and a nerve system of the spinal cord (right; Keith, 1919, 258). Life Sciences Library, McGill University, Montreal.

mode of scientific revelations of today is given by enframing ('*in einem Gestell zurichten*'; *ibid.*, 3). Technological enframings allow the scientist to access the world in such a way that everything seems to 'stand in reserve' ('*bereit stehen*'); consequently, enframing becomes 'the language of modern technology' ('*die Sprache der modernen Technik*'; *ibid.*, 40). For Heidegger, the process of enframing is necessary for that the scientist can see natural objects, because they are presented only through specific instrumental or technological 'frames' ('*Weisen der Anschauung*'; *ibid.*, 3), e.g. through the use of chronometers, micro- and telescopes, or electroencephalograms, etc. In addition, though Heidegger does not directly refer to musical instruments and practices in his discussions on technology, he nevertheless describes and uses modes of acoustic perception to stress his point:

> We never really first perceive a throng of sensations, e.g. tones and noises, in the appearance of things — as this thing-concept alleges; rather we hear the storm whistling in the chimney, we hear the three-motored plane, *we hear the Mercedes in immediate distinction from the Volkswagen.* Much closer to us than all sensations are the things themselves. We hear the door shut in the house and never hear acoustical sensations or even mere sounds. In order to hear a bare sound we have to listen away from things, divert our ear from them, i.e., listen abstractly (Heidegger, Engl. trans. 1971, 21, emph. F.W.S.).

Of course, Heidegger's account of acoustic sensations is still a long way from the use of metaphors in modern neurology. However, I will try to show here that his views on *enframing* deepen our general understanding of why neurological researchers have been preoccupied — in the past and today — with musical instruments and sound technology. As will be more obvious by the end of this chapter, the enframing of scientific observations in technological terms, such as in metaphors derived from acoustical instruments and exemplars of musical practice, has become a necessity and methodological requirement also for the growth of knowledge on the structure and function of the nervous system: The sense of perceptual immediacy in Heidegger's example — i.e. the acoustic differentiation between a Mercedes and a Volkswagen car by someone who belongs to our world of modern technoculture — is also found, for instance, in the parallel phenomenon in musicology when educated orchestra musicians claim to have the immediate capacity to discriminate between a Stradivarius and an ordinary violin.

The role of acoustic learning and practical education for increasing pattern recognition and sound recollection was already highlighted by the Spanish neurohistologist and Nobel Prize laureate of 1906 Santiago Ramón y Cajal (1852–1934). In 1914, he described

Figure 2. Ink illustration representing positive 'collateral neoformations' in the cerebral cortex: 'Section of the motor cortex of a cat, twenty-five days old, which was killed twenty-four hours after the operation. A, D, medium-sized pyramidal cells with hypertrophic arctiform collaterals and a fine atrophic axonic stump (a, b); C, F, G, arctiform pyramidal cells whose peripheral portion of the axon has disappeared; B, pyramidal cell whose axon resolves itself into two recurrent arches; H, wound; c, axons terminating in a club whose cells of origin are probably situated in the zone of small pyramidal cells; e, axon cut above the collaterals and ending in a point of corrosion' (Ramón y Cajal, Engl. trans. 1991, 667). Osler Library, McGill University, Montreal.

a straightforward neuroanatomical substrate (see also Fig. 2) for this phenomenon on the basis of cellular plasticity and neuronal pathway adaptation (see also Pascual-Leone, 2001, 316), which he had theoretically postulated in his influential work *Texture of the nervous system of man and the vertebrates* (*Textura del sistema nervioso del hombre y de los vertrebrados*) earlier on (1904):

> Exercise not always consists in making easy and expeditious what is difficult, but in realising the impossible. *Everybody knows that the ability of a pianist, an*

> *orator, a mathematician, a philosopher, etc.*, results absolutely unreachable for the uneducated person, whose adaptation to the new work, even with the concurrence of favourable organic circumstances, *requires many years of mental and muscular gymnastics*. To understand this important phenomenon, it is necessary to accept that, in addition to the reinforcement of pre-established organic pathways, new pathways are created by the ramification and progressive growth of terminal dendritic and axonal processes. According to this assumption, the acquired capacity [...] would require the existence of primary and secondary memory centers with multiple and complex links between groups of neurons, which are scarce or not present in uneducated brains (Ramón y Cajal, Engl. trans. 2000, vol. 3, 541, emph. F.W.S.).

Hence, it is only a small step from Heidegger's theoretical proposition, about use-dependent acoustical pattern recognition and technological ways of enframing, to explanatory assumptions about the functioning of other technical machines and musical instruments. The analogous view, for example, that a vacuum cleaner acted *as if it were* an inverse hair dryer is explained by metaphorical ways of enframing and so applied to the action of the brain *as if it were* a well-tempered clavier — in which the musical sensations perceived by the brain are represented by the pulls, hammers, and vibrations of the clavier's mechanical intestines.

The existence of this view in the neurological theory of nerve function is testified by the works of Scottish physiologist and physician Robert Whytt (1714–1766). He was the personal physician to King George III (1738–1820), also an influential president of the Royal College of Physicians of Edinburgh, and is regarded by some as 'one of the fathers of modern neurology' (e.g. Rocca, 2007, 87f.):

> Women, in whom *the nervous system is generally more movable* than in men, are more subject to nervous complaints, and have them in a higher degree. On the other hand, old people, in whom the nerves have become less sensible are little afflicted with those disorders. [...]; in grown people, *whose [strings, i.e. the nerves] are less delicate*, this symptom rarely, if ever, happens. On the other hand, people whose solids are less firm, and their nerves more delicate and easily affected, although subject to many complaints, yet are seldom attacked with ardent fevers or violent inflammatory diseases; [...] (Whytt, 1767, 116f., emph. F.W.S.).

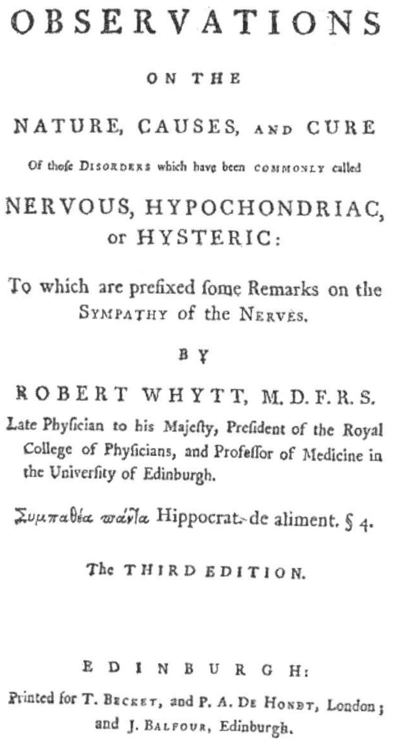

OBSERVATIONS

ON THE

NATURE, CAUSES, AND CURE

Of thofe DISORDERS which have been COMMONLY called

NERVOUS, HYPOCHONDRIAC, or HYSTERIC:

To which are prefixed fome Remarks on the SYMPATHY of the NERVES.

BY

ROBERT WHYTT, M. D. F. R. S.

Late Phyfician to his Majefty, Prefident of the Royal College of Phyficians, and Profeffor of Medicine in the Univerfity of Edinburgh.

Συμπαθέα πάντα Hippocrat. de aliment. § 4.

The THIRD EDITION.

EDINBURGH:

Printed for T. BECKET, and P. A. DE HONDT, London; and J. BALFOUR, Edinburgh.

Figure 3. Front page of Robert Whytt, *Observations on the nature, causes, and cure* [...], 3rd ed. (1767). Redpath Library, McGill University, Montreal.

This citation, which comes from Robert Whytt's *magnum opus* (see Fig. 3), the *Observations on the nature, causes, and cure of those disorders which have been commonly called nervous, hypochondriac, or hysteric* (1767),[3] serves as an initial demonstration of how neurologists equated the structure-function relationship of the nerves with the action of musical instruments. In taking a gendered perspective of the action of the nerves and the nature of brain disease, the anatomical researcher Whytt, with a prime interest in the morphology of the nervous system, assumed that the 'delicate construction' of the female nervous system had quite positive connotations (*ibid.*, 115f.). For him, the mechanical disposition of the chord instrument thoroughly explained why women's nerves are more sensitive, quicker to react, and why women's senses (vision, hearing, smell, etc.) are better than those of men. The flipside of this view, however, was a greater liability of women to have 'nervous-hysterical breakdowns' when the constitution of the nerves no longer complied with the general demands of 'modern cultural life' (*ibid.*, 131f.). Thus, the action of the nerves in both a healthy and a diseased state, was reframed in Robert Whytt's writings with the *technological metaphors* of string instruments, which also accounted for underlying *cultural differences* in the lives of contemporary men and women (see also Stahnisch, 2005, 200–204).

Musical Instruments as Metaphors in Modern Neurology

When trying to understand how metaphors work in neurological science, one realises from its history that those derived from everyday-life experiences and pre-existing technologies have long played an important role in both clinical practice and in the dynamics of basic research investigations. Musical instruments in particular have been used to exemplify the physiological functioning and the disease mechanisms of the nervous system in scientific textbooks, journal articles, and didactic drawings, and also in communicating disease concepts to lay people. The intrinsic purpose of technological metaphors is to make the fabric of the complicated *texture neuronale* more transparent and immediately comprehensible, in other words a *pictorial analogue* serves the important epistemic function of reducing complexity — here with particular regards to the delicate structure of the brain. Consequently, metaphors originating from musicology or the design of musical instruments, such as 'the string chord', 'the harp', 'the clavichord', or 'the well-tempered clavier', etc. (see, e.g., Henson, 1977, 3–10; Welsh 2003, 124–142; or Kursell 2008, 140f.), were frequent in neurological handbooks and articles as the imagination could already envision them. The widespread use of the above mentioned technological enframings in modern neurology will now be further explored, by analysing three historical examples in various areas of neurology and the use of musical metaphors from the succeeding seventeenth, eighteenth and nineteenth centuries.

I will start with the eminent neuroanatomist Thomas Willis (1621–1675), an orthodox, pious, and charitable British physician (Isler, 1965, 13–17), whose interest in the gross morphology of the brain began when he entered a circle of natural philosophers in the royalist milieu of Oxford University during times of religious upheaval and political turmoil (Hughes, 1991, 18–24; Feindel, 1999, 3–11). For him, the pivotal action of the central nervous system hinged upon its delicate anatomical construction of 'ligaments' or 'strings' that were prearranged in the form of a 'celestial harp':

> The reason for the difference is, because the Fibres, thô of the same nature and frame, enter into direct ways of Contractions or wrinklings, from the various strokes of sensible things; *even as the strings of an Harp, from the various strokes of the Musician, give fourth different Sounds, so also the Fibres, which are the Instruments of Touching*, are affected after a different manner, by the various impulse of tangible Things (Willis, Engl. trans. 1683, 61, emph. F.W.S.).

He thus equated the function of the supreme human organ with the most distinguished musical instrument of the time. In *De anima brutorum quae hominis vitalis ac sensitiva est, exercitationes duae prior physiologica ejusdem nataram, partes, potentias and effectiones tradit* [...] (1672) he further notes with greatest respect and admiration:

> This medullary *Epiphysis* reaching above the *Testes* and *Nates*, and going under the Pineal Kernel, tends towards the chambers of the Optick Nerves; approaching which by and by it is Cleft into two Branches, as it were Nervous, one of which, is carried to the Cone of the streaked Body [the Corpora striata], and the other towards its Basis, and in its oblique passage, *sends a shoot into the midst of the Boarder of the streaked Body*: this Branch going to the basis of the streaked Body, behind the root of the *Fornix*, is inserted into an angle of the streaked Body (Willis, Engl. trans. 1683, 26, emph. F.W.S.).

According to Willis, the individual strings of the fornix and the corpora striata (see Fig. 4) did not only act as pulls or levers, which integrated the delicate fabric of the brain.

Apart from their anatomical disposition the nerve strings also provided the physiological explanation for the functions of the brain through their natural capacity to vibrate and conduct nervous functions

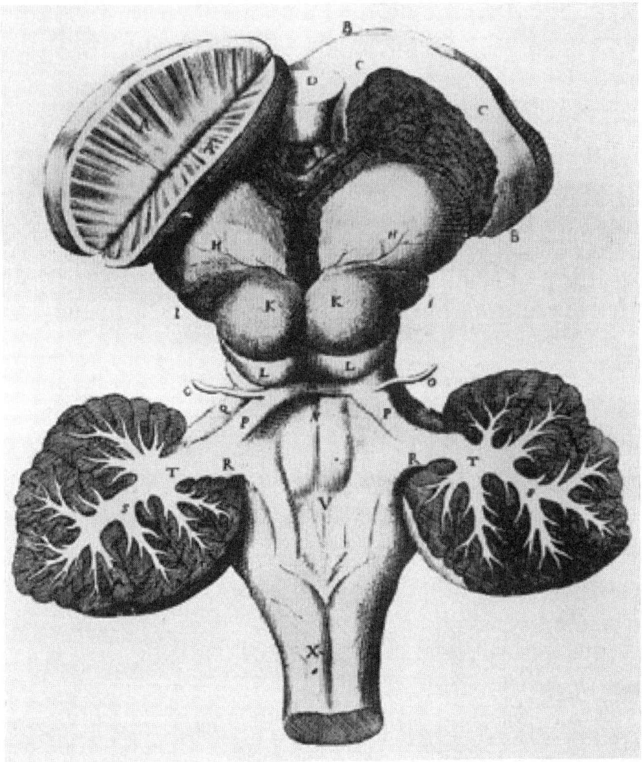

Figure 4. The cerebellar hemispheres partially removed in a sheep, opening the perspective on the dorsal midbrain; in: Willis, *De Cerebri Anatome* [...] (1666), 125, Fig. VIII, by Christopher Wren (1632–1723). Osler Library, McGill University, Montreal.

('*functiones vitales et animales nervorum*') over long distances (ibid., 22f.). Willis, who was himself a Christian believer and devoted lover of spiritual choir songs, went even so far as to anatomically localise the 'action of music' in the 'little brain' (the cerebellum), which he saw as including various areas of the pons and midbrain (Feindel, 1965, 49f.). As a direct expression of his natural theological views, the cerebellum represented the necessary, manifold anatomy of the fine gyri and sulci, which processed the physiological functions of musical tones and harmonies.

This model of the 'Cortical Substances' or 'Cortical Spirits' (Willis, 1683, 24) of the little brain (from the end of the seventeenth century) was markedly different from ancient theories of the workings of the brain (Rousseau, 1973, 141–150): While Willis continued to embrace the Galenic principles of the four humours of the body, blood, phlegm, white and black bile, in explaining the action of the nerves as dependant on the flow or congestion of a humour in what he saw as the hollow interior of the 'narrow channels of the nerves' ('*intra angustos nervorum canales*'; Willis, 1683, 25). Willis clearly applied the knowledge that came out of contemporary technological advances to his understanding of the actions of the nervous system. He viewed the nerves as 'strings' which were influenced externally through the expansion or retraction of the adjacent nerve tubes that mediated individual vibrations and actions. His conceptual assumptions thus further reduced the physiological importance of the ventricles of the brain in earlier theories (see also Finger, 2004). Furthermore, according to this model, the structural integrity of those channels and strings was itself of vital and physiological importance:

> I have observed in many, that when, the Brain being first indisposed, they have been distempr'd with a dullness of mind, and forgetfulness, and afterwards with a stupidity and foolishness, after that, have fallen into a Palsie, which I often did predict; to wit the Morbic matter being by degrees fallen down, and at length being heaped up some where within the Medullar Trunk (where the Marrowy Tracts are more straitned than in the Streaked Body to a stopping fullness. For according as the places obstructed are more or less large, so either an universal *Palsie*, or an half *Palsie* of one side, or else some partial revolutions of members happen. But in either Marrow, and especially the Spinal, an interception or inhibition of the Spirits, creating a *Palsie*, most often happens from a compression, or a breaking of the unity. (Willis, Engl. trans., 1683, 163).

In the ever more mechanised environment of daily working life with its increasing applications of technical apparatus such as driving belts, levers, pumps, and balances, Willis developed into an outstanding advocate of hydraulic and mechanistic theories in neurology, which the Cartesians and iatromechanists had introduced to medicine in general in the seventeenth century:

> While these famous Philosophers suppose Brute Animals to be only certain Machines wonderful made by a Devine Workmanship; to wit, which without any Knowledge, Sense, or Appetite, perform only Corporeal Motions, and the Acts of their Faculties, according to their fitted structure of parts, and the precise direction of the spirits, within Certain measures or bounds of the Animals; yet some of them differ in their Opinions, about the structure and model of the Machine, or moving Engine; to wit, for as much as the figure and property of the Atoms, out of which the same is supposed to be made, are aligned one way by these, and after a diverse way by those. The most illustrious Cartesius, unfolding all things by matter and motion, asserting the souls of brutes to consist altogether of round and highly moveable Atoms, which he Calls the Elements of the first Kind; affirms that nothing else is requisite for all its acts to be performed, than the fibres and nervous parts being struck by a stroke of a sensible thing, they receive a motion after this or that kind of manner, and transfer it by Continued affection of the sensitive parts, as it were by a certain undulation or wavering, into the respective parts [...] (Willis, Engl. trans., 1683, 3).

Mechanical thoughts in physiology and neurology became increasingly popular in the eighteenth century (see Fig. 5). According to the model of mechanical neurology, the utterance of sounds and the perception of harmonies were reduced to the workings and noises of the corporeal strings and ligaments, and the hydraulic action of the blood vessels. What is most evident in the seventeenth century example of the works of Thomas Willis, and his character as a learned engineer and a neuroanatomist, who was loyal to the king, is the frequent employment of musical instruments as metaphors in his theory of brain function — sound evidence of the major influence of the cultural and social contexts on neurological concepts and observations.

Metaphors that were derived from musicology were also of particular significance in the late eighteenth century, when the concept of resonance began also to resonate with the neurologists. Like the formation of waves in water, brain researchers saw sound waves as transmittable in and along the nerves, capable of stimulating vibrations in the soul. A common paradigm of this type was the well-known model

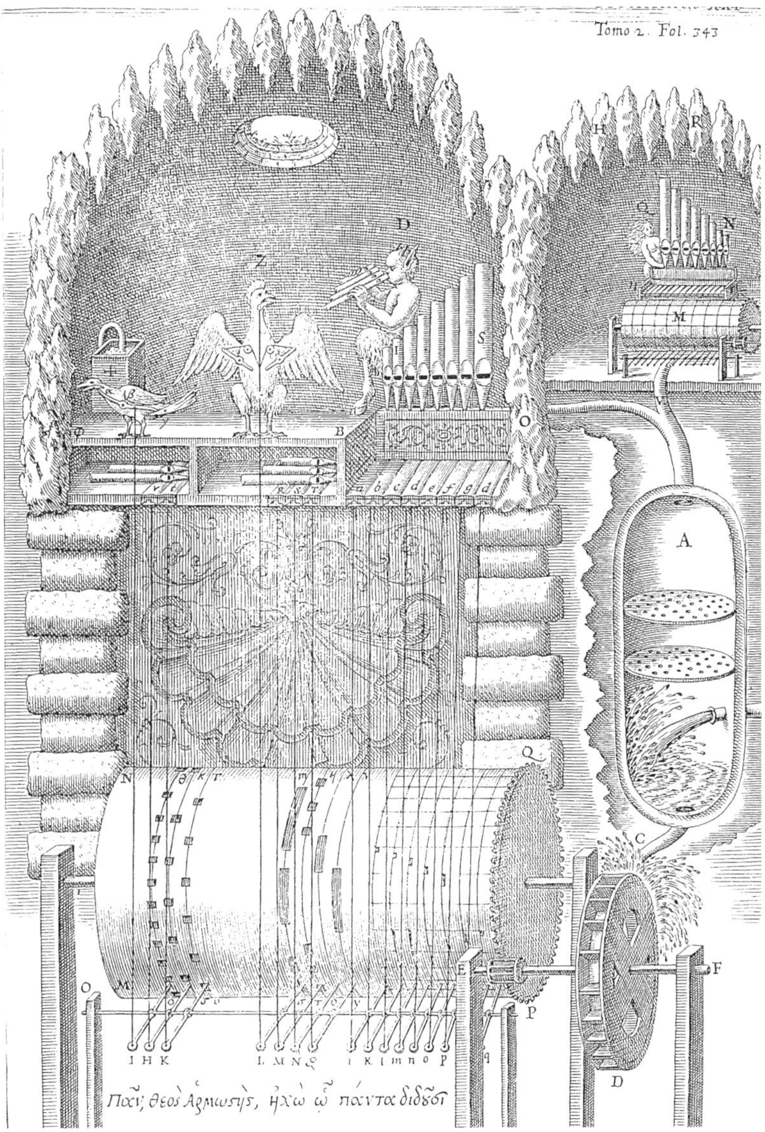

Figure 5. A depiction of the working of an aqueous organ by the German builder Anasthasius Kircher, 1601/02–1661, (Kircher, 1650, vol. 2, Fol. 343, Icon. XXII) like the water-driven apparatus of the French instrument maker Jacques de Vaucanson, 1709–1782 (as in: De Vaucanson, 1738). Duchow Library of Music, McGill University, Montreal.

of 'sound patterns' ('*Klangfiguren*') developed by the Mainz neurologist and anatomist Samuel Thomas von Soemmerring (1755–1830) in Germany. In his *magnum opus* (1796) which analysed *The Organ of the Soul* (*Ueber das Organ der Seele*), he assigned the important physiological role of 'the seat of the human soul' to the cerebral ventricles. In his chapters on 'The Uniting Medium of the Cerebral Terminations of the Nerves is likely the Liquid of the Brain Cavities' ('*Das Medium uniens der Hirnnervenendigungen der Nerven ist offenbar die Fluessigkeit der Hirnhoehlen*'; §§ 28–32), the ventricles of the brain represented the central locus of communication between conscious actions and voluntary motion (Soemmerring, 1796, 31–37).

Soemmerring proposed that the cerebrospinal fluid in the cavities of the brain could be stimulated by the actions of sensory nerves resonating in the liquor-filled cavities of the brain, which then transformed those actions into the faculties of consciousness or visible movements of the body. Though this tied in well with earlier views about the functioning of the brain, his theory of sound patterns developed in a particular local context (see, for example, in: Schott, 1988, 185–189). As a core faculty member of the catholic University of Mainz, Soemmerring had the privilege of living in one of the residential houses, which the medical school provided for its professors in the centre of the city (Dumont, 2005, 3–5). Because he did not have access to a proper scientific laboratory — his course on dissection, for instance, was given in provisional wooden sheds outside the university building — Soemmerring changed the modest room adjacent to the kitchen into his workshop with an extensive anatomical collection of human and animal specimens.

Below his in-house laboratory was the parlour (the '*Gute Stube*') which featured centrally a spinet (the '*Spinett*' has been a small keyboard instrument particularly designed for chamber music). In fact, Soemmerring's house became a primary meeting place for the dignitaries of Mainz and the first point of contact for frequent guests, such as the Goettingen naturalist Georg Forster (1754–1794), Soemmerring's artist friend Wilhelm Heinse (1746–1803) of Aschaffenburg, and the multitalented poet and writer Johann Wolfgang von Goethe (1849–1932) of Weimar, during their visits.

Today, however, we know from his detailed diary that Soemmerring did not play the spinet very often because lectures, clinical and anatomical work, biological preparations, and his demanding undertakings in scientific publishing kept him too busy. However, both Soemmerring and his pupil Jacob Fidelis Ackermann (1765–1815) — who lived for more than five years in the professor's house — frequently mentioned the instrument when exemplifying the workings of the nervous system (Stahnisch, 2007, 424–426).

Nevertheless, the proposed model of 'sound figures' in the ventricles of the brain, as it became developed by the Mainz neuroanatomist, gave rise to heavy criticism from Immanuel Kant (1727–1804), the great Koenigsberg philosopher. Soemmerring had asked him to write a foreword to his book 'The Organ of the Soul' (McLaughlin, 1985, 191–198), which eventually turned into an epilogue of the book, because 'the pride of our times, Kant, was not only keen enough to support the idea of this tractatus, but to also criticise and refine it' (Soemmering, 1796, 81). Kant had been opposed to the neurologist's intention of localising a specific morphological substrate of the soul because, in his view, the soul could not 'show itself' ('*anschaulich machen*') in the body nor project impulses ('*Stoesse auf den Koerper abgeben*') onto it (Kant in Soemmerring, 1796, epilogue, 81–86). Kant further wrote that brain fluid was also 'not organised' and therefore no matter could be reasonably thought of as providing an organ for the soul (Kant, 1800, 117). In his aggrieved reply, Soemmerring answered that if the scientific attention were turned away from the mechanical uniformity of water to its chemical composition there could be 'more scope' for an adequate theory (e.g. Soemmerring, 1811, 5f.).

It is interesting to note, especially from a perspective which explores neurological metaphors, that following Kant's profound criticism and Soemmerring's own preoccupation with telegraphy (see Fig. 6), his explanatory metaphors also changed. In his nearly 20 years at Mainz, Soemmerring promoted the view that nerves acted as musical string instruments. However, later when residing and working in Frankfurt am Main and Munich, Soemmerring began to emphasise 'electricity' as the major explanatory concept for nerve action and

Figure 6. The electrical telegraph, which Samuel Thomas von Soemmerring invented; here the second refined version (1811) is shown after its presentation to the Bavarian Academy of Sciences. Courtesy of the German Museum, Munich.

communication. With his invention of the telegraph, presented to the convocation of the Bavarian Academy of Sciences in 1809, he did indeed observe remarkable action along fibre cables over great distances in the transmission of messages from the sending to the receiving apparatus. Thus the general idea of the vibrating nerve cord had an important after-life even in Soemmerring's telegraph concept, though the change in research interests, his daily life situation, and the professional surroundings of the Bavarian Academy of Sciences in Munich left their imprint on the epistemic value of the musical metaphor.

However, the preoccupation of neurologists with string instruments continued through the late eighteenth and the nineteenth centuries despite the radically changed cultural context after the French Revolution of 1789 — which was increasingly dominated by the bourgeois values of general societal behaviour — and the technological introduction of gas and electricity into modern urban life. Accordingly, concepts that portrayed the temperament, the excitability, and the mechanism of nervous reaction to external stimuli (the '*Reaktionsspannung*') were widely used and had become part of the standard terminology of neurologists. In fact, such terms as 'tenseness' ('*Gespanntheit*'), 'inhibition' ('*Hemmung*'), 'mood' ('*Stimmung*'), and 'irritability' ('*Reizbarkeit*') gave rise to numerous research tropes which clearly transcended the disciplinary boundaries between medicine, music, or fine literature. The terms themselves were certainly not limited to the study of music alone or to that of medical science (Roelcke, 2001, 177–182).

By the middle of the nineteenth century, the concept of musical acoustics was so highly esteemed that those associated with it, were equally highly regarded: because musical acoustics provided access to the higher faculties of the mind, the daily work of the respective neurologists was considered 'learned' and 'philosophical'; consequently some brain researchers were socially elevated over their medical peers. In fact, a number of neurological scientists with an extensive philosophical background themselves taught in philosophy departments of German and other research universities in Europe. Thus, it comes as no surprise that, for example, the neurophysiologist Johannes Mueller (1801–1858) suggested in his famous 1840 collection of student lectures (in his chapter on Voluntary Movements) at the Friedrich-Wilhelms University of Berlin that:

> The fibres of all the motor, cerebral and spinal nerves may be imagined as spread out in the medulla oblongata, and *exposed to the influence of the will like the keys of a piano-forte*. The will acts only on this part of the nervous fibres; but the influence is communicated along the fibres by their action, *just as an elastic chord vibrates in its whole length*, when struck at any one point. It is in the present state of our knowledge–and perhaps always will be–impossible to determine how an exertion of the will in the medulla oblongata the nervous fibres are excited to action. All that we can do is, to consider the fact in its greatest simplicity. (Mueller, 1842, vol. 2, 934, emph. F.W.S.).

While for Mueller the pianoforte was undoubtedly a well-formed and well-founded machine, the aim of truly committed mechanists was to 'shoot the piano player' (i.e. the musician) by replacing him or her with an electromechanical apparatus that could 'play the chords' as well as any pianist. This was not just academic musing as research into the mechanisation of music became a strong and increasing scientific trend — with specific regard to associational and motor phenomena (see also Young, 1970, 116–121).

Another perfect example of this development is the Scottish physicist James Clerk Maxwell (1831–1879). After praising the work-saving properties of the analysing calculator designed by Sir William Thomson, Lord Kelvin (1824–1907), Maxwell marvelled over the range of differing phenomena to which the machine could be applied:

> It would not be devoid of interest, had we opportunity for it, to trace the analogy between these mathematical and mechanical methods of harmonic analysis and the dynamical processes which go on when a compound ray of light is analysed into its simple vibrations by a prism, when a particular over-tone is selected from a complex tune by a resonator, and when the enormously complicated sound-wave of an orchestra, or even the discordant clamours of a crowd, are interpreted into intelligent music or language by *the attentive listener, armed with the harp of three thousand strings*, the resonance of which, as it hangs in the gateway of his ear, discriminates the multifold components of the waves of the aerial ocean (Maxwell, 1890, vol. 2, 800f., emph. F.W.S.).

Maxwell's observations embodied the enormous promise related to the mechanisation of musical acoustics and, following Thomson's apparatus, a great number of similar instruments were soon to be developed (Brain, 2002, 173–175).

However, one neuropathologist — and he is the last example of this historical chapter — the artistically inclined doctor and poet Carl Ludwig Schleich (1859–1922) vehemently opposed the accepted scientific opinion on three counts. He criticised both the bourgeois prestige derived from the practice and research of music from which neurologists benefited, as he found this to be an unwarranted conflation of interests. He also attacked the increasing mechanisation of music through the introduction of early calculating machines, because Schleich found this approach would devalue human creativity which he held in high esteem for his neuroscientific work (Dierig, 2001, 432). In fact, Schleich went so far as to question a fundamental tenet of neurology, that of the passivity of the mediating neuroglia cells *vis-à-vis* the functional activity of the neurons, when he used the analogy of string instruments to negate it (Stahnisch, 2003, 30–33). By the turn of the

Figure 7. Front page of Carl Ludwig Schleich's 1910 essay collection 'On the Soul' ('*Von der Seele*'). Redpath Library, McGill University, Montreal.

century, Schleich had become especially renowned for his collections of essays on popular science, such as the 1910 volume (see Fig. 7) *On the Soul* (*Von der Seele*) and his 1916 treatise *On the Switchboard of Thoughts* (*Ueber das Schaltwerk der Gedanken*), in which he emphasised the neuroglia as the major relay apparatus of the brain.

Even in his first book from 1894, *Painless Operations. Local Anaesthesia with Indifferent Fluids; Psychophysics of Natural and Artificial Sleep (Schmerzlose Operationen. Oertliche Betaeubung mit indifferenten Fluessigkeiten; Psychophysik des natuerlichen und kuenstlichen Schlafes)*, he asserted that an 'active function' and consequently 'psychological importance' were to be ascribed to the interneuronal glia cells of the human brain (Schleich, 1894, 83–87).

As in the case of Thomas Willis or Samuel Thomas von Soemmerring before, Schleich's iconoclastic brain theory came out of a specific cultural context, that of *fin-de-siècle* Berlin. His thoughts about the importance of neuroglia were unquestionably formed beyond the world of established neurological clinics and laboratories, and, instead, in the *milieu* of eccentric bohemians and friends of the arts in Berlin wine cellars, beer gardens, studios, and salons. Theirs was an *avant-garde* laboratory with the Polish physician, writer, and piano player Stanislaw Przybyszewski (1868–1927) at its centre, who invigorated the tight relation between the sciences and the arts. Przybyszewski's piano recitals held at Schleich's home, walking distance from the downtown Charité Hospital, are a perfect example of the creative interaction of cultural and neurological protagonists from which Schleich's new theory on the functioning of the brain emerged (Dierig, 2001, 430–435).

Fascinated by the processes of the diseased brain, Przybyszewski, a student of the renowned Berlin anatomist Wilhelm Waldeyer (1836–1921), made his scientific debut in 1892 with the work *On the Psychology of the Individual* (*Zur Psychologie des Individuums*). In the two-part essay, Przybyszewski asserted that a human who 'lacks general consent for his thoughts and actions' will become ill due to an 'inhibited will'. Faced with 'inhibited outflow' and 'unused nerves', he or she will rather be condemned to lead a life of sorrow (Przybyszewski, 1892, 16). This became a central theme in his writings on psychiatric topics as well as in his musical compositions. Przybyszewski presented numerous of his own piano pieces to the audience in Carl Schleich's private salon, to which the latter regularly invited members of the contemporary Berlin *avant-garde* and befriended colleagues of the medical community of the city. In his memoirs, Przybyszewski refers to Schleich, his home, and in particular his piano:

> Dr. Schleich owned a fine private clinic for surgery which adjoined his elegant apartment. He was not only a brilliant surgeon, inventor of a new medication for local anaesthesia in operations — even in very complicated ones such as the resection of a tumour — he also had a wonderful baritone voice and loved singing above all else. In all of Berlin it would have been

> difficult to find a second instrument as beautiful as Schleich's Steinway. And when we arrived, the whole society had already been assembled. (Przybyszewski, 1965, *posthum.*, 112f., trans. F.W.S.).

Przybyszewski was particularly pleased to see the Steinway pianoforte because of his own enthusiasm for the compositions of his compatriot musician Frédéric Chopin (1810–1849). And hence Schleich's musical salon, in a certain respect, became Przybyszewski's experimental laboratory and working place, where he played his own as well as Chopin's work. The Steinway thus transformed into an instrument which stimulated the inter-disciplinary creativity of Przybyszewski's audience through the mesmerising influences of his piano recitals and life musical play.

Schleich's personal memoirs appeared after the First World War under the title of *A Sunny Past* (*Besonnte Vergangenheit*) in 1923. If these descriptions of his personal experiences and the development of his neurological research are read with Przybyszewski in mind, a striking similarity becomes obvious: The physician as well as the musician and poet both suffered from a sense of intellectual paralysis from the rigid expectations of late Wilhelminian Society in Germany which music did counter:

> Suddenly I [Schleich] jumped up. Stanislaus! I shouted. My dear friend! *The neuroglia is a damper pedal!* An electric sordino, an apparatus for switching registers an inhibition regulator! Eureka! Heavens! And a zillion F sharp majors! Brother, say it again. He has gone crazy. But it is a revelation! (Schleich, 1923, 167, trans. & emph. F.W.S.).

In fact, Schleich regarded his own brain as the site for Przybyszewski's successful piano experiment, when he vigorously played Chopin or presented abstract pieces of early atonal music. According to Przybyszewski, Chopin's piano pieces 'roused correlating feelings' (compare Fig. 8) and music — in this respect — became serious associational 'brain activity' (Schleich, 1916, 41), giving rise to new creative insights and analogies.

For the self-designated scholar-poet Schleich, the clinical laboratory was by no means the only place where scientific knowledge could

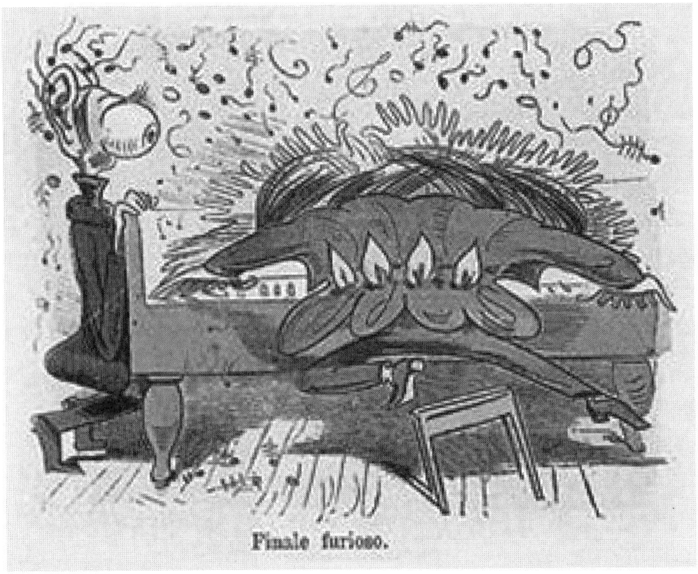

Figure 8. Der Virtuos. Ein Neujahrskonzert, *Finale furioso* (Busch, 1895, vol. 1, 410). University Library, Johannes-Gutenberg University, Mainz.

be gained and developed: microscopes, dissection tools, and staining techniques, etc. — with which he was familiar as a trained neuropathologist — did not have a monopoly on yielding new scientific facts about the brain. In Schleich's writings, the music salon literally became a laboratory where inhibition and excitation in the brain and the theory of the autonomous neuronal cells were unified by the analogous mechanics of the piano.

For Schleich, the piano music literally became the working metaphor for neuronal excitation as modulated by glia cell inhibition, which regulates and checks, disentangles and groups the communication between nerve cells. In a system of nerve strings where — according to the Zurich neurologist and psychiatrist August Forel (1848–1931) — the neurons 'play piano on each other by means of the nerve waves' (Forel, 1894, 16), it appeared as if the neuroglial cells were piano dampers that, when removed from the nerve strings, allowed them to effect excitation; or, if lowered to the strings, as in the famous model of the Vienna neurophysiologist Sigmund Exner,

1846–1927 (see, for example, in Breidbach, 1997, 26–36), inhibit their oscillation.

According to Schleich, instead of the theory of 'action and non-action in the same system' proposed by contemporaneous psychology and physiology (Schleich, 1923, 171–173), there is a double function of the nerves in the brain: inhibition and excitation are linked to different anatomical substrates. His model of neuroglia as piano dampers saw excitation to be modulated by glial inhibition. The neuroglia, in turn, as the mechanism of isolation — in a way like Thomas Willis' earlier interpretation of the vascular substrate — regulated and inhibited the system and checked the nerve conduction pathways, without which brain physiology, psychology, and psychiatry could not manage. These disciplines worked with a thoroughly unreal, purely fantastic inhibition mechanism (*ibid.*, 174f.). Schleich adopted this idea and postulated that the control of neuronal excitation in the brain depended on the type of humour contained in the glial cells: In swollen state, glial cells and their processes insulated the neuronal processes and filled the interneuronal gap. In a non-swollen state, the neuroglia cells were porous elements which permitted neuronal currents. Schleich argued that if neuroglia were in a state of action, which is a state of swelling, the neuronal 'exchange of power suffers' (Schleich, 1894, 90) because glial cells would act like the damper construction of the foot-operated levers at the base of a piano. Furthermore,

> [...] if neuroglia is in a state of imperfect filling, all kinds of psychical motions will be set off, easily and unhindered. In the case of an increased inhibition and glial activity, insulation outweighs excitation; in the case of diminished inhibition and inactivated glial cells, associations will increase and the uninhibited excitability of the neurons mediates a rapid sequence of perception, imagination and action' (Schleich, 1894, 90f., transl. F.W.S.).

Between these two functional states of glia switching, any intermediate neurophysiological condition would be possible.

Though Schleich's theory was inventive and — with hindsight — appeared even compatible with the neuron theory of brain function (Clarke and Jacyna, 1987, 97–99), it nevertheless interfered with contemporary analogies from the science of electricity promoting the view that one and the same element, like the nerve cell, spontaneously alternated between executing inhibition functions and action (Dierig, 1994, 450f.). This thought was, in fact, too *avant-garde* and Schleich's neuroscientific concept shared the fate of his artist friend Przybyszewski, whose writings and piano music were devalued by Wilhelminian cultural and scientific ideology. Schleich's notion of the glia-neuron interaction in the central nervous system was considered as little more than 'a curiosity', in the words of the Berlin anatomist and Charité don, Carl Weigert (1845–1904), which further set the tone for the neuron theory of brain action (Weigert, 1896, 188).

Conclusion: On the Epistemological Value of Metaphors in Modern Neurology

From the historical examples of Thomas Willis, Samuel Thomas von Soemmerring, and Carl Ludwig Schleich, it can be seen how musical metaphors were proposed for in-depth explanation of the structure-function relationship of the human nervous system. It is striking, though, that with their application to modern neurology, the *explanans* (musical perception) has become the *explanandum* (the musical instrument as an epistemological proxy for the brain or individual parts of it). However, the reduction in the overall complexity of the nervous system was not the main epistemological reason why musical instruments were readily accepted as 'good' explanations in neuroscientific thought. As can be seen from the various examples of this chapter, metaphors from musicology and physical acoustics often came out of the lives of contemporary neurologists; they embodied their favourite *passe-temps* and represented additional social prestige of personal and academic distinction. As such, musical metaphors were and are to modern neurologists and brain researchers what office tables, chairs, windows, and pens are to contemporary scholars of the mind-brain relation. The American philosopher Wilfried Sellars has called them the 'medium-size dry goods' which serve as pet objects in serious theoretical explanations (Sellars, 1997, 277).

As a concluding remark, I would like to propose that what is readily available and fits the social context of doctors and medical researchers is also frequently used in scientific explanation: many neurologists either played musical instruments themselves or enjoyed music so much that their research concerns became 'paralleled' by leisure-time activities, thus inverting Friedrich Schiller's (1759–1805) famous *aperçu* (from the '*innere Bildungstrieb als aesthetischer Kunsttrieb*') that 'their aesthetic drive to art virtually became their inner drive to learning' (Schiller, 1795, 153). Therefore, it would be rewarding to further explore the contemporary metaphors that employ 'computers' and neural 'networks' as explanatory models in current-day neurology and brain research (Piszczalski and Galler, 1980, 399–416), and compare them to the historical applications of advanced musical instruments (the 'siren', the 'pianoforte', the 'organ pipe', or the 'electrical piano' etc.) with the same purpose.

This methodological approach allows for more in-depth analysis of the current place of musical and technological metaphors in basic neurology, disease classification, public views on neurological diseases, and ordinary concepts of brain physiology from a historical viewpoint. In drawing on sources from neurological research literature, biographical narratives, high literature, and information pamphlets, from the seventeenth, eighteenth, and nineteenth centuries, I have outlined some of the crucial epistemological and social implications of the historical use of musicological metaphors in modern neurology and the social context from which they originated. In conclusion, I would thus like to place musical metaphors in a more general 'scientific metaphorology', by quoting the German philosopher of metaphors Hans Blumenberg (1920–1996), and give an exploratory answer to the original question 'The Human Nervous System — A Clavichord?':

> [The] metaphor, wherever it is taken from, always represents a self-made bypass through its own 'artificiality' ['*Kuenstlichkeit*']. When viewed as such, the metaphor is necessarily a 'transcendental' element, even when all obligatory prudence is applied against premature idealistic recognition. In themselves, metaphors enable experience without having their own roots in direct experience. [...] Man's metaphorical potency thus enables him to

render a thoroughly alien nature into his own world [‘*eine Eigenwelt zu machen*’]. (Blumenberg, 1986, 175, trans. F.W.S.).

Acknowledgements

The author is grateful to the Alexander von Humboldt-Foundation (Germany) and to the Montreal Neurological Institute (Canada) for their financial support of this project.

References

1. Blumenberg, H. (1986). *Die Lesbarkeit der Welt.* Frankfurt am Main: Suhrkamp.
2. Brain, R. M. (2002). In Clarke, B. and Dalrymple Henderson, L. (Eds.), *Representations in Science and Technology, Art, and Literature*, pp. 155–177. Stanford: Stanford University Press.
3. Breidbach, O. (1997). *Die Materialisierung des Ichs. Zur Geschichte der Hirnforschung im 19. und 20. Jahrhundert.* Frankfurt am Main: Suhrkamp.
4. Busch, W. (1960). In Bohne, F. (Ed.), *Wilhelm Busch: Werke. Historisch-kritische Gesamtausgabe,* 1895, Vols 1–4, pp. 403–411, Wiesbaden: Vollmer.
5. Clarke, E. and Jacyna, L. S. (1987). *Nineteenth-Century Origins of Neuroscientific Concepts.* Los Angeles: University of California Press.
6. De Vaucanson, J. (1979). In Vester, F. (Ed.), *Le mécanisme du fluteur automate: avec la description d'un canard artificiel mangeant* (1738). Reprint: Baren: Frits Knut, Baren.
7. Dierig, S. (1994). *Trends Neurosci.* **17**: 449–452.
8. Dierig, S. (2001). *Configurations* **9**: 413–440.
9. Dierkes, M., Marz, L. and Hoffmann, U. (1992). *Leitbild und Technik. Zur Entstehung und Steuerung technischer Innovationen.* Berlin: Edition Sigma.
10. Dumont, F. (2005). *Samuel Thomas Soemmerring 1755–1830. Mediziner, Naturforscher, Erfinder.* Mainz: Stadt Mainz.
11. Feindel, W. (1965). In Feindel, W. (Ed.), *Thomas Willis. The Anatomy of the Brain and Nerves*, pp. 5–59. Montréal: McGill University Press.
12. Feindel, W. (1999). In Rose, F. C. (Ed.), *A Short History of Neurology. The British Contribution 1660–1910*, pp. 1–18. Oxford: Butterworth Heinemann.
13. Finger, S. (2004). *Minds Behind the Brain. A History of the Pioneers and their Discoveries.* Oxford: Oxford University Press.
14. Forel, A. (1894). *Gehirn und Seele.* Bonn: Strauss.
15. Heidegger, M. (1962). *Die Technik und die Kehre.* Neske: Pfullingen.
16. Heidegger, M. (1971). In *Martin Heidegger: Poetry, Language, Thought*, 1936, pp. 17–87, trans. Hofstadter, A. New York: Harper & Row.

17. Henson, R. A. (1977). In Critchley. M. and Henson, R. A. (Eds.), *Music and the Brain: Studies in the Neurology of Music*, pp. 3–21. London: William Heinemann Medical.
18. Hughes, J. T. (1991). *Thomas Willis (1621–1675): His Life and Work*. London: Royal Society of Medicine.
19. Isler, H. (1965). *Thomas Willis. Ein Wegbereiter der modernen Medizin, 1621–1675*. Stuttgart: Wissenschaftliche Verlagsgesellschaft.
20. Kant, I. (1842). In Rosenkrantz, K. and Schubert, F. W. (Eds.) *Immanuel Kant's Saemmtliche Werke*, 11 vols; Kant's letter no. 871, Koenigsberg in Prussia, August 4th, 1800. Leipzig: Voss, vol. 8: 17.
21. Keith, A. (1919). *The Engines of the Human Body: Being the substance of Christmas lectures given at the Royal Institution of Great Britain, Christmas 1916–1917*. London: Williams and Norgate.
22. Kircher, P. A. (1970). In Scharlau, U. (Ed.), *Musurgia universalis sive Ars Magna Consoni et Dissoni in X libros digestos*, 12 vols, 1664–1678. Reprint: New York: Georg Olms.
23. Kittler, F. (2006). *Theor Cult & Soc* **23**: 39–50.
24. Kursell, J. (2008). *Ber Wiss* **31**: 130–143.
25. Maxwell, J. C. (1890). In Niven, W. D. *The Scientific Papers of James Clerk Maxwell* (2 vols). Cambridge: Cambridge University Press.
26. McLaughlin, P. (1985). In Mann, G. and Dumont, F. (Eds.), *Samuel Thomas Soemmerring und die Gelehrten der Goethe-Zeit*, pp. 191–201. Stuttgart; New York: Gustav Fischer.
27. Mueller, J. (1838–1842). *Elements of Physiology* (1833–40), trans. Baly, W., 2 vols. London: Taylor and Walton.
28. Ott, K. (1994). *Oekologie und Ethik. Ein Versuch praktischer Philosophie*. Bern: Francke.
29. Pascual-Leone, A. (2001). *Ann NY Acad Sci* **930**: 315–329.
30. Pickering, A. (1995). *The Mangle of Practice: Time, Agency, and Science*. Chicago: Chicago University Press.
31. Piszczalski, M. and Galler, B. A. (1980). In Clynes, M. (Ed.), *Music, Mind, and the Brain. The Neuropsychology of Music*, pp. 399–416. New York; London: Plenum Press.
32. Przybyszewski, S. (1892). *Zur Psychologie des Individuums*. Berlin: Fontane.
33. Przybyszewski, S. (1965). *Erinnerungen an das literarische Berlin*. Munich: Winkler-Verlag.
34. Ramón y Cajal, S. R. (2000). *Texture of the Nervous System of Man and the Vertebrates* (1899–1904), trans. and ed. Pasik, P. and Pasik, T., 3 vols. New York: Springer.
35. Ramón y Cajal, S. R. (1991). *Degeneration and Regeneration of the Nervous System* (1913/14), trans. and ed. DeFelipe, J. and Jones, E. J. Oxford: University of Oxford Press.

36. Rheinberger, H.-J. (2007). *Historische Epistemologie zur Einfuehrung.* Hamburg: Junius.
37. Rocca, J. (2007). In Whitaker, H., Smith, C. U. M. and Finger, S. (Eds.), *Brain, Mind and Medicine. Neuroscience in the 18th Century*, pp. 85–98. Heidelberg: Springer.
38. Roelcke, V. (2001). In Gijswijt, M. and Porter, R. (Eds.), *Cultures of Neurasthenia. From Beard to the First World War*, pp. 177–197. Amsterdam; New York: Rodopi.
39. Rousseau, G. S. (1973). In Brissenden, R. F. and Eade, J. C. (Eds.), *Studies in the Eighteenth Century*, pp. 137–157. Toronto: University of Toronto Press.
40. Schiller, F. (1962). In Fricke, G. and Gopfert. H. (Eds.), *Friedrich Schiller: Saemmtliche Werke* (1795), vol. 5. Munich: Carl Hanser.
41. Schirmacher, W. (1983). *Technik und Gelassenheit. Zeitkritik nach Heidegger.* Munich: Karl Alber.
42. Schleich, C. L. (1894). *Schmerzlose Operationen. Oertliche Betaeubung mit indifferenten Fluessigkeiten; Psychophysik des natuerlichen und kuenstlichen Schlafes.* Berlin: Springer.
43. Schleich, C. L. (1910). *Von der Seele. Essays.* Berlin: S. Fischer.
44. Schleich, C. L. (1916). *Vom Schaltwerk der Gedanken. Betrachtungen ueber die Seele.* Berlin: S. Fischer.
45. Schleich, C. L. (1923). *Besonnte Vergangenheit. 1855–1919.* Leipzig: Felix Meiner.
46. Schott, H. (1988). In Mann, G. and Dumont, F. (Eds.), *Gehirn — Nerven — Seele. Anatomie und Physiologie im Umfeld S. Th. Soemmerings*, pp. 183–210. Stuttgart; New York: Gustav Fischer.
47. Searle, J. (1981). In Johnson, M. (Ed.), *Philosophical Perspectives on Metaphor*, pp. 248–285. Minneapolis: University of Minnesota Press.
48. Sellars, W. (1997). *Empiricism and the Philosophy of Mind.* Cambridge, MA: Harvard University Press.
49. Smith, C. U. M. (2004). In Rose, F. C. (Ed.), *Neurology of the Arts: Painting, Music, Literature*, pp. 191–236. London: Imperial College Press.
50. Soemmerring, S. T. (1796). *Ueber das Organ der Seele.* Koenigsberg: Friedrich Nicolovius.
51. Soemmerring, S. T. (1811). *Ueber den Saft, welcher aus den Nerven wieder eingesaugt wird, im gesunden und kranken Zustande des menschlichen Koerpers.* Landshut: Philipp Kruell.
52. Stahnisch, F. W. (2003). *Sitz Ber d Phys-Med Soc Erl, NF* **9**: 29–38.
53. Stahnisch, F. W. (2005). In Stahnisch, F. W. and Steger, F. (Eds.), *Medizin, Geschichte und Geschlecht. Koerperhistorische Rekonstruktionen von Identitaeten und Differenzen*, pp. 197–224. Stuttgart: Franz Steiner.
54. Stahnisch, F. W. (2007). In Schultka, R., Neumann, J. and Weidemann, S. (Eds.), *Anatomie und Anatomische Sammlungen im 18. Jahrhundert. Anlaesslich*

der 250. Wiederkehr des Geburtstages von Philipp Friedrich Theodor Meckel (1755–1803), pp. 421–435. Hamburg: LIT Press.
55. Tresch, J. (2007). *Isis* **98**: 84–99.
56. Weigert, C. W. (1896). *Abh Senckenb Nat-forsch Ges* **19**: 65–209.
57. Welsh, C. (2003). *Hirnhoehlenpoetiken. Theorien zur Wahrnehmung in Wissenschaft, Aesthetik und Literatur um 1800.* Freiburg: Rombach.
58. Whytt, R. (1767). *Observations on the Nature, Causes, and Cure of those Disorders which have been Commonly called Nervous, Hypochondriac, or Hysteric: To which are prefixed some Remarks on the Sympathy of the Nerves* (3rd ed.). London: T. Becket, and P. A. De Hondt; Edinburgh: J. Balfour.
59. Willis, T. (1666). *De cerebri anatome: cui accessit nervorum discriptio et usus* (1664) (2nd ed.) Amsterdam: Gerbandum Schagen.
60. Willis, T. (1683). *Two Discourses concerning The Soul of Brutes, which is that of the Vital and Sensitive of Man. The first is Physiological, shewing the Nature, Parts, Powers, and Affections of the same. The other is Pathological, which unfolds the Diseases which affect it and its primary seat; to wit, The Brain and Nervous Stock, and treats of their cures* (1672), trans. S. Pordage. London: Thomas Ding, C. H. Harper, John Leigh.
61. Young, R. (1970). *Mind, Brain and Adaptation in the Nineteenth Century: Cerebral Localization and its Biological Contexts from Gall to Ferrier.* Oxford: Oxford Clarendon Press.

Chapter 6

The Musician's Brain as a Model for Adaptive and Maladaptive Plasticity

Eckart Altenmüller

For 10,000 years, humans have striven to express and communicate their feelings by singing and playing musical instruments. In order to create new sounds, instruments were invented requiring novel and frequently complex movement patterns. Sensory-motor skills of musicians have some specific qualities: learning begins at an early age in a playful atmosphere. Routines for stereotyped movements are rehearsed for extended periods of time with gradually increasing degrees of complexity. Via auditory feedback, the motor performance is extremely controllable by both performer and audience, and these specific circumstances seem to play an important role for plastic adaptations of the central nervous system. Training-induced changes include both brain function and brain structure and can be observed in sensory-motor and auditory networks. However, in the last two centuries, increasing specialisation and, as a consequence, prolonged training, produced dysfunctional adaptations of the brain, leading to secondary deterioration of movement patterns referred to as musicians' dystonia. This disorder could mark the final point of human evolution of sensory-motor skills.

What is Special About Music Making?

For thousands of years, it has been through the hands of musicians mastering their instruments to have humans communicating and manipulating their emotions. In modern times, we are fascinated by the precise execution of very fast and, in many instances, extremely complex movement patterns which characterise the skills of professional musicians. But when did this faculty develop? Can musicologists provide an answer to these questions concerning the evolution of motor skills by analysing the degree of difficulty required in musical scores? Are there innate limitations of the central nervous system with respect to the mastery of fast and complex movements? Is there evidence of a shift in human motor accomplishments? Or, looking at the problem from another perspective: if a time machine had enough space for a Steinway grand piano and an expert piano teacher, could an inhabitant of ancient Egypt, given that he is properly instructed during childhood and adolescence, learn to perform the Liszt b-minor piano sonata?

Sensory Motor Skills of Musicians in Historical Times

Definite conclusions about manual skills of musicians cannot be made until the advent of musical notation and witnesses on the quality of the execution of music supplied by contemporaries. In Baroque times, outstanding musicians and performers such as Johann Sebastian Bach or Domenico Scarlatti composed extremely demanding music, which in some aspects reaches the limits of technical feasibility, even for highly specialised virtuoso performers of our days. When taking the *Goldberg Variations* of Bach as an example, its execution demands exceptional technical skills, as when concerning the rapidity of trills and passages or the precision of bimanual co-ordination. Later, performing composers such as Liszt — who was an excellent interpreter of Bach's piano music — used these technical refinements as a basis, and added further technical difficulties, such as a novel leap and repetition technique to realise his musical visions. In a very informative article, Lehmann (2006) convincingly demonstrates the increasing demands on manual skills in musicians over the past three centuries. According to this author, the technical challenges are paralleled by the developments of musical instruments, which in turn were often initiated by outstanding performing composers. An example is the extension of range (number of notes) in the piano, which grew from four to six octaves. In the 18th century alone, Beethoven requested larger tonal ranges from his piano maker; the same was true for Franz Liszt, who finally arrived at the 'modern' range of eight octaves in the nineteenth century. A similar extension is documented for other instruments, such as the recorder, violin and flute. Innovations in playing techniques of performing composers added complexity to required manual skills. The 'third-hand' technique, developed by the pianist Siegismund Thalberg in the 1830s, involves distributing the melody notes between the hands in the middle of the keyboard, while the accompaniment is played in scales and patterns to the left and right sides of the melody. This technique destroys not only the classical mapping of hands onto the keyboard with the right hand playing the melody while the left provides the accompaniment, but additionally requires maintenance of dynamic differences between the melody and the accompaniment within one hand, imposing heavy skill requirements on the performer (Lehmann, 2006).

The increasing refinement of musicians' manual skills during the last three centuries is well documented, but the question remains whether this improvement is due to early specialisation and longer cumulative practice times, or whether other factors such as instructional strategies may have had a crucial impact on the acquisition of manual dexterity. When analysing the technical skills of child prodigies performing keyboard music in public from the times of Bach until the twentieth century, Lehmann (2006) comes to the conclusion that during this time span, acceleration in the acquisition of performance skills took place. In other words, there is a significant tendency for prodigies of more recent generations to play technically more difficult pieces after shorter periods of training than earlier prodigies. Several factors contribute to this effect: firstly, over the centuries, there is a tendency towards earlier commencement of musical training. Not uncommonly, outstanding contemporary performers start their systematic training at ages younger than six years. Secondly, accumulated procedural knowledge of the most effective teaching methods handed down from generation to generation of performer/teachers may have resulted in an optimisation of training methods. Thirdly, due to the specialisation of young performers who focus on only one instrument, and neglect other activities, there is increased time spent preparing for performances. Anecdotal evidence for the latter notion is abundant from the nineteenth-century onwards. The pianists Clementi and Czerny are said to have practised at the piano for eight hours daily as children in 'solitary confinement'; Kalkbrenner for 12 hours and Henselt even for 16 hours (Lehmann 2006).

In summary, it is indisputable that the demands on manual skill for the reproduction of composed 'serious' music increased continuously from Baroque times until the middle of the twentieth century. It is not only the complexity of movement patterns, but also the elements of tempo, strength, stamina and the precision of hand and finger movements that constitute this process of increasing perfection over the centuries. Modern society in turn imposes heightened pressures on performers of composed music by comparing the individual live performance in concert with recordings of outstanding peers, easily available on CDs. Additionally, studio recordings with the possibility of obtaining and splicing multiple takes contribute to an illusionary perfection as standard. All of these changes are reflected in the intensification and prolongation of daily practice. However, this development holds only for a relatively small group of musicians, namely the highly specialised classical musicians in Western cultures reproducing classical music. The majority of musicians all over the world are either amateur players, playing their instruments in various social contexts, or professionals relying more on improvisational skills (for example in jazz music) or on the technical developments of instruments and electronic equipment (for example, in rock and pop music).

Brain Adaptations Accompanying Behavioural Pressures

Music, as a sensory stimulus, is highly complex and structured along several dimensions. Moreover, making music requires the integration

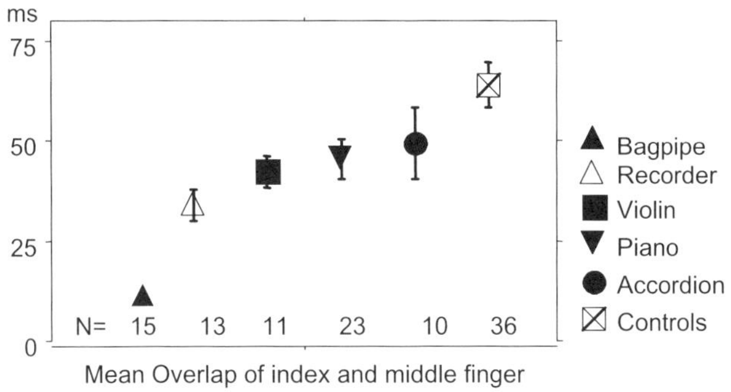

Figure 1. Behavioural adaptations: The role of auditory feedback in a sensory-motor transfer task. Overall averages of a test of synchrony in different groups of professional musicians. The task was to avoid any overlap while touching a metal pad with one index finger and synchronously releasing another finger in a series of trill-like movements, which were executed in a standardised and metronome-paced tempo. The pipers clearly have the smallest amount of undesired overlap, followed by woodwind players. The results demonstrate that motor control in musicians is specifically guided by auditory feedback since avoiding overlap is critical in pipes and woodwind instruments, but not in keyboards and the accordion (bars = ± 1 S.E.M.). (From Walsh *et al.*, 2009, in revision.)

of multimodal sensory and motor information and precise monitoring of the motor performance via auditory feedback (Walsh *et al.*, 2009, Fig. 1).

In the context of Western classical music, musicians are forced to reproduce highly controlled movements almost perfectly and with high reliability. These specialised sensory-motor skills require extensive training periods over many years, starting in early childhood and passing through stages of increasing physical and strategic complexities.

The superior skills of musicians are mirrored in plastic adaptations of the brain on different time scales. At one extreme, years of musical experience, especially in those musicians who begin training early in life, might lead to an increase in grey and white matter volume in several brain regions, including sensory-motor and auditory areas, the cerebellum and the anterior portion of the corpus callosum. These anatomical alterations appear to be confined to a critical period. The fact that in several of the studies a correlation was found between the extent of the anatomical differences and the age at which the musical training commenced strongly argues against the possibility that these differences are pre-existing and the cause for rather than the result of practicing music. At the other extreme, several minutes of training can induce changes in the recruitment of auditory or motor cortex areas, or establish auditory-sensory-motor coupling (Bangert and Altenmüller, 2003, see Fig. 2).

Musicians Dystonia — the Final Point of a Development?

There is a dark side to the increasing specialisation and prolonged training of musicians, namely loss of control and degradation of

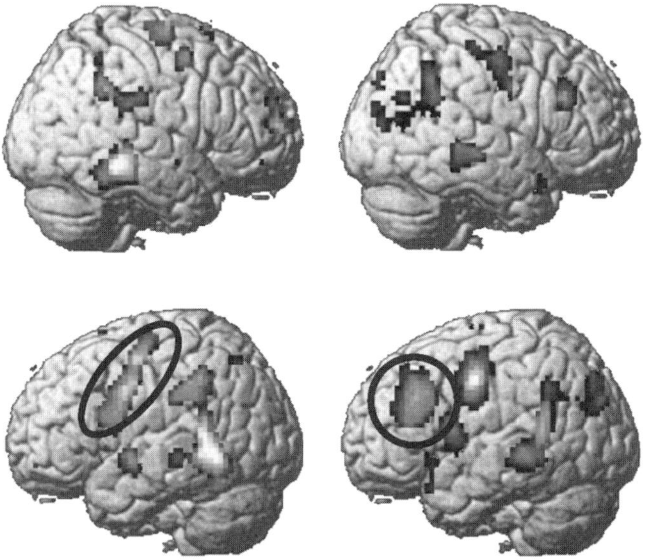

Figure 2. Central nervous adaptations: Auditory-sensory-motor co-representation in a group of seven professional pianists investigated with functional MRI compared to a group of seven non-musicians. The spots indicate increase in brain activation in pianists compared to non-pianists. In the upper row, the right hemisphere is displayed, in the lower the left hemisphere. Listening to simple piano tunes (left side) in professional pianists activates the sensory-motor areas, especially over the left hemisphere (circle on the left). Playing simple piano tunes on a silent keyboard activates the left frontal lobe areas (circle on the right) which are related to gesturing and language processing. (Modified from Bangert *et al.*, 2006, with permission.)

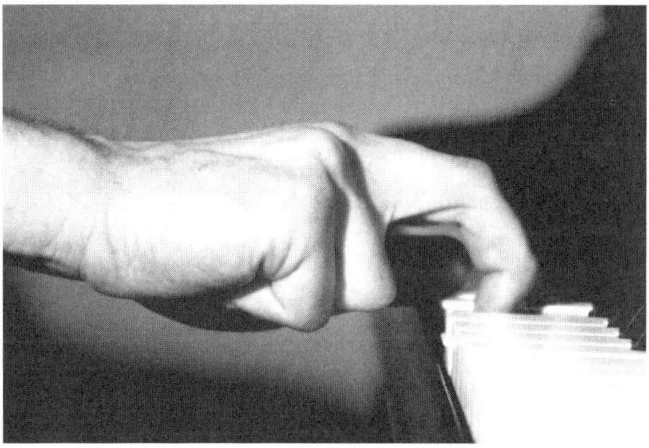

Figure 3. Symptoms of dysfunctional plasticity: focal dystonia in a pianist. Involuntary flexion of the middle, fourth and fifth fingers while attempting to play a C-major scale with the right hand. Typically in dystonia, no pain or sensory symptoms are reported. Dystonia may afflict almost all groups of instrumentalists but is more frequently seen in the right hand of guitarists and pianists and in the left hand of violinists (Altenmüller, 2003).

skilled hand movements, a disorder referred to as musician's cramp or focal dystonia (Fig. 3). The first historical record, from 1830, appears in the diaries of the ambitious pianist and composer Robert Schumann. As was probably the case for Schumann, prolonged practice and pain syndromes due to overuse can precipitate dystonia, which is developed by about 1% of professional musicians and in many cases ends their career (Jabusch and Altenmüller, 2006). Neuroimaging studies point to dysfunctional (or maladaptive) neuroplasticity as one of the relevant pathomechanisms.

Support for this theory comes from a functional brain-imaging study performed in musicians with focal dystonia. Compared to healthy musicians, the patients showed a fusion of the digital representations in the somatosensory cortex, reflected in the decreased distance between the representation of the index finger and the little finger when compared to healthy control musicians (Fig. 4). Such a fusion and blurring of receptive fields of the digits may well result in a loss of control, since skilled motor actions are necessarily bound to intact somatosensory feedback input.

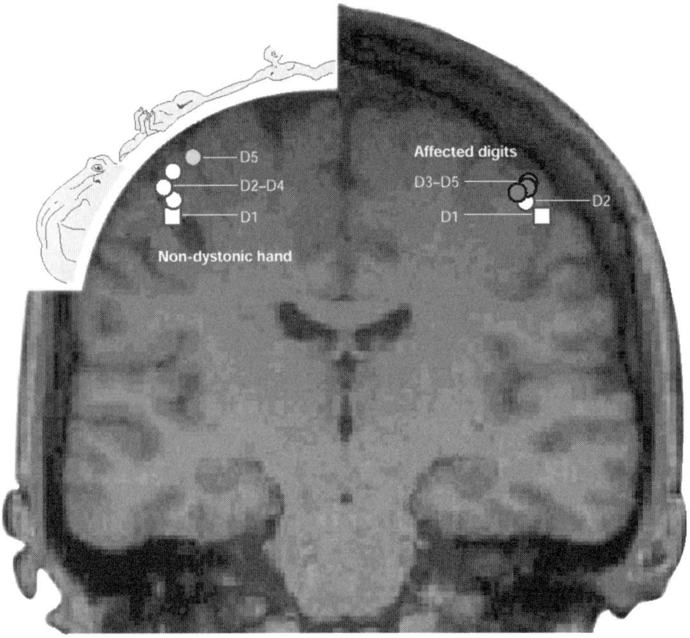

Figure 4. Neuronal correlates of dysfunctional plasticity. Fusion of the somatosensory representation of single digits of the hand in a musician suffering from focal dystonia. The best fitting dipoles explaining the evoked magnetic fields following sensory stimulation of single fingers are shown projected on the individual's MRI. Whereas for the non-affected hand the typical homuncular organisation (see inset) reveals a distance of about 2.5 cm between the sources for the thumb and the little finger (white square and circle on the left), the somatosensory representations of the fingers on the dystonic side are blurred, resulting from a fusion of the neural networks which process incoming sensory stimuli from different fingers (circles on the right). (Modified from Elbert *et al.*, 1998 and Münte *et al.*, 2002.)

Considering (a) the historical advent of the disorder in the nineteenth century with rapidly increasing technical demands imposed on musicians, (b) the epidemiological data with rapid and repetitive finger movements as a risk factor, and (c) the above mentioned neurobiological findings of the blurring of hand representations, one is tempted to state that focal dystonia finally marks the natural limits of a process of refinement of manual dexterity over many years, although, according to a very recent study hereditary factors with a certain predisposition to develop this condition may also play a role (Schmidt *et al.*, 2006).

From Neandertal to Carnegie — and Future Developments?

Finally, we have to answer the questions raised in the first paragraph. The time machine with the expert piano teacher from a renowned German music academy arrives with the Steinway D grand piano 3500 years ago in the valley of the river Nile. I have no doubt that the young son of the Pharaoh, if taught to play from age four by our German professor, could have learned the Liszt Sonata, given that he was endowed with enough passionate motivation for the approximately 10,000 hours of training required, and given that he had access to a protein-rich diet to develop large bones, robust muscles, and hands flexible enough to span the tenths.

According to paleo-neurological findings, the brain's structure has not changed significantly in the last 100,000 years. And it is highly probable that humans of the upper Paleolithic period — around 30,000 years B.C. — were able to execute demanding movement patterns when playing on their bone flutes. The true 'revolution' which enabled *Homo sapiens* to master novel tasks occurred much earlier in human evolution. It was enabled by the development of **neural plasticity**, the potential to adapt to new environmental stimulation and to new challenges by modification of neural networks — and this potential was most likely present long before the first musical instruments were developed.

As has been demonstrated in the preceding paragraphs, the musician's brain is an excellent paradigm to study the short-term and long-term effects of neural plasticity and long-term adaptation in sensorimotor systems, even in macroscopical brain structures. Furthermore, in the case of focal dystonia, neuroscientists have become aware of the limits of these adaptations under certain stressful conditions.

It is beyond doubt that creative innovations will continue to be made, but as far as manual skill of the independent use of the fingers is concerned, it seems that a final point in a million-year long development was reached somewhere between the beginning and the middle of the twentieth century. For the piano, the works of late romantic performing composers such as Rachmaninoff, Godowski, Albeniz and Alcan with their extension of the Lisztean technique mark an end point for the age of virtuosi. When comparing the available recordings, contemporary performers do not seem to be essentially superior to the previous generation.

The new challenge in performing compositions of the 'classical' modern composers, for example Messiaen, Boulez, Ligeti, all extremely difficult to master, is not based in new demands on manual skill, but rather in their complex musical structure and novelty of rhythmic and harmonic patterns. Since these patterns are as yet usually not integrated in the systematic training of music students, they seem extremely demanding, but they do not present any new qualitative or quantitative challenges in respect to manual dexterity. Many contemporary composers try to overcome the natural limitations of hand skills by exploiting unusual ways of producing sound, such as plucking the strings in the piano or using the open holes for glissandi in the flute. These new techniques challenge manual dexterity in a new way, but do not add new complexity to the independent use of fingers as was for example the case with Thalberg's third-hand technique. Another aspect is important concerning future developments: a majority of music enthusiasts feel uncomfortable when exposed to contemporary music and many of them feel unable to judge the quality of the performance. As a consequence they may have difficulties in perceiving outstanding perfection and will not reward extraordinary manual skills, as was the case in earlier times. In other words, society will cease to offer appropriate incentives for performers to study these pieces for months or even years.

After all, I suggest that manual dexterity has reached the end of adaptation — at least in the conventional style of music making. The advent of disorders such as musicians' cramp may well be a warning sign of biological limitations in individuals, who are especially susceptible to disturbances in neural plasticity. But it is not only pathology or the maximal available accumulative time of optimal training which limits the 'artistic' aspect of musical performance — it is the fact that society wants to feel **the need to communicate behind the fingers**, the original personal expression of emotional experience. The latter, of course, has to be collected somewhere outside the practising room, limiting the time assigned for manual exercises in a natural way.

Acknowledgments

I am very grateful to Hans-Christian Jabusch and Marc Bangert for their most significant contributions to the experimental data reported in this chapter and for comments on an earlier version of the paper. This work was supported by a grant of the DFG (SPP 1001, Al 269/1-3). A modified version of this article has been published in Williamon A., Coimbra D., (Eds.), *Proceedings of the International Symposium on Performance Science 2007.* European Association of Conservatoires (AEC).

References

1. Altenmüller, E. (2003). Focal dystonia: Advances in brain imaging and understanding of fine motor control in musicians. *Hand Clinics* **19**: 1–16.
2. Bangert, M. and Altenmüller, E. (2003). Mapping perception to action in piano practice: A longitudinal DC-EEG-study. *BMC Neuroscience* **4**: 26–36.
3. Bangert, M., Peschel, T., Rotte, M., Drescher, D., Hinrichs, H., Schlaug, G., Heinze, H. J. and Altenmüller, E. (2006). Shared networks for auditory and motor processing in professional pianists: Evidence from fMRI conjunction. *Neuroimage* **15**: 917–926.
4. Elbert, T., Candia, V., Altenmüller, E., Rau, H., Rockstroh, B., Pantev, C. and Taub, E. (1998). Alteration of digital representations in somatosensory cortex in focal hand dystonia. *NeuroReport* **16**: 3571–3575.
5. Jabusch, H. C. and Altenmüller, E. (2006). Epidemiology, phenomenology and therapy of musician's cramp. In Altenmüller, E., Wiesendanger, M. and Kesselring, J. (Eds.), *Music, Motor Control, and the Brain*, pp. 265–282. Oxford: Oxford University Press.
6. Lehmann, A. C. (2006). Historical increases in expert music performance skills: Optimizing instruments, playing techniques, and training. In Altenmüller, E.,

Wiesendanger, M. and Kesselring, J. (Eds.), *Music, Motor Control, and the Brain*, pp. 3–22. Oxford: Oxford University Press.

7. Münte, T. F., Altenmüller, E. and Jäncke, L. (2002). The musician's brain as a model of neuroplasticity. *Nat Rev Neurosci* **3**: 473–478.
8. Schmid, A., Jabusch, H. C., Altenmüller, E., Hagenah, J., Brüggemann, N., Hedrich, K., Saunders-Pullman, R., Bressman, S., Kramer, P. L. and Klein, C. (2006). Dominantly transmitted focal dystonia in families of patients with musician's cramp. *Neurology* **6**: 691–693.
9. Walsh, G., Jabusch, H. C. and Altenmüller, E. (2009). Synchronization of contrary finger movements in pipers, woodwind players, violinists, pianists, accordionists and non-musicians. *Acta Acoustica* (in revision).

Chapter 7

Temporal Co-ordination of the Two Hands in Playing the Violin

Mario Wiesendanger

This chapter reports on the remarkable co-ordination of the two hands while playing the violin. The two hands function very differently, i.e. in asymmetric movements: first of all fingering of the left hand and bowing with the right hand. In addition, the left arm is also involved in postural functions including the left shoulder and hand. Left elbow movements are frequently used for position changes along the fingerboard. The right bowing arm needs to control the dynamics, i.e. pressure changes on the string and securing the grip-force at the frog end of the bow. All of this derives from basic sensorimotor mechanisms. But on top of it, the player has to add the genuine musical ingredients — to create the musical emotion.

Uni- and Bi-manual Skills in Daily Life Bear Substantially on Playing Music

Music belongs to culture, like objects of art, rock paintings and skillful, nonperishable tools, dating back thousands of years. Although we can admire art in museums and all over the world, unfortunately, we know only little about the origin of music (although it was recently claimed that flutes made of leg bones had been detected at a Neanderthal site[1]). One may speculate that hunters had also plucked their stringed bow in rhythmic ceremonies. But the existence of string instruments has only been documented since the ninth century.[2]

Bimanual co-ordination and the use of the *precision grip* was an enormous step forward in many everyday occupations; movements are constrained by the environment and centered toward goals. Grasping an object with the left hand and manipulating it with the right hand is a typical manoeuvre learned early in childhood. By experience, grasping forces have to be judged adequately in a feed-forward mode in order not to lose or damage the object. Division of labour is a typical situation whereby one hand (usually the left) has a grasping-holding function whereas the right hand is manipulating the object.[3] In dynamic situations of holding objects, the safety margin has to adapt to the moving hand carrying a tool, e.g. a hammer. Subtle co-ordination of forces is amazingly well controlled. Johansson and his co-workers have, over many years, developed a most interesting story on the fine control of grip forces in everyday manual behaviour.[4] My interest in hand actions and particularly of bimanual co-ordination in natural everyday manipulations led me to venture into music, to investigate the issue of bimanual co-ordination in violinists.

A New Step into a Brain Circuit for the Use of the Hands — the Work in Physiology at Oxford

In August 1969, I received a stipend to work with Professor Charles Phillips. So, I left the Brain Research Institute in Zurich to work for a year at the Physiology Institute in Oxford. We travelled — in two packed cars including my wife and four daughters — through France and over to Oxford. The plan was to investigate the functional characteristics of the primate's cortical area 3a, situated deeply in the fundus and the posterior bank of the central sulcus of baboons. Professor Tom Powell from Anatomy joined us also to localise the fast-green dye spots which had been deposited *via* the recording glass-pipettes during the experiment. The histology convinced us that the 'green-spots' were indeed in area 3a. In brief, the results of these experiments were as follows: Short current pulses of very low-threshold were applied to the deep radial nerve (which contains fast conducting fibres from muscle spindles and Golgi tendon organs). The response volleys travelled rapidly to the root-entry zone of the spinal cord and, on the way up, excited the dorsal column nuclei, the ventral-basal thalamus and finally the cortical area 3a, deep in the central sulcus. At all these stations the synaptic transmission was powerful up to the cortex.

This old story seems far-fetched from the theme of music! But some of the 3a-story at the Oxford laboratory was mentioned by Charles Phillips in 1972 at the Music Conference in Vienna (the *Danube Conference*). In its following publication (1977), *Music and the Brain — Studies in the Neurology of Music*,[5] the article by Phillips mentioned the 3a story briefly, published in the *Journal of Physiology*:

> Modern work has discovered the cortical area 3a which receives signals from Ia muscle spindles. It lies immediately adjacent to the motor cortical area and partly overlaps the adjacent receiving areas for skin and joints. All these areas are mutually interconnected by short neural linkages.[6]

At the same conference, Phillips, having mentioned a series of statements relevant for subhuman primates, then turned to the human:

> Use of the hands in man has progressed to include music-making which go far beyond the function of prehension, tactile exploration and simple manipulation and construction... and the modern musician shapes his hands in prehensive patterns and grips, like those of other people; at the keyboard he shapes them into chords, that is into patterns representing complex sounds, before they touch the keys.

Of course Phillips also mentioned the prerogative of monosynaptic cortico-motoneuronal connections in higher primates — a major achievement, serving for quick and direct commands to motoneurones. In his Ferrier lecture, Phillips had already summarised beautifully his work on the primate's cortico-motoneuronal system and its role in fine-control of the hand.[7]

Creation of Neurophysiology in Fribourg: Cortical Loops and Bi-manual Co-ordination

In 1975, I moved to the University of Fribourg. The theme of fast *transcortical loops* was *en vogue* and also extensively explored in human subjects.[8] My team was then ready to concentrate on the issue of *bimanual co-ordination* in subhuman and human subjects. Many complicated tasks are already learned in early childhood, and used intensively. It is remarkable how fast and precise the two hands co-operate naturally, achieving many manipulative goals of everyday life. We chose a natural drawer-pulling task which involved both hands differently: the left for opening the drawer and the right for lifting and re-inserting a peg to the drawer. The manoeuvre was repeated ten or more times. This is a natural, goal-directed bi-manual action. The results were clear-cut: whereas the timing of reaching to the handle and opening the drawer (left arm) was clearly different from the timing of the right arm, which had to pick up a peg from the opened drawer and to re-insert it. But the two hand components came together at the goal in perfect synchrony (drawer fully opened and peg picked up). This fits well with Lashley's *principle of constant goal with variable means*.[9] Later on, further experiments gave us a clue that messages from hand muscle- and skin-receptors are likely to adjust the matching timing.[10]

I mentioned already the extraordinary manual precision, notably the fine regulation of the *grasping hand*. The grip force is automatically adjusted, including the safety margin (grip-force: load-force); the latter is automatically controlled, depending on the variable dynamics, as for example in hammering. Young children learn fast from experience. Going back to the drawer manipulandum, we observed that children at a young age who, for the first time, pulled the drawer to the mechanical stop, mostly lost the drawer handle at the impact. We noticed that all healthy subjects (except small children) gradually increased their grasp force *in anticipation* of the impact at the mechanical stop.

Together with Deborah Serrien, we had the opportunity, at the Neurological Department of the University of Berne, to test patients with various neurological deficits of their hand motility and/or sensibility.[11,12] Most chronic neurological patients, with mainly cortical and cerebellar lesions, slowly recovered to some degree. This recovery was, however, non-adaptive, but to some extent useful. For example, the grip force was not dynamically adjusted in anticipation as described in healthy subjects, but changed instead to a *non-adaptive and high force plateau.*

With Allen Wing (Cambridge) and Roland Johansson (Umeå, Sweden), we organised in 1995 an international meeting at the Monte Verità near Locarno (Switzerland). Allen Wing and his collaborators were in charge of the publication *Hand and Brain* with Academic Press.[13] Three years later, a similar conference took place again at the Monte Verità. The papers were collected in a special issue of *Experimental Brain Research*, with the title *Neural basis of hand dexterity*, by Eric M. Rouiller, Marie-Claude Hepp-Reymond and me as guest editors.[14]

Motor Skills for Playing the Violin

Professional musicians begin their career at a young age and soon acquire all the tricks of the highest grade in motor control: speed of fingering, precision of finger-stops and -lifts, the non-linear tonality of strings along the space of the fingerboard, exact intervals of double or triple finger-stops. To use the entire length of the fingerboard, the left hand needs to change position in short or long steps (called *position changes*). All of this is assigned to the left-fingering hand with the need to determine the correct pitch and speed. The left hand is as important as the right: the bow trajectories should be straight, parallel to the bridge. It requires multiarticular changes of the elbow and wrist angles, in such a way that the bow is moved in a straight line, parallel to the bridge.[15] Another twist of bow movements is to elevate the elbow to reach a lower string or, *vice-versa*, to a higher string. The reversal points of the bow movements can be soft or abrupt; the bow stroke can be long or short, or 'jumping' from the string. The dynamics of bowing, controlled by the pressure of middle finger, has to change all the time, first to compensate the change in the length of the couple (variable distance of the hand from the string) as well for steering the musical intensity, from *fortissimo* to *pianissimo*. Rarely, bowing actions are also replaced by plucking the strings, often in fast succession. All of these manoeuvres (described here more in physical terms) are often very difficult, controlled movements of both hands; last not least, playing music may also transmit deep emotional signs, perhaps typical for an individual player, but hardly a measureable item!

There are of course many books with recipes of how to 'handle' the bow, the fingers, to catch the precise intonation, especially with combined position and string changes, the timing, i.e. keeping the tempo (with allowance of intended emotional signs, such as 'grace notes'). If you think of Bach's solo sonata, one is faced with many double and more simultaneous finger-stops; or of Paganini with his incredible tempo and multiple finger-stops and bowing acrobatics. I like the instructive and richly illustrated book by Galamian about finger-arm movements and postures in playing the violin. It contains good advice about specific finger movements bowing movements and postures required for the difficult instrument.[16]

I discovered also a book, published in 1924 (with later editions up to 1970) of the German physiologist Wilhelm Trendelenburg (1877–1946). He occupied successively the German University chairs of Physiology in Fribourg, Giessen, Tübingen and Berlin. He was an all-round physiologist, most interested in the physiology of the senses and of motor control. But he was also a melodist, playing the cello and the violin, also in ensembles of professional musicians, particularly in string quartets. I discovered only recently his publication (in German) about *The Basis of Playing String Instruments.*[17] The book of 300 pages is well illustrated, and was still available from a 1970 reprint in a second-hand book shop. Apart from physical details, he also described with many illustrations the motor aspect of fingering and bowing, and also the postures of the player. One of his observations was particularly interesting: he investigated the initiation of string vibration in relation to the initiation of the bow movement. On the rotating black-paper drum he recorded and measured the delay between bow-movement onset and the initiation of the string vibration. The book illustrates him in a photograph: playing the cello himself with the setup of the string-galvanometer, recording the bow movement onset and the delayed onset of the string oscillations; the difference amounted to about 50ms. Obviously, the 'unprecison' does not interfere with the precison of hearing (see also[18–20]).

A New Laboratory to Embark on Studies of Motor Skills in Violin Playing

Being retired, I ventured to create a laboratory for testing motor skills in amateur and professional violinists. Having had the experience of bi-manual co-ordination and bi-manual goal achievement with the drawer paradigm, it was natural to choose how the right bowing arm and the left fingering arm are coordinated in time.

I needed some hard and software: four special video cameras (Vicon system) instrumented with infrared strobes (200 measures/s), and some other equipment that allowed me to record movements of interest in the subjects. A set of reflecting markers were assigned to the four fingernails of the left hand (index, middle, ring and little); additional three markers were fixed on the bow, and the three last markers were placed in an asymmetric triangle of the violin roof. Thus, the room-coordinates could be transformed into the violin-centered coordinates which eliminated the whole body movements (for details see Ref. 19). First we had to digitise the trajectories of the four fingertips (left hand) and of the bowing (right-hand), as derived from violin-centered 3D-coordinates. In order to have sufficient data, a melody was first played with bow reversals for each tone. Our strategy was to repeat the simple tune ten times. The results were clear: finger-stops and -lifts were displayed on four lines, corresponding separately to the actions of the index-, middle-, ring-, and little-fingers as displacement profiles. The fingers move sharply down with an impact at the string (a short horizontal line), followed by the sharp finger-lift. The coordination could now be exactly measured, i.e. to measure the timing between finger-impact and the timing of the sharp bow reversal. Our measures of bimanual coordination varied from near 0 to 70 ms.[19,20] This fits also with observations of synchronisations in professional string players: Rasch[21] reports on measured asynchronisations of a professional string trio that amounted to 37–49 ms. As also reported by Rasch, '… the onset of string tones is a more gradual process resulting in long rise times (30–100 ms)'.

My previous co-worker, Oleg Kazennikov from the Russian Academy in Moscow came several times for short periods to Fribourg, with his skillful handling and knowledge of how to extract numerical data from the played data of a number of complex recordings. His programming skill was an immense help to me. I tested amateurs and professional players again, to measure bi-manual co-ordination under the condition of many position- and string-changes in selected exercises of the played melodies (in preparation). Both of the above-mentioned mechanisms need additional time: position changes with an opening or closing angle of the left elbow, bringing the hand higher up or lower down along the fingerboard. String changes need additional changes of the right elbow, up to reach the lower string and down to reach the upper string. The up- and down- movements of the bowing elbow result in rotations of the bow strokes around the strings. The bi-manual, more complex co-ordination needs a measureable increase; however, the co-ordination interval was still below the 70 ms limit. Such physical deviations from zero do not obviously perturb the audition of the musical flow. Interestingly, there was no difference between professionals and amateurs in this respect.

Another twist was to play as fast as possible with the left fingering hand during repetitive bowing at 1 Hz. One finger after the other was selected to play the trills under four bow strokes. Our best results (in professionals) amounted to just about 10 Hz (ten up and ten down movements, or five finger-stops). A piano artist was said to generate, with both hands and all fingers, up to 80 notes.[22]

Another theme of our studies was the consistency of the recorded finger traces of a given violinist. It is not the place here to describe *in extenso* how we can measure the movement characteristics of the violinist, since this has been published before.[18,19] In short, four lines corresponded to the four strings and the profiles of the left-hand fingering were attributed to the correct string-line over the fingerboard. The player was asked to play a short piece of a Bach concerto. Fast finger-stops and -lifts were produced and later displayed on the four string-lines, corresponding respectively to the actions of the index (first line), middle finger (second line), ring finger (third line) and little finger (fourth line), as mentioned previously. Thus, one can easily identify the sequence of the finger actions; fingers which are not acting on the string, frequently have similar profiles as the finger-stops, but are located above the string line. These are mostly (passive) mirror movements (as when you make a willed index flexion, the middle finger makes a similar, but weaker flexion). When we superimposed the recorded profiles of the Bach-fragment (played five to ten times), we observed that the displacement profiles of the finger- and bow-movements (corresponding to the notation) were highly congruent. The traces were overlayed at the middle of the display. There was only a small jitter among the profiles toward the start and the end. It demonstrates that, for a given subject, a well-rehearsed piece of music uses strict time constraints for the four finger displays and for the tempo.

Conclusions

Having played the violin since primary school, curiosity was probably my main drive to investigate some basic characteristics of motor control in violin playing; but this was only possible after retirement. I had the fortune to have three wonderful teachers and found colleagues to play with in quartets, trios and to play in the school and the University orchestra in Zurich. During my stays in Paris, Oxford, London (Canada) and finally in Fribourg and in Berne, the fiddle always accompanied me. While my group did our academic research in motor control, the thought of '*motor control of music*' was often in my mind, but to deviate from the classic problems was too risky. In my final 'music-lab', I was free to study a few problems that interested me: the bi-manual co-ordination, with or without simultaneous position and/or string changes; the effect of delayed auditory feedback during playing; measuring the anticipation of sight-reading the notes; the dynamics of bowing by means of recording the grip-force at the frog of the bow and the pressure exerted on the bow by the middle finger; and finally the maximum speed of finger-stops in trills.

Acknowledgements

The greatest help I received for the music project came from Dr. Oleg Kazennikov of the Academy of Sciences in Moscow, who came four times for short periods to help me in the project. He also gave me a lot of help through email. Notably he developed Matlab programs that I could handle, which were specific for the project and which helped me enormously to manage the large data flow, including the graphic representations. I am also grateful for the support of Professor Eric Rouiller who is now head of Neurophysiology and who was most supportive of my 'exotic' work.

References

1. Kunej, D. and Turk, I. (2000). New perspectives on the beginnings of music: Archeological and musicological analysis of a middle Paleolithic bone "flute". In Wallin, N. L., Merker, B. and Brown, S. (Eds.), *The Origins of Music*, pp. 235–268. Cambridge, MA: MIT Press.
2. Valentin, E. (2004). *Handbuch der Musikinstrumente*, pp. 1–417. Kassel: Gustav Bosse.
3. MacNeilage, P. F. (1987). The evolution of hemispheric specialization for manual function and language. In Wise, S. P. (Ed.), *Higher Brain Functions*, pp. 285–309. New York: John Wiley.
4. Johansson, R. S. (1996). Sensory control of dexterous manipulation in humans. In Wing, A. M., Haggard, P. and Flanagan, J. R. (Eds.), *Hand and Brain — the Neurophysiology and Psychology of Hand Movements*, pp. 381–414. San Diego: Academic Press.
5. Critchley, M. and Henson, R. A. (1977). *Music and the Brain — Studies in the Neurology of Music*. London: Heinemann.
6. Phillips, C. G. (1977). Brains and hands. In Critchley, M. and Henson, R. A. (Eds.), *Music and the Brain — Studies in the Neurology of Music*, pp. 48–58. London: Heinemann.
7. Phillips, C. G. (1969). The Ferrier Lecture 1968: Motor apparatus of the baboon's hand. *Proc R Soc London Ser B* **173**: 141–174.
8. Wiesendanger, M. and Miles, T. S. (1982). Ascending pathways of low-threshold muscle afferents to the cerebral cortex and its possible role in motor control. *Physiol Rev* **62**: 1234–1270.
9. Lashley, K. S. (1930). Basic neural mechanisms in behavior. *Psych Rev* **37**: 1–24.
10. Kazennikov, O. V. and Wiesendanger, M. (2005). Goal synchronization of bimanual skills depends on proprioception. *Neurosci Lett* **388**: 153–156.

11. Wiesendanger, M. and Serrien, D. J. (2005). Bimanual coordination and its motor disorders. In Freund, H. -J., Jeannerod, M., Hallett, M. and Leiguarda, R. (Eds.), *Higher-Order Motor Disorders — from Neuroanatomy and Neurobiology to Clinical Neurology*, pp. 193–236. Oxford: Oxford University Press.
12. Wiesendanger, M. and Serrien, D. J. (2001). Neurological problems affecting hand dexterity. *Brain Res Rev* **36**: 161–168.
13. Wing, A. M., Haggard, P. and Flanagan, J. R. (1996). *Hand and Brain: Neurophysiology and Psychology of Hand Movements,* pp. 1–513. London; New York: Academic Press.
14. Rouiller, E. M., Hepp-Reymond, M. -C. and Wiesendanger, M. (1999). *Neural Basis of Hand Dexterity*, vol. 128, pp. 1–261. Heidelberg: Springer.
15. Braitenberg, V. (1986). The cerebellum and the physics of movement: Some speculations. In Glickstein, M., Yeo, C. and Stein, J. (Eds.), *Cerebellum and Neuronal Plasticity*, pp. 193–207. New York: Plenum.
16. Galamian, I. (1983). *Principles of Violin Playing and Teaching.* New York: Prentice-Hall Inc.
17. Trendelenburg, W. (1925). *Die natürlichen Grundlagen der Kunst des Streichinstrumentenspiels.* Berlin: Springer.
18. Wiesendanger, M., Baader, A. P. and Kazennikov, O. (2006). Fingering and bowing in violinists: A motor control approach. In Altenmüller, E., Wiesendanger, M. and Kesselring, J. (Eds.), *Music, Motor Control and the Brain*, pp. 109–123. Oxford: Oxford University Press.
19. Baader, A. P., Kazennikov, O. and Wiesendanger, M. (2005). Coordination of bowing and fingering in violin playing. *Cognitive Brain Research* **23**: 436–443.
20. Wiesendanger, M., Baader, A. P. and Kazennikov, O. (2003). Vom Steinwerkzeug zum Geigenspiel. In Stulz, P. u. L. A. (Ed.), *Musik und Medizin — Zwei Künste im Dialog*, pp. 95–109. Zürich: Chronos Verlag.
21. Rasch, R. A. (1988). Timing and synchronization in ensemble performance. In Sloboda, J. A. (Ed.), *Generative Processes in Music*, pp. 70–90. Oxford: Clarendon Press.
22. Smith, H. W. (1959). *From Fish to Philosopher*, pp. 1–304. CIBA.

Chapter 8

Music as a Calibrator of Time: Auditory Processing

Steve Jones

The biological significance of music and the reasons for its specific neurological representation in the human brain remain topics of intense speculation. I propose here that music helps us understand and practise using the sub-divisions of time, on a scale of milliseconds to minutes or even hours. Using non-invasive auditory evoked potentials and magnetic fields (more suitable than fMRI or other brain imaging methods for studying temporally detailed mechanisms) we are able to demonstrate and evaluate the participation of large neuronal populations in the analysis of temporal acoustic regularities which may be interpreted as 'pitch' or 'rhythm'. Some possible clinical applications are discussed.

Introduction

A universally agreed and satisfying definition of 'music' is probably unattainable, even (perhaps 'especially'?) among skilled practitioners of the art. This leaves an inviting gap for opinionated amateurs. My own definition of music is 'a period of time filled by sound patterns'. Everyone knows, and possibly even agrees, what is meant by 'a period of time', but what is a 'sound pattern'? It doesn't seem too controversial to propose that, while visual patterns are our perceptions of regular and irregular groupings of light energy across external space (with temporal variations optional), so sound patterns represent the grouping of acoustic energy *across time*, with spatial variations optional. Music cannot be divorced from time, and with few exceptions a particular piece of music only 'works' within circumscribed limits of temporal variation (tempo and duration). Many definitions of music would insist on sounds possessing 'pitch' as an essential element. It seems to be one of the central assumptions of auditory neurophysiology that complex sound waveforms must first be broken down into their frequency components, and that this is the basis for our perception of pitch. Such, however, is by no means the whole story, and it is arguable that much of the important *temporal* information conveyed by sound, including that species of pitch known as 'periodicity' or 'residue' pitch that is essential for musical melody and harmony, could be extracted without such a breakdown.

When we consider the possible reasons for Homo sapiens' evolutionary success, and the disproportionate size of our brains, we tend not to assume any great superiority for our senses, but invariably think of the possession of language and prolific tool use as being two attributes that elevate us above other terrestrial mammals. Equally important, I would suggest, is our unique ability to perceive and analyse the passage of time. Only humans possess such a wide repertoire of motor skills or 'programs', ranging from a fraction of a second to hours or longer. It appears to be only humans that construct detailed internal models of their own hypothetical future (or futures), calibrated on a scale of minutes to years. Other species may be capable of organising a communal hunt or conserving their food supplies for times of lower abundance, but only humans (I believe) ever wake up in the morning with a detailed plan of what they intend to do with the day.

Music as an aid to understanding the passage of time

Does music assist us in this? Ubiquitous in human cultures and with a specific representation in the human brain, music hardly seems likely to be just a by-product of linguistic evolution that most of us fortuitously happen to find pleasurable. The elements of my argument outlined above combine in the postulate that music is helpful to humans as a means of understanding and calibrating the passage of time.

For the last few centuries we have had excellent temporal calibration devices known as 'clocks'. It is hard for us today to conceive of our existence without clocks, but on reflection, the difficulties soon become clear. For how long should we boil an egg? At what time will the tribe gather to go hunting (on a cloudy day)? When should I expect my repaired shoe to be ready? Even clocks are of little help to us in learning how to co-ordinate our activities second-by-second with those of other humans — surely a major factor in our evolutionary success. Since music is unshakeably grounded in time, I believe its evolution alongside that of language has been a significant factor in how humans co-ordinate their activities, particularly over periods of seconds to minutes. In a limited sense, music makes a model for events of the real world. Better musicians who apply the understanding acquired through their art to the temporal dynamics of human society may indeed make better hunters, home-makers, leaders and followers. An orchestral conductor of my acquaintance believes that musicians also make better drivers — himself, presumably, included.

Music represents 24 octaves of sound

Excluding music-drama, where the visual element and literal storytelling clearly assume an importance equal to or greater than that of the sound, musical 'objects' in Western culture vary in duration from less than a minute to an hour or more. In binary terms, this amounts to about eight factors of two — in musical terms eight 'octaves'. Musical structures of less than a minute but more than two or three seconds are termed 'phrases' or 'melodies' — a further four octaves. Musical shapes of less than a second or two tend to be repetitive and to form rhythmic structures whose individual elements, notes or beats, go down to around one-tenth of a second in length — four more octaves. The fastest musical pieces I can think of have note rates of about 12/s, which may be a constraint imposed by motor performance limitations. Above 20/s, periodic sounds start to acquire a different musical attribute, that of 'pitch' which, within the confines of music, spans about eight more octaves. We see, therefore, that music performs a kind of temporal calibration of sound that is effectively continuous from an hour to a fraction of a millisecond — 24 octaves in all.

Studying auditory processing in the laboratory and the clinic

Functional imaging methods such as fMRI and PET are rather poorly suited to examine processes of the human brain in which temporal variations are fundamental. The BOLD response indirectly reflects variations in metabolic activity over periods of a few seconds. The electroencephalogram (EEG), however, and its magnetic counterpart the MEG, are more direct manifestations of events occurring synchronously in large populations of neurons, with a resolution of a millisecond or less. Such methods are therefore well suited to the study of auditory processing. In this chapter, I will describe three studies demonstrating neurophysiological correlates of temporal sound processing, the first two at resolutions of less than a millisecond and the third over longer time epochs. The second and third examples seem to point specifically to temporal calibration processes occurring while listening to music.

Interaural Time Difference Processing in the Brainstem

The first example is provided by the brainstem auditory evoked potential (BAEP) to binaural as compared with monaural sounds. The BAEP to monaural clicks consists of a sequence of peaks, alternately positive and negative in polarity, reflecting the propagation of neuronal activity from the eighth cranial nerve to the midbrain. The response to simultaneous binaural stimulation is almost equal to the algebraic sum of the left and right monaural responses, indicating that the pathways from the two ears are substantially independent of one another. However, when the binaural response is subtracted from the

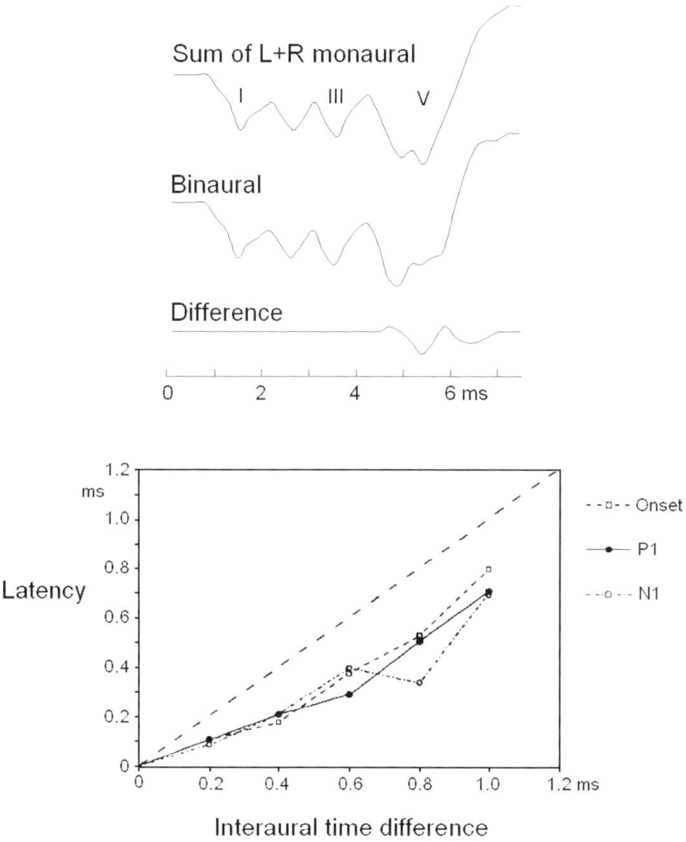

Figure 1. Upper panel: Derivation of the binaural interaction component from the difference between binaural BAEPs and the algebraic sum of left and right monaural BAEPs. Lower panel: The latency of the binaural interaction component increases by approximately half the interaural time difference up to 1.0 ms, and is usually unrecordable at longer differences. From Jones and van der Poel (1990).

sum of the monaural responses (Fig. 1), a residual potential remains which represents the binaural interaction occurring at or above the level of the superior olivary complex (for example, Furst *et al.*, 1985; Jones and van der Poel, 1990).

This 'binaural interaction component' of the BAEP, usually comprising positive (down-going in our convention) and negative peaks termed P1 and N1, has a latency of 5–7 ms. This is consistent with the activity of third order neurons arising in the superior olivary complex, the first nucleus of the auditory pathway receiving input from both ears and within which binaural interaction is known to occur. The most important function of the binaural interaction is to compare input from the two sides for differences in the intensity and arrival time of input from the two ears. Such differences are interpreted in terms of variation in the location of the sound source in the horizontal plane; a source located substantially to the left or right of the midline causes sound to arrive later at the contralateral ear (owing to the extra distance travelled by sound waves around the head) and also at a lower intensity (owing to the sound 'shadow' cast by the head).

When binaural clicks are presented, not simultaneously but with an interaural time difference of between about 0.05 and 1ms, the subjective impression is that the sound source is laterally located toward the side of the leading clicks. A binaural interaction component can be recorded with interaural time differences up to about 1.0ms, but is delayed as compared with that to simultaneous clicks, by an amount approximately equal to half the interaural time difference. This is what would be expected from the output of a binaural coincidence detector such as that proposed by Jeffress (1948). Here a hypothetical column of neurons receives input from both ears, via presynaptic axons whose respective lengths (and conduction velocities) accurately mirror the range of naturally occurring interaural stimulus delays. If each neuron responds only to the simultaneous arrival of presynaptic input from both sides, it is effectively 'tuned' to a particular interaural delay.

Although the mechanisms of acoustic source location and 'auditory scene analysis' have nothing specifically to do with music, this example does serve to indicate how potential fields recorded from the surface of the scalp can accurately reflect the temporal structure of acoustic energy and the processes concerned with analysing that structure.

The 'Pitch Onset' Response

Much more closely related to the neurological representation of music is a phenomenon first described in detail by Krumbholz *et al.* (2003).

These authors studied the long-latency evoked auditory magnetic fields (representing substantially the same processes as evoked potential fields) occurring when continuous random noise changes to a 'regular interval' noise (RIN) in which the fine temporal structure tends to repeat itself at regular intervals. RIN has a similar overall spectral composition to random noise but sounds very different, with a strong pitch equivalent to that of a sinusoid whose temporal period is equal to the repetition interval of the fine structure. The first main component of the evoked magnetic field to the transition from random to regular interval noise peaks at about 100 ms, comparable to the well-known N1 component of the AEP (not to be confused with the N1 of the binaural interaction component!) and therefore termed 'N1m'. This is followed by a peak of the opposite polarity, peaking at 150–200 ms (P2m). When compared with the superficially similar N1m to the onset of random noise, the N1m to the transition to RIN (the 'pitch onset response') was found to be significantly more anteriorly distributed over the scalp. Source location analysis suggested that both responses were likely to be generated in the superior temporal cortex of both hemispheres, but that the neuronal population responsible for the pitch onset response must be located a few millimetres anterior and inferior to that responsible for the noise onset response.

Repeating these studies in the EEG domain (Jones, 2006), it was possible to confirm the more anterior distribution of the pitch onset response, peaking at the mid-frontal electrode (Fz), several centimetres anterior to the central electrode (Cz) where the noise onset response was maximally recorded (Fig. 2). It was also possible to show by low- and high-pass filtering the noise that the pitch onset response was virtually independent of the frequency content of the sounds. This conforms with the subjective impression, that regular internal noise with a periodicity of, say, 5 ms has a pitch equivalent to that of a 200 Hz sinusoid, irrespective of whether the noise contains any significant energy at 200 Hz ('periodicity' or 'residue' pitch). In subsequent studies, we (Subrati and Jones, in preparation) have found that a pitch onset response can be recorded to regular interval noise whose period ranges over at least six octaves (repetition rates of 43 Hz

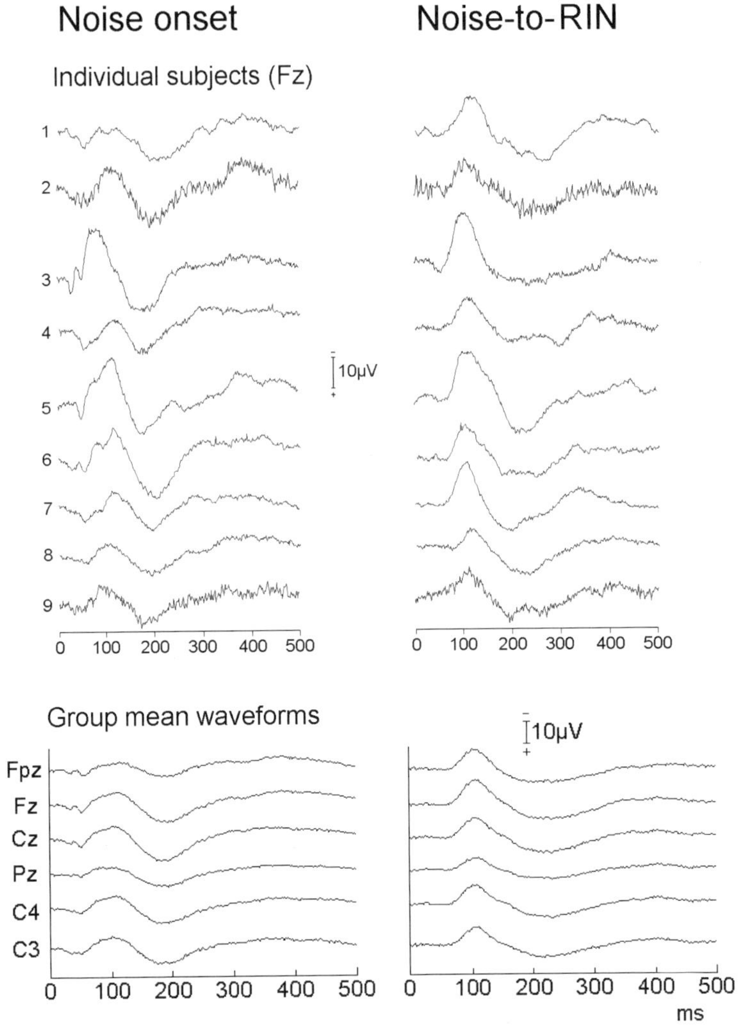

Figure 2. Evoked potentials to the onset of random noise and the transition to regular interval noise (RIN, the 'pitch onset response') in nine individual subjects and the mean of all subjects, showing the tendency of the pitch onset response to be larger at frontal electrodes Fpz and Fz. From Jones (2006).

to 2765 Hz) — a very substantial proportion of that commonly occurring in music. There was evidence, moreover, that the magnitude of the pitch onset response tended to decrease at the upper and lower extremes of the repetition range giving rise to a sense of musical pitch.

Krumbholz *et al.*, (2003) concluded from the change in the latency of the N1m as the periodicity of regular interval noise was increased, that four cycles of the latter were required in order to establish its periodic nature as distinct from random noise, and entrain the processes responsible for generating the response. Our own data, however, suggested that as little as two cycles may be sufficient, particularly at the lower repetition rates.

We concluded from these studies that a substantial population of neurons in the auditory cortices of both hemispheres is strongly sensitive to periodic sounds that are associated with a sense of pitch. This kind of 'periodicity' or 'residue' pitch, which is likely also to be the basis of musical melody and harmony, is distinct from the cruder kind of pitch (sometimes known as pitch 'height') attributable to the spectral breakdown performed by the cochlea. Since the cochlea is temporally uncalibrated, there seems to be no way in which the breakdown performed here can accurately reflect harmonic relationships between sound frequencies. For this information, the brain must depend on the periodic structure of sound (irrespective of its spectral content) being accurately transmitted by the timing of neuronal activity in the eighth nerve, and subsequently in the auditory pathways of the brain. To be able to recognise the periodic nature of an otherwise random noise waveform after as little as two cycles implies a remarkable ability to retain and compare the fine temporal structure, presumably by a process analogous to autocorrelation.

Evoked Potentials to Disturbance of Rhythmic Patterns

When a continuous tone abruptly changes in frequency, N1 and P2 potentials are generated which are similar to those occurring at the onset of random noise. We (Jones *et al.*, 1998) compared the responses obtained when complex harmonic tones — synthesised musical instrument sounds containing a large number of higher partials in addition to the fundamental — changed so as to produce the impression either of a change in pitch (in which case all the partials changed frequency by the same percentage) or a change in timbre (in which case the partials changed their relative energy levels but not their frequencies). The responses were found to be essentially similar to one another, suggesting an underlying process of 'spectral profile analysis' apparently performed in the auditory cortices of both hemispheres (Jones and Perez, 2002).

When spectral energy changes are made at progressively faster rates, the N1 and P2 become progressively refractory, such that at rates greater than about 8/s the waveform resembles a low-amplitude sinusoid. At still higher rates of 16/s or more the response virtually disappears altogether. However, if the changes abruptly cease and the frequencies of the tone become steady again, a large N1/P2 complex is once again generated (Fig. 3; Vaz Pato *et al.*, 1999).

We argued that this complex could not be due to the same spectral profile analysis process as generates the N1/P2 to individual frequency changes. Once the process has been made refractory by a rapid rate of changes, how can it know that the next change is to be the last one and suddenly become dishabituated? We found, moreover, that the N1/P2 at the end of a period of frequency changes was more anteriorly distributed on the scalp, in a manner very similar to the difference between the responses to the onset of random noise and the pitch onset response. A second important difference concerned the latency of the N1, which (following frequency changes made at different rates) was virtually constant, not with respect to the last change that actually occurred, but the *next* change which might have been anticipated but which failed to occur. This response appears to be generated by the sudden, unexpected occurrence of 'no change', and implies a process whereby the preceding temporal pattern of changes is extrapolated in order to predict when the next one should occur.

In subsequent studies, we examined the stimulus variables affecting this waveform. As well as being generated at the end of a period of oscillatory frequency changes, on resumption of steady frequencies, it was also produced after rapid sequences of three, four or five notes, presented regularly or in random order (Vaz Pato *et al.*, 2002). The largest response, however, was recorded at the end of regularly repeated, rhythmic note patterns. Only a very short period of steady frequency was necessary in order to generate the response, which could also be produced by single 'deviant' notes in a regular sequence

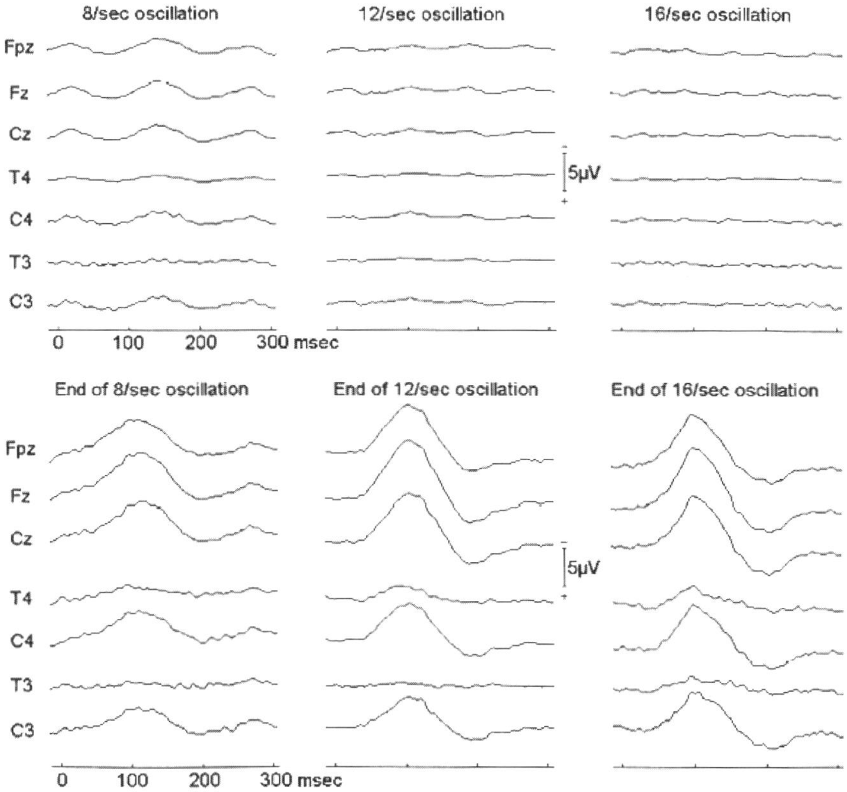

Figure 3. Low amplitude, sinusoidal evoked potentials to frequency oscillations of a complex tone occurring at eight changes/s, further attenuated at 12/s and virtually abolished at 16 changes/s. A large response is once again generated on resumption of steady frequencies. From Vaz Pato and Jones (1999).

presented at 16 notes/s (at this rate there was virtually no detectable response to each individual note). Larger responses were produced by a greater degree of deviance (11 semitones rather than seven semitones higher than the 'expected' note), and smaller responses were obtained when the sequence of repeated notes was presented in pseudo-random order (Fig. 4).

When two deviant notes occurred consecutively in a regular pattern, a broader and rounder response was recorded. If, however, the

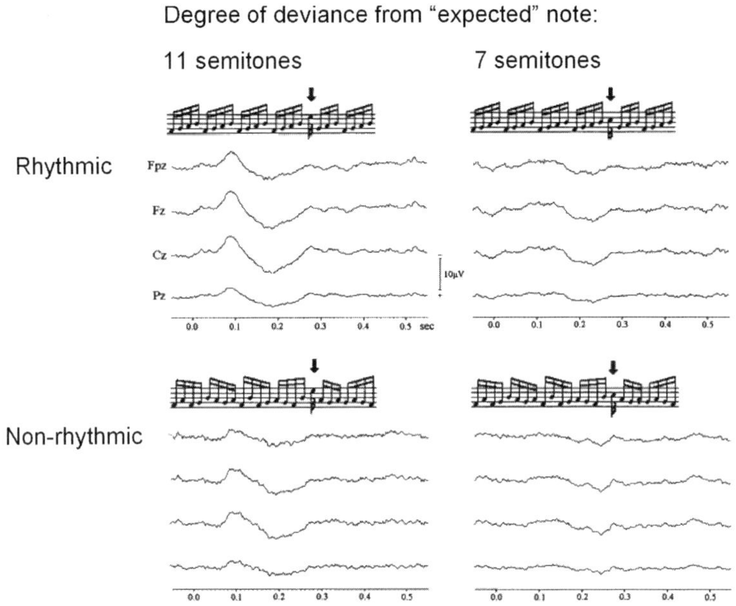

Figure 4. Evoked potentials to a 'deviant' note in a regular (upper) or irregular (lower) four-note pattern presented at 16 notes/s. In both the rhythmic and non-rhythmic conditions the deviant note was outside the range of the ongoing pattern, although in the non-rhythmic conditions the 'expected' note was of course not fully specified. From Vaz Pato *et al.* (2002).

response to a single deviant was subtracted from the double deviant response, a smaller N1/P2 was revealed which appeared to be a distinct response to the second unexpected note.

Many of these properties resemble those of the well known 'mismatch negativity' (MMN) produced when an infrequent 'deviant' sound occurs amidst a sequence of 'standards'. Generally speaking however, the MMN is reckoned to represent a distinct process from the N1, producing a peak at a substantially longer latency than the 100 ms typical of our studies. However, the similarities caused us to label this component 'mismatch-type' N1, or MN1, as distinct from the 'change-type' N1 produced by individually occurring changes in spectral energy.

The fact that the MN1 is produced most effectively by the unexpected occurrence of 'no change', or of a deviant note in a regular, rhythmic note pattern, suggests that the underlying process is one which analyses the rhythmic structure of sounds in order to detect when a change occurs. Rhythm is of course one of the most prevalent elements of music. Recalling the '24 octaves of sound' discussed in my introduction, rhythm represents periods of sound longer than those of residue pitch (the sensation of which is lost above about 50 ms), occupying the four octaves between about 100 ms and one or two seconds.

We do not know exactly where the proposed process of rhythmic analysis takes place in the brain, but the close similarity of these AEPs to the 'pitch onset response' suggests that both may be due to the activity of neurons in the anterior part of the superior temporal lobes, closely adjacent to the primary auditory cortex in Heschl's gyrus. No significant tendency has yet been reported for either response to be more strongly represented in one hemisphere than the other.

Implications for Clinical Neurology

Evoked potentials are a non-invasive methodology which has been fruitfully employed in clinical neurology for more than 30 years. BAEPs are still found useful in screening patients for tumours affecting the eighth nerve, and in demonstrating involvement of the brainstem in multifocal conditions such as multiple sclerosis.

Multiple sclerosis patients are remarkable in that they exhibit a high incidence of abnormal BAEPs, although very few complain of auditory symptoms. This may be because the complexity of the auditory pathways confers a degree of functional redundancy. In a group of patients who had normal BAEPs, we (Jones *et al.*, 2002) found the cortical responses to spectral energy change of complex tones also to be normal, but those occurring at the end of a period of 16/s frequency changes to be significantly delayed. This suggests a subtle, and functionally probably unimportant, disturbance in the processes of rhythmic sound analysis, consistent with the kind of effect that might be expected of diffuse demyelination of the auditory pathways.

Long-latency AEPs, particularly the mismatch negativity, have been used to try to ascertain the prognosis of comatose patients (e.g. Fischer *et al.*, 1999). Our own experience of patients who had awoken from coma, but were in various states of impaired consciousness, were that AEPs to complex tone modulation tended usually to reflect the patients' behavioural level of responsiveness (Jones *et al.*, 2000). There was a small minority, however, in whom well preserved cortical AEPs to spectral energy and rhythm changes raised the possibility of otherwise unsuspected powers of auditory perception and discrimination. No particular tendency was noted for the responses to rhythmic changes to be more often affected than those to changes in spectral energy, so it seems temporal auditory processes may not be especially vulnerable to diffuse traumatic brain injury.

A hypothesis for possible future investigation is that some patients with auditory processing disorders (complaints of poor or distorted hearing in the absence of any demonstrable peripheral hearing deficit) might specifically show abnormalities to processing in the temporal domain. Our evidence so far is purely anecdotal, but does suggest that minor disturbance in the timing of eighth nerve activity may sometimes cause distortion of pitch perception, even when the pure tone audiogram is normal.

References

1. Fischer, C. *et al.* (1999). *Clin Neurophysiol* **110**: 1601–1610.
2. Furst, M., Levine, R. A. and McGaffigan, P. M. (1985). *J Acoust Soc Am* **78**: 1644–1651.
3. Jeffress, L. A. (1948). *J Comp Physiol Psychol* **41**: 35–39.
4. Jones, S. J. and van der Poel, J. C. (1990). *Electroenceph Clin Neurophysiol* **77**: 214–224.
5. Jones, S. J., Longe, O. and Vaz Pato, M. (1998). *Electroenceph. Clin. Neurophysiol.* **108**: 131–142.
6. Jones, S. J. *et al.* (2000). *Brain* **123**: 1007–1016.
7. Jones, S. J. and Perez, N. (2002). *Clin Neurophysiol* **113**: 1558–1565.
8. Jones, S. J., Sprague, L. and Vaz Pato, M. (2002). *J Neurol Neurosurg Psychiatry* **73**: 561–567.

9. Jones, S. J. (2006). *Hear Res* **221**: 65–72.
10. Krumbholz, K. *et al.* (2003). *Cereb Cortex* **13**: 765–772.
11. Vaz Pato, M. and Jones, S. J. (1999). *Cogn Brain Res* 7: 295–306.
12. Vaz Pato, M., Jones, S. J., Perez, N. and Sprague, L. (2002). *Clin Neurophysiol* **113**: 519–527.

Chapter 9

Musical Reading and Writing

John Brust

Music's relationship to language, including reading and writing, has long been the subject of debate, with implications as to whether music is an adaptive mental facility. Clinical studies often reveal striking dissociations between alexia and agraphia for language and for music, and functional brain imaging shows that brain regions underlying musical literacy are more widely distributed than those underlying language literacy. There is also evidence that learning how to read music confers cognitive advantages outside the musical realm.

The question of music's relationship to language is inextricably linked to music's survival value. Darwin believed that musical 'calling systems' evolved into speech, whereas Herbert Spencer believed that music evolved as a stylised form of speech.[1,2] How the brain processes music overlaps with how it processes language,[3,4,5] but there are obvious differences. Language is 'about things in the world': music 'cannot be used to send a message about an object or event outside itself'.[6] The survival value of language is obvious, whereas the survival value of music has generated so many different hypotheses as to call into question whether it is adaptive at all.[7]

Compared with language, musical aptitude has a much broader range across a population. Mental retardation sufficient to prevent language acquisition must be severe. Four per cent of people, however, are 'tone deaf' (selective abnormality of pitch perception),[8] whereas fewer than one in a thousand people has absolute pitch.[9] Children are not taught how to talk or to sing. They do have to be taught how to read and write either language or music.

Neurological phenomena involving music can be positive or negative. Positive phenomena include musicogenic epilepsy,[10] musical seizures,[11] and release hallucinations.[12] Negative phenomena include a large variety of amusias. Like aphasia, amusia is a disorder of cognitive processing unexplained by motor or sensory deficit. Amusia and aphasia can occur together, or each can occur in the absence of the other. Unlike the subtypes of aphasia, which in over 95% of right-handers predict lesion localisation in the left cerebral hemisphere, usually involving opercular cortex, amusia can follow lesions in either hemisphere, and the variety of phenomena reported are enormous. There are disorders of pitch, interval, tonal structure ('contour'), timbre, rhythm, meter, melody recognition, or emotional response.[13] There are also disorders of musical writing and reading.

Clinically significant aphasia is nearly always accompanied by agraphia; co-existing alexia is less predictable. By contrast, musical agraphia and alexia may or may not be present in patients with amusia or in patients with language agraphia and alexia. Such unpredictability was reflected in two patients, both professional musicians, one of whom had aphasia and amusia and the other aphasia with preserved musical processing.[14] The first patient, with conduction aphasia following a small left temporoparietal infarct, had expressive and receptive amusia and musical agraphia and alexia affecting pitch and rhythm to an equal degree. The second patient, with transcortical aphasia following a much larger left cerebral infarct, had preserved musical expression and reception (including pitch, melody, harmony, intervals, timbre, duration, rhythm, and loudness), but musical agraphia and alexia, worse for pitch than for rhythm. It was unclear from the size and location of the lesions why the patients' musical agraphia and alexia had such different features or why amusia followed the smaller infarct but not the larger one.

A number of other reports have described preserved musical aptitude in the presence of aphasia and musical agraphia and alexia. A famous example is Maurice Ravel, who had Wernicke's aphasia, agraphia and alexia for both language and music, and ideomotor apraxia, but whose musical thinking remained intact. He recognised errors in the performance of his own music but could neither write nor dictate music or play the piano, declaring, '[The music] is here, in my head; I hear it, but I will never write it.'[15] Similarly, an orchestral conductor and composer was unable to write music properly or play the piano while declaring, 'I have the music in me.'[16]

Additional forms of musical agraphia and alexia are described in case reports. A woman with Wernicke's aphasia had musical alexia restricted to unfamiliar melodies; rhythm reading was severely impaired but pitch reading was intact.[17] Following left temporoparietal infarction a professional musician could not discriminate or reproduce auditorily presented rhythms although metric judgment and melodic processing were preserved, and he could reproduce rhythmic patterns reading from a score.[18] An aphasic patient could not name musical notes on a score or play notes after hearing their names spoken but could sight-read normally and match written letters and notes.[19]

While there have been reports of amusia and musical agraphia and alexia in the absence of aphasia or language agraphia and alexia, isolated musical alexia has not been described. A case that nearly fit such a description was the self-report of the British neurologist Ian McDonald, who following infarction of the right inferior parietal lobule had a left inferior homonymous quadrantanopia, topographical disorientation, disturbed perception of the velocity of moving objects, dyscalculia, and impaired sight-reading at the piano.[20] There was no aphasia or language agraphia or alexia. His major difficulty reading music involved identifying pitch by placement on the stave; letters designating pitch were correctly interpreted. There were also reading errors of timing and rhythm, and he had a milder degree of musical agraphia. Music perception was nearly normal, yet the music seemed to have lost its emotional content. Dr. McDonald attributed his musical alexia to spatial disruption resulting from damage to his non-language hemisphere.

Strikingly different from cases such as these are reports of aphasics who not only had preserved musical ability but continued to read and write music. The Russian composer Vissarion Shebalin had severe Wernicke's aphasia with alexia and agraphia for language, yet he continued to compose songs, quartets, and a symphony which other musicians considered up to his usual standard.[21] Following a left occipital hemorrhagic infarct, the American composer Randall Thompson had alexia with agraphia, with inability to read more than single letters; although he made errors writing music, his composing skills remained intact.[22]

A professional pianist with Wernicke's aphasia continued to play publicly and 'could compose brief melodies whose writing corresponded to what he hummed.'[23] He read note pitches correctly but had difficulty with their 'rhythmic value.' A professor of piano with Wernicke's aphasia, mild alexia, and severe agraphia had preserved musical reading and writing.[24] An orchestra conductor with global aphasia, alexia, agraphia, and ideomotor apraxia could read but not write music, and he continued to conduct.[25] A blind organist and composer who used Braille for both language and music had Wernicke's aphasia and verbal alexia and agraphia in Braille; he did not have amusia, and he continued to read, write, and compose in Braille.[26] A musicologist with a right temporal lobe infarct could not recognise familiar melodies or compositions when he heard them, yet he could easily sing melodies from memory and interpret written scores.[27]

Reading and writing music involves more heterogeneous symbols than reading and writing language. There are real words, often abbreviated (e.g. largo, DC, f, pp), notes designated by letter (e.g. A, G) but represented spatially with duration indicated by appearance (e.g. whole, half, eighth) and by intervallic distance, and purely musical symbols (e.g. sharps, flats, rests, fermata, clef, crescendo, staccato, repeat). Both simultaneous and sequential events are indicated. Such heterogeneity might predict that the circuitry underlying musical reading and writing would be more widely distributed than that underlying language reading and writing.

Functional imaging confirms this prediction. In a study using positron emission tomography (PET), subjects sequentially listened to scales, played scales on a keyboard and listened, read a score silently, read a score while hearing it played, and sight-read a score, playing and listening.[28] With each task signals from the previous task were subtracted from the image. The final images then revealed activation of the left inferior frontal and premotor cortex (reflecting 'patterning of motor sequences required for the right manual execution of the piece'), the left supramarginal gyrus (reflecting 'mapping between musical notation and its corresponding sounds or melody'), and the occipitoparietal and superior parietal lobule bilaterally (reflecting 'sensorimotor transformations of visually guided skilled movements'). This network is distinct from that used for verbal processing. For example, music sight-reading does not activate the lingual and fusiform gyri as reading language does. Sight-reading reflects analysis of the spatial location of the notes, mapping between musical notation and its corresponding sounds, and mapping between musical notation and visually guided skilled actions. Subsequent studies using functional magnetic resonance imaging have drawn similar interpretations.[29,30]

A study measuring reaction time and evoked potentials while reading music concluded that pitch and duration are processed independently.[31] An evoked potential study of musicians found that diatonic violations of familiar music produced different responses when the music was read compared with when it was heard.[32]

Compared with non-musicians, musicians demonstrated superior performance on a non-musical task involving vertical-to-horizontal stimulus-response mapping. The findings suggested that learning to read music aids the development of visuomotor mapping skills 'that generalise outside the musical context.'[33] A number of other studies have suggested that musical training that includes standard notation leads to greater improvement in spatial skills than musical training that does not.[34] Such findings have evoked comparisons to the 'Mozart effect,' in which listening to Mozart produced temporary improvement in visuospatial skills.[35] The Mozart study, however, did not involve reading music; moreover, the transient benefit might have been a non-specific effect of arousal and attentiveness rather than a result of specific musical experience.[36] Such interpretation does not rule out the possibility that musical literacy might confer cognitive advantages outside the musical realm.

References

1. Darwin, C. (1871). *The Descent of Man, and Selection in Relation to Sex.* New York: Appleton.
2. Spencer, H. (1857/1951). The origin and function of music. In *Literary Style and Music.* New York: Philosophical Library.
3. Patel, A. D. (2008). *Music, Language and the Brain.* Oxford: Oxford University Press.
4. Levitin, D. J. (2007). *This Is Your Brain on Music.* New York: Pluma.
5. Sacks, O. (2007). *Musicophilia.* New York: Alfred A. Knopf.
6. Warren, J. (2008). Another musical mystery tour. *Brain* **131**: 890–894.
7. Pinker, S. (1997). *How the Mind Works.* New York: WW Norton & Co.
8. Foxton, J. M, Dean, J. L., Gee, R., *et al.* (2004). Characterization of deficits in pitch perception underlying 'tone deafness.' *Brain* **127**: 801–810.
9. Athos, E. A, Levinson, B., Kistler, A., *et al.* (2007). Dichotomy and perceptual distortions in absolute pitch. *Proc Natl Acad Sci USA* **104**: 14795–14800.
10. Cho, J.-W., Seo, D. W, Joo, E. Y., *et al.* (2007). Neural correlates of musicogenic epilepsy: SISCOM and FDG-PET. *Epilepsy Res* 77: 169–173.
11. Penfield, W. and Perot, P. (1963). The brain's record of auditory and visual experience. *Brain* **86**: 595–696.
12. Warren, J. D. and Schott, G. D. (2006). Musical hallucinations in a musician. *J Neurol* **253**: 1097–1099.
13. Brust, J. C. M. (2001). Music and the neurologist: A historical perspective. *Ann NY Acad Sci* **930**: 143–152.
14. Brust, J. C. M. (1980). Music and language: Musical alexia and agraphia. *Brain* **103**: 367–392.
15. Sergent, J. (1993). Music, the brain, and Ravel. *Trends Neurosci* **16**: 168–171.
16. Critchley, M. (1953). *Aphasiology and Other Aspects of Language.* London: Edward Arnold, pp. 256–257.
17. Midorikawa, A., Kawamura, M. and Kezuka, M. (2003). Musical alexia for rhythm notation: A discrepancy between pitch and rhythm. *Neurocase* **9**: 232–238.
18. DiPietro, M., Laganaro, M., Leemann, B., *et al.* (2004). Receptive amusia: Temporal auditory processing deficit in a professional musician following a left temporo-parietal lesion. *Neuropsychologia* **42**: 868–877.
19. Bevan, A., Robinson, G., Butterworth, B., *et al.* (2003). To play 'B' but not to say 'B': Selective loss of letter names. *Neurocase* **9**: 118–128.
20. McDonald, I. (2006). Musical alexia with recovery: A personal account. *Brain* **129**: 2554–2561.
21. Luria, A. R., Tsvetkova, L. S. and Futer, D. S. (1965). Aphasia in a composer. *J Neurol Sci* **2**: 288–292.
22. Judd, T., Gardner, H. and Geschwind, N. (1983). Alexia without agraphia in a composer. *Brain* **106**: 435–457.

23. Assal, G. (1973). Aphasie de Wernicke sans amusie chez un pianiste. *Rev Neurol* **129**: 251–255.
24. Assal, G., Buffet, J. (1983). Agraphie et conservation de écriture musicale chez un professeur de piano bilingue. *Rev Neurol* **139**: 569–574.
25. Basso, A., Capitani, E. (1985). Spared musical abilities in a conductor with global aphasia and ideomotor apraxia. *J Neurol Neurosurg Psychiatry* **48**: 407–412.
26. Signoret, J. L., van Eeckhout, P., Poncet, M., *et al.* (1987). Aphasie sans amusie chez un organiste aveugle. *Rev Neurol* **143**: 172–181.
27. Sparr, S. A. (2002). Receptive amelodia in a trained musician. *Neurology* **59**: 1659–1660.
28. Sergent, J., Zuck, E., Terriah, S., *et al.* (1992). Distributed neural network underlying musical sight-reading and keyboard performance. *Science* **257**: 106–109.
29. Schon, D., Anton, J. L., Roth, M., *et al.* (2002). An fMRI study of music sight-reading. *Neuroreport* **13**: 2285–2289.
30. Stewart, L., Henson, R., Kampke, K., *et al.* (2003). Brain changes after learning to read and play music. *Neuroimage* **20**: 71–83.
31. Schon, D., Besson, M. (2002). Processing pitch and duration in music reading: A RT-ERP study. *Neuropsychologia* **40**: 868–878.
32. Gunter, T. C., Schmidt, B. H. and Besson, M. (2003). Let's face the music: A behavioral and electrophysiological exploration of score reading. *Psychophysiology* **40**: 742–751.
33. Stewart, L., Walsh, V. and Frith, V. (2004). Reading music modifies spatial mapping in pianists. *Percept Psychophysic* **66**: 183–195.
34. Hetland, L. (2000). Learning to make music enhances spatial reasoning. *J Aesthetic Educ* **34**: 179–238.
35. Rauscher, F. H. and Shaw, G. L. (1998). Key components of the Mozart effect. *Percept Motor Skills* **86**: 835–841.
36. Schellenberg, E. G. (2001). Music and nonmusical abilities. *Ann NY Acad Sci* **930**: 355–371.

Chapter 10

'Fools at Musick' — Thomas Willis (1621–1675) on Congenital Amusia

Marjorie Lorch

A brief passage interrupts Thomas Willis' discussion of audition with respect to the cranial nerves and memory in his *Cerebri Anatome* (1664). This digression describes the anatomical underpinnings of what would now be referred to as tone deafness, or more recently, congenital amusia. Willis discusses variation in musical talent, the human exclusivity of this ability, and the nature of its development. Willis' inclusion of a consideration of the neurology of musical ability is speculative with regard to matters that are not clearly connected to the surrounding text. The interest Willis had in the physical substrate of musical ability is understandable in light of certain philosophical and theological issues regarding the nature of sense perception, the passions and the human soul and the research activities of Willis' colleagues in the Royal Society.

Introduction

This chapter will consider the ideas of Thomas Willis (1621–1675) on how musical talent may be understood in terms of the structure, action, and use of the human brain. It will begin with a biographical introduction to Willis and the intellectual context of the English Reformation and Restoration. The main objective of this chapter is to consider Willis' neurological texts with respect to notions about hearing, auditory perception and memory, and music. His first major work on the anatomy of the brain, *Cerebri Anatome* was published in 1664. This book contains a little noted passage which considers how music is instantiated in the brain and discusses individual differences in musical ability in development. His later work *De Anima Brutorum* (1672) contains further descriptions of the auditory system and the differences between humans and animals with regard to higher mental functions. Willis' treatment of the topic of musical talent is remarkable for both its speculative nature and its original insight. This chapter will conclude with a consideration of the contemporaneous interests of his colleagues, focusing particularly on the work of Robert Hooke (1635–1703), and the intellectual, political and theological issues in which Willis' work was situated, in order to fully understand Willis' proposals for a neurology of music.

Biographical background

Thomas Willis has been the focus of much historical scholarship[1–10] over the past 50 years which includes a number of publications[11–13] marking the tercentenary of the publication of what is considered the first major textbook of neurology, his *Cerebri Anatome* in 1664. He was born in Great Bedwyn, Wiltshire and received private schooling from Edward Sylvester in Oxford, and from Dr. Thomas Iles, Canon of Christ Church. He received his BA in 1639 and MA 1642 from Christ Church, Oxford and began to practice medicine in 1646. Willis' life was affected greatly by the English Civil War throughout the early part of his medical career (1644–1665). He was a Royalist supporter and devout Anglican, with a large medical practice based in Oxford throughout the Reformation. Upon the restoration of Charles II to the throne in 1660, Willis was made Sedleian Professor of Natural Philosophy at Oxford, a post which he retained until his death 15 years later.

As Sedleian Professor, he was required to give weekly lectures on Aristotle and 'comments on the Offices of the Senses, both external and also internal as of the Faculties and Affections of the Soul, as also of the Organs and various provisions of these'. Willis 'resolved to unlock the secret places in Man's Mind' and devote himself 'wholly to inquire into the offices and uses of the Brain.'[14]

His first book, *Diatribae Duae Medico-Philosophicae*[15] published in 1659, was a collection of three essays on fermentation, fevers and urines. It drew on the experience he had gained from his patients and his research in chemistry. His clinical contributions include distinguishing between *diabetes mellitus* and *diabetes insipidus*[16]; and describing *myasthenia gravis.*[17,18] He gave the first account of typhoid fever as he met with it among the troops during the Civil War,[19] and was the first to describe and name puerperal fever.[20]

Willis' contributions to neurology

Willis made major contributions to neurology which have been the focus of recent researchers in the history of neuroscience.[21] He is credited with creating the term *'neurologie'* and the first use of the notion of 'reflex action.' His most well-known discovery is the eponymous Circle of Willis which explained the functional significance of anastomosis against occlusive deficiency.

Willis produced three major texts of neurological significance over an eight year period: *Cerebri Anatome* (1664),[22] *Pathologiae Cerebri* (1667)[23], and *De Anima Brutorum* (1672).[24] His work is also known through the notes of his Oxford lectures[25] as taken down by his student John Locke (1632–1704) who went on to write *An Essay Concerning Human Understanding* in 1690.[26]

In his *Cerebri Anatome,* Willis reclassified and renumerated the cranial nerves into nine pairs, clearly illustrated in his Figure 1, drawn by his assistant Christopher Wren (1632–1723). In Britain, Willis' system of nine pairs was retained up to the nineteenth century.[27] Willis also gave the first account of the intercostal nerves which he termed 'reins of the soul.'[28]

The anatomy of the brain and the human soul

Cerebri Anatome, translated into English as *The Anatomy of the Brain* by Samuel Pordage in 1681,[29] was dedicated to his patron,

Gilbert Sheldon (1598–1677). The book was groundbreaking in its use of extremely detailed anatomical dissections in conjunction with years of observational experience as a practitioner to form original insights into function. Willis went beyond the teachings of Aristotle, Galen, and Descartes and Gassendi to develop a new view of brain function.

One of the central concerns for any new scheme linking neuroanatomy and physiology to behaviour and mind was the conceptualisation of the human soul. Willis' major innovation was to introduce the idea of the 'sensitive soul' and 'rational soul.' In chapter ten of *Cerebri Anatome*, Willis follows his teacher William Harvey's (1578–1657) idea that a 'vital soul,' the *flamma vitalis*, acted within the blood. The sensitive soul arose from the vital soul, formed by 'the procreation of spirits in a 'double fountain' of arteries supplying the cerebrum and cerebellum in a parallel neural circulation of *spiritus*.'[30]

In Willis' system, the vital soul and sensitive soul controlled sensation and motion, as well as some higher functions including knowledge and simple reasoning. These capacities were found in both animals and humans. However, only human beings had an immortal 'rational soul' for higher thought, will and judgment. Though the rational soul was immaterial, in Willis' scheme it operated on the body. Willis claimed that the 'rational soul variously moves the sensitive,' using it as a vehicle.[30] The structure which served to link the passion and the body to the rational soul was identified by Willis as the intercostal nerves. Through them the bodily actions in humans could be driven by rational judgement.[28]

The theological context of Willis' work

Willis was by all accounts a deeply religious person. Martensen remarks that: 'Throughout his career contemporaries commented on Willis' piety and devotion to traditionally formalist religious services.'[31] During the period of Puritan Reformation, formalist liturgy was banned, but Willis maintained highly observant Anglican practices. Illegal Latin services were regularly held in secret in his home in

Oxford. Willis was married to the daughter of Samuel Fell (1584–1649) the Anglican Dean of Christ Church and Vice-Chancellor of the University and his patron was Gilbert Sheldon, the Archbishop of Canterbury, who was a leading spiritual and political figure of the day. *Cerebri Anatome* was dedicated to Sheldon, and thus, to the Church and God.

In his prefatory remarks, Willis set out his research goal as being to 'unlock the secret places of Man's Mind, and to look into the living and breathing Chapel of the Deity.' This statement confirms that Willis undertook his work on the anatomy of the brain in a deeply religious spirit. In his dedication to Sheldon in *Cerebri Anatome*, he states:

> It hath been a long while accounted as a certain Mystery and School-house of Atheism to search into Nature, as if whatever Reasons we grant to Philosophy should derogate from Religion. But truly he doth too much abuse the Name of Philosophy who considers... all the make and provision of a Clock, by which invented Machine the course of the Time may be exactly known and measured, if that at length, when by his search and consideration he hath profited himself so much, he should not acknowledge the Artist to whose Labour and Wit he owes all those things.

In subsequent passages of his dedication, Willis repeats his belief that the study of the human brain is closely related to the revelation of the 'Divine Word'.[5]

Willis' multiple sources of neurological evidence

Willis' understanding of the function of the nervous system was a result of a unique synthesis. He had by all accounts an exceptionally large medical practice. Indeed, Martensen[31] states that the income from this practice was the largest in Oxford (£300 per annum by 1667). He was an acute observer of patients' signs, symptoms and behaviour. In addition he actively pursued research in chemistry and carried out experimental manipulations. At the same time, he spent huge amounts of time dedicated to the dissection of the human body and a variety of animals in order to understand what was and was not unique to the human anatomy. Willis' neurological ideas, as expressed in *Cerebri Anatome*, arise out of his experiences drawn from this large body of clinical observations, pathological confirmations, and experimental inquires. These different sources of evidence were then presented for each part of the nervous system considering in turn their structure, action and use according to the programme of the 'old anatomy.'[28]

Another significant factor that uniquely contributed to Willis' work was the team of friends, junior colleagues and student he collaborated with in his empirical pursuits. Willis' work was enriched by an incredibly talented and multidisciplinary group of individuals. For the work which served as the foundation for *Cerebri Anatome,* Willis acknowledged by name the contributions of Richard Lower (1631–1690), Christopher Wren and Thomas Millington (1628–1704). There were, in addition, a larger group of associates who may have also played a part in developing Willis' ideas, including his young assistants Robert Hooke, Robert Boyle (1627–1691) and John Locke.

Parts of the nervous system described

After a methodological discussion of dissection techniques at the beginning of *Cerebri Anatome,* Willis goes on to describe the 'oblong marrow', the 'cerebel', the skull, and 'pia mater.' Each structure is considered in turn with respect to his dissection findings along with a discussion of the actions and uses of each. Chapter 17 is entitled 'of the Nerves, which receiving the stores or companies of the Spirits from the Cerebel, bestow them on the Acts of involuntary Function.'

Willis begins this section with a consideration of the 'fifth and sixth conjugation of Nerves.' He then proceeds to the seventh pair which he refers to as 'the hearing nerves'. He details his description as follows: 'We have shown before the Processes (which in a manner may be called distinct Nerves) of the seventh pair to be two fold on either side: one, the softer of these, serves only for the sense, but the other harder seems to perform some motions. This latter Nerve, being carried without the Skull, is divided into three branches, all which serve to pathetick motions, or at least to such as are performed without consulting the Brain.' This description of the facial and vestibulocochlear nerves is consistent with that of Galen's fifth pair. (It was not until the 1790s that Soemmerring separated the two and renumbered the nerves so that the vestibulocochlear became the eighth nerve, as today.[27])

A Digression on Musical Ability and Brain Anatomy

The musical text from Cerebri Anatome

After his discussion of the cranial nerves, Willis departs from this theme to offer a digression on musical talent. Surprisingly, while there has been a great deal of recent commentary on the significance of various aspects of Willis' work, his discussion of musical talent has been completely overlooked. Owing to the exceptional nature of Willis' comments on this topic, the brief passage on musical ability is reproduced here in full:

> But here (if I may digress a little) we should inquire in what part of the Head the Ideas of sounds are left: whether only in the Brain, which is the Chest of Memory acquired as it were artificial, or whether not also in the Cerebel, which is the place of natural memory? Truly we suppose, that sounds belong to both these, as it were to distinct Store-houses. Every audible impulse being struck against the Ear, it is presently carried by the passage of the auditory Process to the annulary Protuberance; but from thence it is carried, as other sensible Species, to the chamsered bodies or the common Sensory; (which way it passes thither, shall be shewed afterwards) this impression tending from thence farther, and being also delivered to the Brain, stirs up the Imagination, and so leaves in its Cortex an image or private mark of it self for the Memory. Further also, as the auditory Process depends on the Cerebel, and receives from it the provision of the animal Spirits: so it is most likely, that by the recess of the same Spirits the Ideas of the Sounds are conveyed also to the Cerebel; which forming there footsteps or tracts, impress a remembrance of themselves, from whence when afterwards the Species there laid up are drawn forth by the help of the vocal process, voices, like the sounds before admitted, and breaking forth in a certain ordained series, come to be made.
>
> Hence it is usual, that musick or melody is soon learnt by some men, which afterwards they bring forth with exact Symphony, without any meditation or labour of the Brain; to wit, from the distinct accents of the heard harmony, the Spirits moving within the Cerebel are disposed into peculiar

Schemes; according to which, when they flow on both sides into the vocal process of the auditory Nerve, they render as it were with a certain spontaneous voice, and like a Machine or Clock with the succession of Species, the measures or Tunes of the Instrument which they had drunk in at the ears. Without doubt hence the reason may be sought, why some men learn Musick without any trouble, and others hardly or not at all. For it is observed, that some Children, before they can speak distinctly, quickly sing, and remember certain Tunes; whilst others, though very ingenious men and of excellent memory, are very Fools at Musick, and become uncapable, as an Ass for the Harp; wherefore 'tis commonly said, that some have musical ears, and others are wholly destitute of that faculty. In the mean time, 'tis to be confessed, that in these the Organs of the Voice are not defective; but all the fault, though wrongfully, is cast on the hearing.

But in truth the genuine cause of this defect seems to consist in this, that when in all, the audible Species go to the Cerebel sooner and more immediately than the Brain; yet in some the Cerebel being harder, and not easily yielding to the received impressions, those Species, because they could impress nothing of themselves in their passing to the Cerebel, being carried towards the common Sensory, leave their Types or Ideas chiefly and almost wholly in the Brain: which part being still busied with disturbed motions, is less apt to keep distinctly the composures of Harmony. But in the mean time, in others the Species of audible things, besides that they are carried to the common Sensory and to the Brain, do also affect the Cerebel, especially if they are harmonically figured (forasmuch as in them there is a softer capacity of the impressions) with a peculiar order and Scheme of the animal Spirits: where, as the Species of the Harmony being disposed in convenient little places and cells are kept, afterwards they flow out from thence, almost unthought of, without any endeavour or labour of remembrance, but in a distinct series, and as it were in composed modes and figures, and so by blowing up the vocal processes, they constitute sweet Tunes and vocal Musick.

If that the divers ways of passage are inquired into, to wit, whereby the audible Species, being carried into the annular Protuberance, do get both to the Brain and Cerebel; I say it is not improbable, but out of that Protuberance both a passage lyes open into the underlying tract of the oblong Marrow, and as it were the high road as also another passage is opened into the Cerebel through the medullar processes of the same Ring. But lest there should perchance be a confusion of the animal Spirits and the sensible Species, (which indeed can hardly be avoided) if the way made for their passage should lye open into various passages and manifold apertures; therefore concerning this it may well be supposed, that the Ideas of the Sounds pass through the Cerebel, when they are carried to the common Sensory; which region being first past, they are at length brought by a by-path, *viz.* through the orbicular Prominences to the chamfered Bodies: which perhaps is partly the reason, that in the Hearing the perception of the sense succeeds so late, and the impulse of the object, in respect of sight, follows so slowly.

Whilst therefore the audible Species passes through the Cerebel, in some men, it leaves in this region (for that it is of a soft temper, and fit for the receiving impressions) and tracts and marks of it self, and so they obtain musical ears. But in others who have a harder frame of the Cerebel, they produce no tracts of the same Sounds, and therefore such are wholly distitute of the faculty of Musick.

As therefore we suppose the audible Species to pass through the Cerebel after this manner, a reason may be given from hence, wherefore Musick does not only affect the Phantasie with a certain delight, but besides chears a sad and sorrowful Heart; yea, allays all turbulent Passions excited in the Breast from an immoderate heat and fluctuation of the blood. For since the animal Spirits, serving for the motion of the *Praecordia*, are derived from the Cerebel; as the perturbations conceived in the Brain, the influence being transmitted hither by moving these Spirits, in the Fountain itself, transfer the force of their Affections on the Breast; so the Melody introduced to the Ears, and diffused through this Province, does as it were inchant with a gentle breath the Spirits there inhabiting, and composes them, called off from their fury, to numbers and measures of dancing, and so appeases all tumults and inordinations therein excited.

From these may in some measure be known the reason of the difference, why the hearing Nerves are after a different manner in man and in four-footed beasts: for, because in these there is little need that the audible Species should pass through the Cerebel, either for the reciprocations of the sound heard, by the voice, or for the impressing there the Tunes of the Harmony (for neither is Musick required, what ever Poets feign, to the taming the Affections which move the breasts of beasts) therefore in these (I mean in four-footed beasts) the annular Protuberance dispensing the animal Spirits to the auditory Nerves, and receiving from them the sensible Species, requires not so strict an affinity with the Cerebel: yea, whenas in may suffice, that those Nerves arise from the oblong Marrow, yet the annular Protuberance, as it were a common Porch, ought to be prefixed to them; to wit, in which both the Spirits going out from either side, and the sensible Species to be carried to either, ought first to be mixed and united together, lest otherwise every sound should become double.

Among the Nerves which are seen to belong to the Cerebel, and to perform its offices, lastly follow the eighth or wandering pair, which indeed hath its rise out of the common Trunk of the oblong Marrow, near the place where the last process of the Cerebel is terminated, and over against where the

> pyramidal bodies, being produced from the annular Protuberance, end: so that we think these Nerves also, by that process coming between on either side, and also perhaps in some measure through the passage of the pyramidal bodies, do derive all manner of influence of the animal Spirit from the Cerebel.
>
> The beginning of these consists of very many fibres and filaments or little threads presently distinct one from another; to which belongs, from the very beginning of every Nerve, a noted Trunk arising out of the spinal Marrow. The description of the wandring pair of Nerves, and its protension into the *Praecordia* and some *Viscera*, are added hereafter. For the present it shall suffice, that we take notice, for that as much as this Nerve is bestowed chiefly on the *Praecordia*, the acts whereof are involuntary, and are performed without our care or knowledge in sleep as well as waking; and for that the same Nerve seems to receive the forces of the Spirits wholly from the nearer fountain of the Cerebel; from hence it may certainly be well concluded, that the government or oeconomy of the Cerebel regards only the involuntary Function.

Willis' original observations about music and the brain

Willis begins by asking 'we should inquire in what part of the Head the Ideas of sounds are left?' This is a question which is in line with the programme of anatomy and physiology from the time of Galen onwards.[32] He considers the pathway for auditory perception from the ear up to the Cortex. At this point, he raises the issue of auditory memories. Interestingly, from this he goes on to consider the relative ease or difficulty of learning a melody by certain individuals: 'Without doubt hence the reason may be sought, why some men learn Musick without any trouble, and others hardly or not at all.'

Willis goes on to offer his observation that skilled singing is sometimes acquired before speech in young children. It is not just the vocal production that is of interest to him, but the memory of the tunes as well as instrumental musical performance. In this way, Willis broadens his view from auditory perception and recognition to vocal production and fine motor performance. He considers musical talent to be of a piece. Moreover, Willis observes that musical ability appears to be unrelated to other aspects of memory, reasoning or intellect generally. A physiological difference between individuals is sought as an explanation for this variation of those who have 'musical ears' and those who are 'wholly destitute of the faculty of Musick'.

Finally, Willis returns from his quite original speculations to a discussion of the more traditional topic of the effect of music on 'the passions.' He suggests that this is another point where the mental abilities of humans are distinct in their anatomical composition from animals: 'From these may in some measure be known the reason of the difference, why the hearing Nerves are after a different manner in man and in four-footed beasts...'

Willis' Later Work on the Auditory System

In Willis' major work on comparative neuroanatomy, *De Anima Brutorum* (1672), Willis describes many aspects of higher cognitive functions which are distinct in humans. The agenda is made clearer in the title given to the English translation of the book (1683): 'Two discourses concerning the soul of brutes which is that of the vital and sensitive of man'. Willis offered his own account of how sound was constituted, a question of active research at the time:

> That some Sonorifick Particles, or Causing sounds, are diffused thorow the Air, and as they are more subtil than the little Bodies of the Air, and are indued with a more rapid Motion, the Transmission or Propagation of the sound, depends upon the peculiar motion and waving of these, made apart from the inclination of the whole Air... Propogation of sounds 'whil'st the Sonorific Particles leap back from a solid Body, they cause the audible Species to be every where represented, according to the stroke there made upon them, in the whole Sphear of Vibration, whether by a like Contortion, or Gyration, or any other ways of Conformation in Motion, of the symbolar Particles.

Willis also extended the understanding of the anatomy of hearing and the function of the eardrum. 'Nigh to the most intimate recess of this Den [inside the ear canal], a thin Membrane is placed, with a Circular Bone, fitted to the same, which wholly shuts up the Cavity of the Ear, and distinguishes the Interior Cloyster from the Exterior; so that the Impulse of the sound, shaking this Membrane like a Drum, delivers the Impression to the Sonorifick Particles planted beyond, and they being moved, affect the Fibres, with the Auditory or Hearing Nerve'. Next he describes the anatomy of the cochlea and its function: 'use of the Shell, no less noted, to wit, that the audible Species may be impressed on the Fibres and the ends of the sensible Nerves, inserted in this place, not at once or at large, but by little and little, and as it were in a just proportion and dimension'. This final comment suggests the possibility of tonotopic mapping in the basilar membrane.

In the midst of this discussion of the function of the peripheral organs in audition, Willis presents the case of a woman who was hard of hearing, whose ability to understand conversations improved if a servant repeatedly banged on a drum. This is the first description of what is now referred to as the paracusis of Willis. As is typical for Willis, this anatomical description is followed by speculation on the possible function of these structures. Here Willis discusses various aspects of auditory processing including selective attention and what is now referred to as the 'cocktail party effect':

> That we may give our Conjecture concerning these, perhaps there is need for the audible Species, to be carried toward the common Sensory, that its passage may be the more certain, and that the perception of the sensible thing, may be put out of doubt; but we rather think, that this Sensory is made double, that when oftentimes the Idea's of sounds ought to be heard and perceived together, some might pass this way, and others that way, without Confusion. For it is observ'd, that the Hearing, not only as the other Senses, receives many objects together; and by and by whether united or confused, comprehends them, by the same act of the Sense; but moreover, this faculty in the time of Hearing, so distinguishes things, often divers, admitted together at the Ears, that it seems to hear one after another: It ordinarily happen'd, that in a confused multitude of voices and sounds, that I have my self taken notice to have heard the peculiar voice of a certain Man, and then a little after, I have known that I have heard, at the same time, some other words of another Man, that I did not perceive before; the reason of which is, that this sound, being received together with that, reached not at the same instant to the Common Sensory: wherefore, we may believe, that the sensible Species of the former sound, passing thorow only one Shell, is by and by conveyed, by the first branch of the auditory Nerve, sooner to the Sensory, but the other sensible Species, because it could not be carried with it together by the same Nerve; therefore it is carried by a winding about thorow the second Shell, and at length

to the second branch of the auditory Nerve, and so coming later to the Common Sensory, is afterwards perceived.

The Early Modern English Context

In order to fully appreciate why Willis thought to include this digression on musical talent in his account of the anatomy and physiology of the brain in *Cerebri Anatome*, we must take into account the context in which he worked, and the work of his close associates. There are several key issues that may be argued to have contributed to the interest Willis had in this topic. First, the status of music theory as part of natural philosophy and, in turn, the program of empirical enquiry which was then developing in England in the middle of the seventeenth century. Second, the theological and linked political issues at this time regarding Anglican and Puritan ideas about music in relation to God and the soul, and how those issues played out for Royalist and Parliamentarian supporters.

The place of music in natural philosophy

In his English Dictionary *Glossographia* (1656), Thomas Blount defined 'Philosophie' as being comprised of three parts: 1) 'Rationall', 2) 'Naturall' and 3) 'Morall.' Natural Philosophy was further described as concerning the study of Arithmetic, Music, Geometry, and Astronomy. Thus, the study of music theory was viewed as part of a larger programme to understand the properties of the physical universe. At the same time, musical instruments in the seventeenth century were more technologically complex than most other mechanical devices, such as fountains or windmills, and their production required mathematical knowledge of harmonics and physics.[33]

With respect to music theory, the seventeenth century was a time of redefinition of the conceptualisation of pitch from two- to three-dimensional space. This is referred to as the 'problem of consonance.'[34] Up to the early seventeenth century, theories of harmony had been confined to musical sounds heard melodically, as, for example, in church modes, and consonance was defined through the ratios of the Pythagorean scale. By the middle of the century, an alternative way of conceptualising harmony was beginning to develop, based on a tonal system where musical intervals are heard simultaneously, that is to say, harmonically.[34] At the same time there was a strong link between music and medicine. Music was seen as having a therapeutic action on the body. This is discussed in detail by Robert Burton (1577–1640) in his Anatomy of Melancholy (1621). Gouk suggests that: 'In the seventeenth century not just court physicians, but anyone versed in Scripture knew that psalmody was a means of raising spirits and restoring souls.'[35]

Acoustics, music and the Royal Society

Meetings were held in both Oxford and London from the mid 1640s of 'divers worthy persons inquisitive into natural philosophy and particularly what hath been called the New Philosophy or Experimental Philosophy'.[20] Willis was actively involved in the original meetings of the Oxford Philosophical Club, and both of these two groups had as members all of Willis' circle of collaborators — i.e., Wren, Lower, Hooke, Boyle, as well as other friends who interested in music theory in Oxford such as John Wallis (1616–1703) and John Wilkins (1614–1672). The two Societies kept up close relations and were soon merged, meeting in London at Gresham College. These formed the foundation of what was to become 'The Royal Society of London for Improving Natural Knowledge' when King Charles II granted a charter in 1662.[20]

One of the key figures in the Royal Society was Robert Hooke who had trained with Willis. It was Willis who recommended Hooke as an assistant to Robert Boyle. Hooke became the 'Curator of Experiments' for the newly founded Society. Like research into the physical nature of light, there was an active programme of investigation into how sound was propagated at this time. These topics were being actively pursued by both of Willis' own young colleagues, Robert Hooke and Robert Boyle, and others in the Royal Society circle.

Hooke carried out experiments on various aspects of the production and propagation of sound at the Royal Society in 1664, at the same time Willis was writing of the *Cerebri Anatome*. For Hooke, stringed percussion and wind instruments served as models for conceptualising both the physiological and ethical aspects of internal character.[36]

Hooke conceptualised brain function in terms of a sounding bell, its mode of vibrations different from a string's vibration, with the soul playing on this instrument. There were a number of aspects to Hooke's interest in music. First, Hooke was a musical practitioner: as a student he had been a 'singing man' (i.e., had a music scholarship) and played the organ at Oxford University. He also made musical instruments. He was known to have had tinnitus (possibly from diabetes). In addition, Kassler[36] links the fact that Hooke had shaking palsy to his interest in muscle 'tone.' Thus, Kassler contends that Hooke's use of musical metaphors in his writings on Natural Philosophy were tacitly grounded in experience, and were not simply literary metaphors.[36] In contrast, although Willis, like Hooke, used the metaphor of the organ in his own writings on brain physiology, he believed that the nerves were not hollow but filled with fluid, and so more often used analogies to water courses, ponds, tides, seas and light rays.[37]

The Conceptual History of Music and the Brain

Willis' digression on this topic appears to have a primarily local and contemporaneous resonance. It was cited by a several authors in the second half of the century. One noted example of Willis' work cited in the context of music theory, brought to my attention by David Cram, was by Thomas Salmon (1647–1706).

Salmon was Rector of Mepsal, Bedfordshire, and a music theorist who proposed a 'universal character' for music notation in which G was always the first line on the staff. He later attempted to construct a viol which could produce mathematically pure intervals.[38] In his vindication of *An Essay to the Advancement of Music*[39] written in 1672 states: '... the four first pages of my *Essay*, where I so largely treated of the divine institution of Musick: *Which* (I said) *needed nothing else, nor could have any thing greater to command acceptance, than a challenge of its Institution from divine Providence it self, who had provided a peculiar faculty for its reception*. (Which I since find the eminent Dr. *Willis* place in some peculiar *Schematismi* of the *Cerebellum, Anat. Cer. Cap.* 17.)...'

In addition, Willis' neurological theories were used in the context of theological debates. Robert Sharrock (1630–1684) was an Oxford colleague of both Willis and Boyle. In his book *De Finibus Virtutis Christainanae* (1673), Sharrock discusses the Resurrection with reference to Willis' theories about the function of the corpus striatum and corpus callosum with respect to his conception of the rational soul.[40]

There appears to have been little consideration of the neurological instantiation of music *per se* in subsequent periods until the work of Gall and his phrenological followers in the early nineteenth century.[41] Gall included music as one of the faculties to be identified with a particular location on the skull adjacent to that dedicated to language. There were a number of subsequent discussions in the literature on whether musical ability was regulated solely by the faculty of tune or if there were additional contributions by other faculties to produce this complex set of abilities.[42]

Our modern history of music and the brain has a relatively shallow time depth. In his 1977 review of the subject, Henson cites the work of the German surgeon, Billroth (1894) as the first attempt to define the neurological substrate of musical ability.[43] Advancement of research in this area was aided by the work of Carl Emil Seashore (1866–1949) on the psychology of musical talent and the educational aspects of its development in the early twentieth century.[44] Research was also facilitated by Seashore's development of standardised test materials separating out various components of musical skill.

With the advent of clinical–pathological correlations studies of cortical function in the second half of the twentieth century, music again became a focus of interest as a complex non-linguistic skill of auditory processing which was subject to impairment from focal neurological damage.[45] Typically, the work by Critchley and Henson,[46] nearly 40 years ago, is considered to have pioneered the kind of interdisciplinary consideration of music and the brain to which the present volume aims to contribute.

It is worth pointing out how much the developments in our understanding of complex skilled auditory processing have lagged behind our understanding of the visual system. As detailed above, Willis himself made several contributions to the understanding of the peripheral auditory system. In contrast, central mechanisms involving cortical processing of auditory stimuli are still little understood. There is yet to be a full account of many aspects of auditory behaviour in humans, including such things as perfect or absolute pitch. Both congenital and acquired amusia have become active areas of research interest in recent decades.[47–57]

Willis' interest in the relation between music and the passions is also mirrored in current research. The changes in neurophysiological activity brought about by listening to particular types of music (such as Mozart) is currently being studied with respect to both beneficial effects in healthy individuals and in reducing epileptiform activity.[58]

With respect to Willis' observations on individual differences in the perception of music, present-day research suggests they may be generated by experience in early life: several studies have documented that musical experience modifies the tonotopic organisation of the primary auditory cortex and that the earlier in life this musical experience begins the more difference there is to non-musicians.[59] Other researchers have identified asymmetries in the *planum temporale* in musicians versus non-musicians.[60]

An early, but little cited, case study of an individual with difficulty in musically processing was published by Grant Allen in 1878.[61] He likened what he termed 'note deafness' to the work of Dalton on colour-blindness. Allen reported the case of a 30-year-old man with a solid education and without neurological lesion, who suffers from a severe musical handicap. The man was unable to discriminate the pitch of two successive tones, failed to recognise familiar melodies and could not carry a tune. He exhibited an overall indifference towards music. Yet, the musical defect could not be explained by a lack of exposure to music, since the man had received musical lessons during childhood. From the observations Willis makes on individuals lacking musical talent, it seems he would have been unsurprised by this case description.

Conclusion

In this chapter, I have considered a passage in Thomas Willis' *Cerebri Anatome* (1664) which has hitherto received little attention by scholars of seventeenth century neurology. Willis uniquely considers the variation in musical talent with respect to brain function. He discusses individual variation in musical ability and its development; the role of perceptual and expressive aspects of music production; and the independence of this ability from general intellect or memory. There is also a discussion of the difference between humans and animals with respect to such aspects of auditory processing, and the effect of music on the passions and the soul.

Consideration of the conceptual landscape of Reformation England leads to an understanding of why Willis devoted a section of *Cerebri Anatome* to an exposition on musical talent. It should be pointed out that there is no equivalent treatment of art or dance or mathematical ability. I suggest that the reason for Willis' particular interest in the neurophysiological underpinnings of music is that he viewed music to be a real part of the human rational soul and its instantiation in the anatomy of the brain through the sensitive soul. In Willis' world, there was a strong connection between music and spirituality with a direct action on the body from the soul through religious ecstasy.[62]

Although appreciation of these issues is gained by considering the broader intellectual context and contribution of those in his social network, it can not be denied that throughout his writings there is evidence of highly original thought and a unique ability to synthesise evidence from a diverse set of observations which led to his major innovations in neurology. Keele asserts, 'Willis loved speculation for its own sake. He believed that, given certain data as a basis, truth could be evolved therefrom by the process of reasoning.'[5] Willis' consideration of the neurological basis of musical talent is an example of the most fruitful type of speculation.

Acknowledgements

Dr. David Cram, Oxford University, initiated this project by bringing the passage by Willis on the music faculty to my attention. I am grateful to him for the many discussions and exchanges we have had over the years which have been instrumental to the development of this paper. I would also like to thank Prof. Robert J. Frank, Jr. UCLA for the invaluable discussions we had on this project. Any remaining errors or limitations in this paper are my own.

References

1. Eadie, M. J. (2003). *J Clin Neurosci* **10**: 14–29.
2. Eadie, M. J. (2003). *J Clin Neurosci* **10**: 146–157.
3. Hierons, R. (1966). *J Neurol Sci* **4**: 1–13.
4. Isler, H. (1968). *Thomas Willis 1621–1675: Doctor and Scientist.* New York: Hafner.
5. Keele, K. (1967). *Medical History* **11**: 194–200.
6. Martensen, R. L. (2004). *The Brain Takes Shape.* Oxford: Oxford University Press.
7. Meyer, A. and Hierons, R. (1965). *Med Hist* **9**: 142–155.
8. Meyer, A. and Hierons, R. (1965). *Med Hist* **9**: 1–15.
9. Molnar, Z. (2004). *Nat Rev Neurosci* **5**: 329–335.
10. Zimmer, C. (2004). *Soul Made Flesh: The Discovery of the Brain and How it Changed the World.* Portsmouth, NH: Heinemann.
11. Feindel, W. (1964). *Thomas Willis: The Anatomy of the Brain and Nerves.* Montréal: McGill University Press.
12. Dewhurst, K. (1964). *Proc R Soc Med* **57**: 682–687.
13. Dewhurst, K. (1964). *Thomas Willis as a Physician.* Los Angeles: William Andrews Clark Memorial Library, University of California.
14. Feindel, W. (1995). In Awad, I. A. (Ed.), *Philosophy of Neurological Surgery,* pp. 49–66. Thieme.
15. Willis, T. (1660). *Diatribæ duæ medico-philosophicæ, quarum prior agit de fermentatione ... altera de febribus. ... His accessit dissertatio epistolica de urinis. Editio secunda, ab authore ... locupletata,* p. 12. Londini.
16. Allan, F. N. (1953). *Diabetes* **2**: 74–77.
17. Toyokura, Y. (1973). *Nippon Rinsho* **31**: 241–246.
18. Alajouanine, T. and Bourguignon, A. (1954). *Presse Med* **62**: 519–520.
19. Aronson, S. M. (2007). *Med Health R I* **90**: 203.
20. Nicholls, A. (1941). *Can Med Assoc J* **44**: 66–71.
21. Hierons, R. A. (1964). Myer. *Proc R Soc Med* **57**: 687–692.
22. Willis, T. (1664). *Cerebri anatome, cui accessit nervorum descriptio et usus,* p. 4. Londini.
23. Willis, T. (1667). *Pathologiae cerebri et nervosi generis specimen. In quo agitur de morbis convulsivis et de scorbuto,* p. 4. Oxonii.

24. Willis, T. (1672). *De Anima brutorum quæ hominis vitalis ac sensitiva est, exercitationes duæ: Prior physiologica ... altera pathologica, etc.* 2 pt, p. 8. Londini.
25. Dewhurst, K. (1980). *Thomas Willis's Oxford lectures.* Oxford: Sandford Publishing.
26. Locke, J. (1690). *An Essay Concerning Human Understanding.* London.
27. Shaw, J. B. (1992). *Clinical Anatomy* **5**: 466–484.
28. Knoeff, R. (2004). *J Hist Med Allied Sci* **59**: 413–440.
29. Willis, T. Feindel, W. and Pordage, S. (1965). *The Anatomy of the Brain and Nerves.* Montréal: McGill University Press, printed in U.S.A. p fol.
30. O'Connor, J. P. (2003). *J R Soc Med* **96**: 139–143.
31. Martensen, R. L. (2004). In *Dictionary of National Biography.* Oxford: Oxford University Press.
32. Cunningham, A. (2002). *Stud Hist Phil Biol Biomed Sci* **33**: 631–666.
33. Kassler, J. C. (1984). *History of Science* **22**: 59–62.
34. Gouk, P. (2007). *Senses and Society* **2**: 303–328.
35. Gouk, P. (2005). In Pelling, M. and Mandelbrote, S. (Eds.), *The Practice of Reform in Health, Medicine and Science, 1500–2000: Essays for Charles Webster.* Aldershot: Ashgate.
36. Kassler, J. C. (1995). *Inner Music; Hobbes, Hooke and North on Internal Character.* London: Athlone Press.
37. Smith, C. U. M. (2004). In Rose, F. C. (Ed.), *Neurology and the Arts*, pp. 191–206. London: Imperial College Press.
38. Harley, J. (1956). *The Musical Times* **97**: 191–192.
39. Salmon, T. (1672). *A Vindication of an Essay to the Advancement of Musick, from Mr M. Lock's Observations, by Enquiring Into the Real Nature and Most Convenient Practise of that Science. [With a letter to T. Salmon on the same subject subscribed N. E.] Few MS. notes.* London, p. 8.
40. Frank, R. G. (1980). *Harvey and the Oxford Physiologists.* Los Angeles: UCLA Press.
41. Gall, F. J. (1835). *On the Functions of the Brain and of Each of its Parts: With Observations on the Possibility of Determining the Instincts, Propensities, and Talents ... of Men and Animals, by the Configuration of the Brain and Head.* Translated from the French by W. Lewis.
42. Combe, G. (1836). *Elements of Phrenology.* Edinburgh: MacLachan & Stewart.
43. Henson, R. A. (1977). In Critchley, M. and Henson, R. A. (Eds.), *Music and the Brain*, pp. 3–21. London: Heinemann Medical Books.
44. Seashore, C. E. (1919). *The Psychology of Musical Talent.* Boston: Silver Burdett Company.
45. Milner, B. (1962). In Mountcastle, V. B. (Ed.), *Interhemispheric Relations and Cerebral Dominance*, pp. 177–195. Baltimore: Johns Hopkins Press.
46. Critchley, M. and Henson, R. A. (Eds.), (1977). *Music and the Brain.* London: Heinemann Medical Books.

47. Stewart, L. von Kriegstein, K. Warren, J. D. and Griffiths, T. D. (2006). *Brain* **129**: 2533–2553.
48. Ayotte, J. Peretz, I. and Hyde, K. (2002). *Brain* **125**: 238–251.
49. Zatorre, R. J. and Gandour, J. T. (2008). *Philos Trans R Soc Lond B Biol Sci* **363**: 1087–1104.
50. Bautista, R. E. and Ciampetti, M. Z. (2003). *Epilepsia* **44**: 466–467.
51. Botez, M. I. and Wertheim, N. (1959). *Brain* **82**: 186–202.
52. Di Pietro, M. Laganaro, M. Leemann, B. and Schnider, A. (2004). *Neuropsychologia* **42**: 868–877.
53. Mavlov, L. (1980). *Cortex* **16**: 331–338.
54. McFarland, H. R. and Fortin, D. (1982). *Arch Neurol* **39**: 725–727.
55. Bella, S. D. and Peretz, I. (2003). *Ann NY Acad Sci* **999**: 166–169.
56. Hyde, K. L. Lerch, J. P. Zatorre, R. J. Griffiths, T. D. Evans, A. C. and I. Peretz. (2007). *J Neurosci* **27**: 13028–13032.
57. Mandell, J. Schulze, K. and Schlaug, G. (2007). *Restor Neurol Neurosci* **25**: 323–334.
58. Hughes, J. R. and Fino, J. J. In F. C. Rose (Ed.), (2004). *Neurology of the Arts*, pp. 237–273. London: Imperial College Press.
59. Platel, H. Eustache, F. and Baron, J.-C. In Rose F. C. (Ed.), *Neurology and the Arts*, pp. 175–189. London: Imperial College Press.
60. Charness, M. E. and Schlaug, G. (2004). In Rose F. C. (Ed.), *Neurology and the Arts*, pp. 153–173. London: Imperial College Press.
61. Allen, G. (1878). *Mind* **10**: 157–167.
62. Finney, G. L. (1947). *Journal of the History of Ideas* **8**: 153–186.

Chapter 11

Musicogenic Epilepsy

Jock Murray

Musicogenic epilepsy was described by Critchley in 1937 and to date almost 100 cases have been described in the literature. The nature of the musical stimulus is specific to the patient and can be a tone, any music, a specific singer, song or musical instrument, a musical form or musical group. Musicogenic epilepsy has been classified as a form of reflex epilepsy, but it differs in significant ways from most of the reflex epilepsies. The epileptic focus is usually in the temporal lobes and more often on the right side, but it can be on either side or bilateral and in many probably involves complex interconnecting networks. Usually the patients have unprovoked seizures as well as those precipitated by music. If a unilateral focus can be determined these patients may be candidates for neurosurgical resection.

Introduction

The profound effect of music on the brain and emotions has long been noted by poets and writers. More recently there has been interest in the remarkable cases where music precipitated epileptic seizures, but even here there are some earlier instances.[1] Scallinger in the sixteenth century described seizures precipitated by listening to the lyre.[2] The Chinese poet Kung Tzu Chen said he became 'absent minded' and 'fell sick' whenever he heard the sound of a street vendor's flute.[3] Shakespeare in *The Merchant of Venice* spoke of 'others, when the bagpipe sings i' the nose, Cannot contain their urine.' (Act IV, Scene I).[4] Joan of Arc was said to have her visions precipitated by the sound of church bells. There have also been reports of musical hallucinations, humming and singing as part of epileptic seizures. Another well-documented early case was the music critic Nikonov, presented to the 1884 St. Petersburg meeting of the Russian Psychiatric Association by Merzheevsky[5], and later published by von Bechterev.[6] The music critic's tendency to have seizures to music was further described by him in a pamphlet *Fear of Music* and he mentions running down a back street covering his ears to escape the sound of a military band.[1]

Critchley in 1937 used the term musicogenic epilepsy to describe seizures induced by some aspect of music.[7] He first noted the phenomenon in a hospital ward maid he was caring for, and she was found to have seizures to Tchaikovsky's *Valse des Fleurs* played by the Berlin State Opera Orchestra. In his paper, Critchley analysed 20 cases, 11 of his own and nine from the literature, and noted that the seizures to music occurred somewhat later in age than idiopathic seizures, but equally in men and women. The musical stimulus varied and was characteristic for the individual patient. There was also variability in the patient's musical sophistication. The usual type of seizure was partial complex or grand mal seizures, but with a good prognosis.

Many reports note an emotional component of the music or sound that forms the stimulus. The seizures are of simple or complex partial type, with interictal and ictal epileptiform activity recorded from either temporal region,[8] usually the right.[3] Most patients also have spontaneous seizures and reflex seizures often begin over a year, after the onset of spontaneous attacks.[3]

The Quandary of Classification

Musicogenic epilepsy is classed as a form of reflex epilepsy, a term put forward by Merlis in 1974 to cover the array of unusual sensory stimuli that could provoke seizures.[9] The concepts and the terms musicogenic epilepsy and reflex epilepsy have engendered discussion and controversy. Music is defined by the OED as 'sounds in melodic or harmonic combination', but some cases published as musicogenic epilepsy have been to a single sound, and some series of cases include machinery sounds that would not normally be regarded as musical.

The most common reflex epilepsy is photosensitive epilepsy, accounting for over half the cases in the Western world, but eating epilepsy and hot water epilepsy are the most common forms in India.[10] Reflex epilepsy includes eating epilepsy, reading epilepsy, television epilepsy; hot water epilepsy; startle epilepsy; movement induced epilepsy; somatosensory forms (touching, tapping the skin); tooth-brushing seizures, seizures induced by sexual activity; and epilepsy precipitated by non-linguistic cerebral functions such as doing arithmetic, card games, checkers, working the Rubik's Cube, or thinking certain thoughts.

Animal models for reflex epilepsy have been studied since 1929 and include applying strychnine to the visual cortex of dogs to study hyperexcitability and photic seizures. Another approach has been to study animals that are genetically susceptible, such as photosensitive epileptic chickens, rodents susceptible to sound-induced convulsions, the E1 mouse sensitive to vestibular stimulation, and the Mongolian gerbil sensitive to a variety of stimuli. The animal most closely resembling the reflex seizures and EEG findings of humans is the baboon Papio papio.[11]

Physicians have been aware of reflex seizures for many centuries. In biblical times, a slave offered for sale would be placed facing a potter's wheel with a shiny vessel spinning to see if they had seizures, attempting to evoke photosensitive epilepsy. In fact, the list of precipitating sensations to be avoided by epileptics in Roman times up to the middle ages was very long and indicated recognition of various factors that were likely to bring on a seizure. Because the seizures evoked were similar to other seizures, reflex epilepsies were named for the precipitating factor such as light or music. Marshall Hall in 1833 discussed the nature of epilepsy precipitated by various stimuli.[12] Attempts to classify reflex epilepsies within the overall classification of epilepsy were made by Forster (1972) and the Commission on Classification and

Terminology of the International League Against Epilepsy (1985, 1989).[13,14] More recently, Engel defined reflex epilepsy syndromes as those '...in which all epileptic seizures are precipitated by sensory stimuli. Reflex seizures that occur in focal and generalised epilepsy syndromes that are also associated with spontaneous seizures are listed as seizure types'.[15] Thus, few reflex epilepsy syndromes are recognised as separate entities.

Complex reflex epilepsies are seizures precipitated by elaborate stimuli whose specific pattern is the determining factor. The seizures are precipitated by integration of higher cortical function, and even by the anticipation of the specific stimulus. The Commission on Classification and Terminology of the International League against Epilepsy (1989) classify the reflex epilepsies under both localisation-related epilepsy, and generalised categories, but Senanayake argues for them to be classified only as location related epilepsy.[10] Merlis divided the syndrome into two types: the more common group with an affective associative response necessary for the precipitation of the seizure, and a smaller group in which a simple sound brought on the seizure.[9] Intermediate cases also occurred.

Others object to the term 'reflex' as no recognised reflex is involved.[9] As well, many of the patients can develop methods of aborting the onset of a seizure when they perceive the music is producing effects they identify with the onset of a seizure. The term 'sensory precipitation epilepsy' was suggested instead of reflex epilepsy, but doesn't cover those who have seizures just thinking certain thoughts or performing mental mathematics, as no external sensory input is involved. The term reflex epilepsy continues to be used as it pictures a stimulus and an epileptic response.

As pointed out by Gastaut, most cases of musicogenic epilepsy do not fall into the strict definition of reflex epilepsy because of the variable latency, as there is often a prolonged time of musical stimulus before the seizure, and many developed techniques to abort the seizure.[3,16] Avanzini comments that the term musicogenic epilepsy is problematic as the patients usually have spontaneous seizures as well and it is inappropriate to use the occasional precipitating factor as the criterion for defining their syndrome.[17] It is also unclear if the concept of musicogenic epilepsy should include seizures to a single sound, to non-musical sounds or just thinking about music.

Vizioli investigated a case who had seizures with secondary generalisation precipitated by any type of music. He did not agree with classifying such cases as reflex epilepsy as too many structures are involved and the stimulus cannot be reduced to crude perception of rhythm and melodies.[18] It is considered that three levels of integration are involved in music: a sensory level (the lowest one) an emotional and waking level (the second one) and finally a highest level, allowing aesthetic appreciation. The author concluded that musicogenic epilepsy has to do mostly with the limbic brain, that is, with the second level of integration.

Review of Cases

Cases in the literature have been periodically reviewed[1,3,4,7,17,19] and there are now almost 100 published cases, and others are known by personal reports.

In general, about 5% of seizures seem to have some precipitating event or stimulus but ranges of 2.6% to 18.2% have been reported.[10] Musicogenic epilepsy is a rare epilepsy syndrome in which seizures are precipitated by music with a population prevalence of 1:10,000,000. Because the seizures are evoked by specific stimuli, musicogenic seizures represent a valuable tool in the investigation of the mechanisms of seizure initiation and of the human brain's processing of music.[20]

Wieser *et al.* reviewed the 76 cases in the literature up to 1995 and added seven of their own for a total of 83 cases. They found that 17% had seizures exclusively to music while in 78% it was the only precipitating factor in their epilepsy. Information was missing from many reports but in 36 cases where it was clear, seizures occurred that were both precipitated by music and without music. In 31%, the provoked and unprovoked seizures developed at the same time. In 58% where the information was known, unprovoked seizures to music preceded the music-provoked seizures, and in 11%, the music-precipitated seizures developed first. Where the focus was clearly defined, 75% were in the temporal lobe, and more often on the right (61% vs 39%).

Most had frequent seizures and had a mean age of onset of 27.7 years (±12.5 years).There was a slight preponderance of women (54% vs 46%). In 78% music was the only precipitating stimulus. For 4% it was tones; 14% to all kinds of music and sounds; and in the remaining 34 there was stimulus with a description of the features. In 17 it was specific and in 17 nonspecific. In 15, the affective content of the sound or music was important. When specific instruments were the stimulus, they were piano and organ 11 times, jazz instruments twice, string instruments once and wind instruments once. There is often a long latency period up to several minutes before the stimulus precipitated the seizure and during this time there is a buildup of emotional feeling. They noted that in 35 patients where there was information, four were professional musicians (11%); 11 were talented amateurs (31%); five were 'music fans' (14%); seven had above average interest in music (20%) and only eight (23%) said they were not especially interested in music, suggesting a greater involvement or interest in music predisposed to musicogenic epilepsy. They recommended that future cases have more detailed analysis of the patient's involvement, interest and emotional response to music by a musical test battery. Avanzini reviewed the 83 cases of Wieser and added four more from the literature and a personal case for a total of 88 cases (17). Individual cases are often so interesting and idiosyncratic that single cases merit publication in an age when editors usually treat case reports with disdain.

Mechanisms in Musicogenic Epilepsy

The pathophysiology of musicogenic epilepsy is unclear. A conditioned response has been suggested,[21] but this view is not now generally accepted.[22] Penfield noted that electrical brain stimulation that evoked musical hallucinations, often in very vivid experiential form, were most often on the right side of the brain.[23] As pointed out by Wieser,[3] the temporal lobes are crucial in most phenomena related to music, particularly the right temporal lobe and acoustic cortex, in pitch discrimination; musical hallucinations during seizures and in musicogenic seizures. Pleasurable experience with music relates to the mesiolimbic temporal lobe structures. Wieser *et al.*[3] suggest there is a right temporal predominance and documented right anterior and mesial hyperperfusion during ictal SPECT as did Genc *et al.*[24] The case of Lin[25] in a six-month-old child showed hyperperfusion on the left as did the case originally published by Brien and Murray in 1984,[26] and this case has been studied recently with SPECT by Sadler.[27] Chronic temporal lobe depth electrode studies in epileptic subjects without musicogenic epilepsy suggest different lateralisation for different components of a musical stimulus.[28] Creutzfeldt and Ojemann confirmed that musical stimuli may have widespread effects on neuronal activity in the temporal lobes extending well beyond the more restricted primary auditory area that different components of music have different effects possibly with specialised lateralisation and localisation, and that the effects of music are different from those of speech.[29] PET studies show predominant involvement of right hemisphere structures in networks involved in processing musical information, extending well beyond the classical auditory cortex of Heschl's gyrus.[30] Studies in subjects with musical hallucinations show that the primary auditory cortex is not 'a sufficient substrate for higher-order pattern perception'.[31] A lot of musical functions such as tonal patterns, some aspects of pitch, processing complex harmonic structure and timbre discrimination are more related to right temporal lobe. The primate auditory cortex is considered to consist of a central core of primary cortex which receives thalamic projections and which is linked to several 'belt' areas. Primary cortex has multiple tonotopically organised sections and is especially sensitive to pure tones. Belt regions show more sensitivity to complex stimuli and are less tonotopically organised.[30] Zifkin and Zatorre also note that more complex musical processing tasks activate more cortical and subcortical territory bilaterally but with right hemisphere predominance.[32] Thus, hyperexcitable cortical areas could be stimulated to different degrees and extents by different musical stimuli in patients sensitive to musical triggers. Gloor suggested that responses to limbic stimulation in epileptic subjects depend on widespread neuronal matrices linked through connections which have become strengthened through repeated use, of interest in considering the delay from seizure onset to the development of sensitivity to music.[33] It has questioned whether music is a unique stimulus requiring specialised neural mechanisms, or whether music is just one of the many complex sounds processed by the auditory nervous system.[34] Studies of amusia after stroke with no other auditory deficit suggest the former. It is likely there are neural mechanisms dedicated to processing musical features that are functionally separate enough to be affected by bilateral temporal lesions. These complex neural systems would also be triggered in musicogenic epilepsy.[34] As mentioned, both temporal lobes are probably involved, even though more of the cases seem to have a focus on the right side. Others, however, had a left-sided focus and some bilateral. The notion of cerebral dominance for music has been replaced by a concept of modular but interconnected networks that have wide bilateral localisation in the brain and that are molded both by genetics and experience.

Types of Music/Sound

Critchley listed a number of precipitating types of music in his cases, including piano played by the patient, bass notes on brass instruments, rhythms, tap dancing, and certain songs and orchestral pieces.[7] He also noted one patient had only to think about music, and another had to consciously concentrate on the music for it to precipitate a seizure.

Individual cases of musicogenic epilepsy often showed a specificity for the stimulus peculiar to each patient. Interesting examples were certain frequency bands of church bells, singers who used their throats to project sound (microphone singers), a certain hymn or anthem, a particular musical instrument such as the piano or organ, certain types of music (Thai female rock and roll singers), certain composers or specific pieces of music. In musicogenic epilepsy the stimulus is not consistent and published reports have included various musical instruments, church bells, singers, certain composers, and certain pieces of music (the French anthem Marseillaise). So it is interesting to study the nature of the specific stimulus in each patient but there is little similarity in the stimuli except that these have all been grouped together as they all are related to some sounds regarded as musical.[1]

There are cases of seizures precipitated by organ and hillbilly music,[19] listening to the specific hymn, *Now Thank We All Our God*,[35] to certain strong beat or rhythms, the rhythms of the group Alabama,[20] music of The Beatles,[25] classical music, various instruments and sometimes a simple tone.[1] For some, loudness is important; for others concentrating on the music rather than having it in the background makes a difference. The stimulus can be very specific as shown by Poskanzer whose patient had seizures to church bells in the frequency between 290 and 1120 c/s or narrower band.[36] The impact of the bell sound seemed important as a tape recording of the bell sound played backwards did not precipitate the seizure. The left temporal lobe was the area of the seizure development on the EEG. The case of Brien and Murray had seizures regularly precipitated by singers who used their mouth and throat to produce the voice (microphone singers), but did not have seizures to trained singers and choirs or to instrumental music.[26,27]

The case of Newman and Saunders shows the complexity of the music form to some patients. A 39-year-old woman had seizures when listening to light popular music. On testing various musical forms, they could not induce seizures with photic stimulation, hyperventilation or by listening to piano music of Handel's *Messiah*, or Beethoven's *Choral Symphony* or Gilbert and Sullivan, but she would consistently have a seizure with a focus in the left temporal area when listening to a song, *I Think I'm Gonna Fall In Love With You* by a singing group called The Dooleys.[37] When they electronically separated the rhythm background, the melody and alternately separated the high and the low frequencies, none of these played separately precipitated a seizure, but when the completed song was again played she had a generalised major convulsion.

Many cases report an emotional component to the seizure onset and Daly and Barry presented three cases emphasising the strong emotional and psychological overtones.[38]

Investigations

Shaw and Hill first recorded the EEG features in a case in 1947.[21] In some cases, the interictal EEG is normal, but foci on right, left and bilateral temporal lobes have been reported. In some there was a delay of minutes listening to the music before the epileptic activity appeared on the EEG. Joynt *et al.* found a delay of 441 and 897 seconds after Bach's organ music was played.[19]

To localise the neural correlates of musicogenic epilepsy, Cho in 2007 used subtraction ictal SPECT coregistered with MRI (SISCOM) and[18] F-fluorodeoxy glucose positron emission tomography (FDG-PET) in a woman who had suffered from frequent musicogenic seizures.[39] She had complex partial seizures consisting of palpitation and an unpleasant feeling, which were followed by staring and oroalimentary automatisms. Ictal EEG showed rhythmic theta waves originated from the right temporal lobe, and SISCOM showed ictal hyperperfusion on right insula, amygdala, hippocampal head, and anterior temporal lobe, whereas interictal FDG-PET showed interictal hypometabolism in the same brain regions, suggesting dysfunction and abnormal activation of right temporo-limbic structures related to an emotional response to music. Others have used SPECT[3,25,27] functional MRI (fMRI).[40] Wieser *et al.* used SPECT in a 32-year-old woman with musicogenic seizures precipitated by Italian songs and found in the preictal phase a discrete asymmetry of the temporal lobes with a slight hypometabolism of the right temporal and to a lesser degree of the right frontal lobe.[3] In the ictal phase, there was a right neocortical posterior temporal hypometabolism at the height of the seizure.

The heterogeneity of the syndrome is also evidenced by the various epileptic foci found in the cases. For instance, Tayah and colleagues showed that in three patients with intractable musicogenic seizures, different foci were found.[20] They were initially studied by non-invasive methods including, in one patient, ictal magnetoencephalography (MEG) and magnetic resonance spectroscopy (MRS). The ictal-onset zones in these patients were then localised with intracranial EEG monitoring, and the outcomes of the two patients who underwent epilepsy surgery were determined. In the first patient, the musicogenic seizures were localised to the right lateral temporal lobe, in the second to the right mesial temporal lobe, and the third arose independently from both mesial temporal lobes. The first two patients with unilateral foci underwent resective epilepsy surgery and are seizure free. They concluded that musicogenic epilepsy is a heterogeneous syndrome with seizures that can arise from multiple temporal lobe foci, and if they have unilateral ictal onset are good candidates for resective epileptic surgery.

Genc *et al.* reported the case of a 48-year-old woman whose complex partial seizures began at the age of 32.[24] Cranial computed tomography, magnetic resonance imaging (MRI) and interictal SPECT showed no abnormality. Interictal EEG showed paroxysmal bitemporal sharp wave discharges predominant on the right side. Epileptic activity began after four to five minutes and showed a combination of high voltage sharp and slow sharp waves and spikes that originated from the right temporal leads and then became generalised. Ictal SPECT study demonstrated a right anterior and mesial temporal hyperperfusion. They concluded that this supports the dominant role of the right temporal lobe and the possible relation of mesial temporal structures to the affective content of music in musicogenic epilepsy.

Although the onset of musicogenic epilepsy is somewhat later than in idiopathic epilepsy, the case of Lin was in a six-month-old infant,[25] a child noted to have seizures to loud music, particularly The Beatles. The interictal EEG was normal but the ictal EEG showed spikes throughout the left temporal area. The MRI was normal but single-photon emission computed tomography of the brain revealed hypoperfusion of the left temporal area.

The case published by Brien and Murray in 1984 was more recently studied by Dr. Mark Sadler (personal communication) and found to have a focal discharge in the left temporal lobe, with hyperperfusion of the left temporal lobe on SPECT.[26,27]

PET studies in normal individuals showed increased blood flow in the right temporal lobe in normals listening to music and suggests functional networks dedicated to processing pitch and perhaps other aspects of music.[34] Listening to more complex musical forms showed activation of both temporal lobes, but more on the right. Shibata *et al.* studied a 49-year-old Japanese male who had musicogenic partial and simple seizures using the dipole tracing method (DTM), which showed equivalent current dipoles in the posterior transverse temporal gyrus.[41]

Conclusion

Musicogenic epilepsy is a rare form of reflex epilepsy in which some aspect of music is a specific triggering stimulus for at least some of the patient's seizures. There is localisation to the temporal lobes in most, more often the right, and if a unilateral temporal lobe focus is found, the person may be a candidate for epilepsy resection surgery.

References

1. Kaplan, P. W. (2003). Musicogenic epilepsy and epileptic music: A seizure's song. *Epilepsy and Behavior* **4**: 464–473.
2. Scallinger, J. J. (1605). *Le Loirier's Treatise of Spectres.*
3. Wieser, H. G., Hungerbühler, H., Siegel, A. M. and Buck, A. (1997). Musicogenic epilepsy: Review of the literature and case report with ictal single photon emission computed tomography. *Epilepsia* **38**: 200–207.
4. Critchley, M. (1977). Musicogenic epilepsy. (I) The beginnings. In Critchley, M. and Henson, R. A. (Eds.), *Music and the Brain*, pp. 344–353. London: William Heinemann.
5. Merzheevsky, I. P. (1965). *Slochai epliepsi pripedki kotorsi vizibayootsya kekotopemi misikelnimi tonomi.* Minutes of the meeting of the St. Petersbourg Soc Psichiatr (quoted by Titeca).
6. Von Bechterev, V. O. (1914). Reflectornoi epliepsi pod oliyanienevyoskovich razdrazlenye. *Obozrenie Psychiatr* **19**: 513–520.
7. Critchley, M. (1937). Musicogenic epilepsy. *Brain* **60**: 13–27.
8. Scott, D. F. (1977). Musicogenic epilepsy. (2) The later story: Its relation to auditory hallucinatory phenomena. In Critchley, M. and Henson, R. A. (Eds.), *Music and the Brain*, pp. 354–364. London: William Heinemann.
9. Merlis, J. K. (1974). Reflex epilepsy. In Vinken, P. J. and Bruyn, G. W. (Eds.), *Handbook of Clinical Neurology*, Vol. 15, pp. 452–456. Amsterdam: Elsevier.
10. Senanayake, N. (2000). Epilepsies with seizures precipitated by specific modes of activation. In Vinken, P. J. and Bruyn, G. W. (Eds.), *Handbook of Clinical Neurology*, Vol. 29, revised series. Amsterdam: Elsevier.
11. Menini, C. and Silva-Barrat, C. (1998). The photosensitive epilepsy of the baboon. A model of generalized reflex epilepsy. In Zifkin, B. G. *et al.* (Eds.), *Reflex Epilepsies and Reflex Seizures. Advances in Neurology*, Vol. 75, pp. 29–47. Philadelphia: Lippincott-Raven Press.

12. Hall, M. (1833). Epilepsy. *Philosophical Transactions* **123**: 635.
13. Commission on Classification and Terminology of the International League Against Epilepsy. (1985). Proposal for classification of epilepsies and epileptic syndromes. *Epilepsia* **26**: 268–278.
14. International League Against Epilepsy. International Classification of Functioning and Disability. Beta-2 draft. Full version. Geneva. World Health Organization. July 1999.
15. Engel, J. Jr. (2001). A proposed diagnostic scheme for people with epileptic seizures and epilepsy: Report of the ILAE Task Force on Classification and Terminology. *Epilepsia* **42**: 796–803.
16. Gastaut, H. and Poirier, F. (1964). Experimental, or "reflex", induction of seizures. Report of a case of abdominal (enteric) epilepsy. *Epilepsia* **5**: 256–270.
17. Avanzini, G. (2003). Musicogenic seizures. *Annals NY Acad Sci* **105**: 95–102.
18. Vizioli, R. (1962). The problem of human reflex epilepsy and the possible role of masked epileptic factors. *Epilepsia* **3**: 293–302.
19. Joynt, R. J., Green, D. and Green, R. (1962). Musicogenic epilepsy. *JAMA* **179**: 501–504.
20. Tayah, T. F., Abou-Khalil, B., Gilliam, F. G., Kowlton, R. C., Wushensky, C. A. and Gallagher, M. J. (2006). Musicogenic seizures can arise from multiple temporal lobe foci: Intracranial EEG analyses of three patients. *Epilepsia* **47**: 1402–1406.
21. Shaw, D. and Hill, D. (1947). A case of musicogenic epilepsy. *J Neurol Neurosurg Psychiatry* **10**: 107–117.
22. Forster, F. M. (1972). The classification and conditioning treatment of the reflex epilepsies. *Int J Neurol* **9**: 73–86.
23. Penfield, W. and Perot, P. (1963). The brain's record of auditory and visual experience: A final summary and discussion. *Brain* **86**: 596–696.
24. Genc, B. O., Genc, E., Tastekin, G. and Iihan, N. (2001). Musicogenic epilepsy with ictal single photon emission computed tomography (SPECT): Could these cases contribute to our knowledge of music processing? *Eur J Neurol* **8**: 191–194.
25. Lin, K.-L., Wang, H.-S. and Kao, P.-F. (2003). A young infant with musicogenic epilepsy. *Pediatric Neurology* **28**: 379–381.
26. Brien, S. E. and Murray, T. J. (1984). Musicogenic epilepsy, *CMAJ* **131**: 1255–1258.
27. Sadler, M. (personal communication).
28. Wieser, H. G. and Mazzola, G. (1986). Musical consonances and dissonances: Are they distinguished independently by the right and left hippocampi? *Neuropsychologia* **24**: 805–812.
29. Creutzfeldt, O. and Ojemann G. (1989). Neuronal activity in the human lateral temporal lobe. III. Activity changes during music. *Exp Brain Res* **77**: 490–498.
30. Johnsrude, I. S., Giraud, A. L. and Frackowiak, R. S. J. (2002). Functional imaging of the auditory system: The use of positron emission tomography. *Audiol Neurootol* **7**: 251–276.

31. Griffiths, T. D. (2000). Musical hallucinosis in acquired deafness. Phenomenology and brain substrate. *Brain* **123**: 2065–2076.
32. Zifkin, B. G. and Zatorre, R. (1998). Musicogenic epilepsy. In Zifkin, B. G., Andermann, F., Beaumanoir, A. and Rowan, A. J. (Eds.), *Reflex Epilepsies and Reflex Seizures. Advances in Neurology*, Vol. 75, pp. 273–281. Philadelphia: Lippincott-Raven.
33. Gloor, P. (1990). Experiential phenomena of temporal lobe epilepsy. Facts and hypotheses. *Brain* **113**: 1673–1694.
34. Zifkin, B. G. and Kasteleijn-Nolst Trenité, D. (2000). Reflex epilepsy and reflex seizures of the visual system: A clinical review. *Epileptic Disord* **2**: 129–136.
35. Sutherling, W. W., Hershman, L. M., Miller, J. Q. and Lee, S. I. (1980). Seizures induced by playing music. *Neurology* **30**: 1001–1004.
36. Poskanzer, D. C., Brown, A. E. and Miller, E. (1962). Muscicogenic epilepsy caused by a discrete frequency band of church bells. *Brain* **85**: 77–92.
37. Newman, P. and Saunders, M. (1980). A unique case of musicogenic epilepsy. *Arch Neurol* **37**: 244–245.
38. Daly, D. D. and Barry, M. J. (1957). Musicogenic epilepsy: Report on three cases. *Psychosom Med* **19**: 399–408.
39. Cho, J. W., Seo, D. W., Joo, E. Y., Tae, W. S., Lee, J. and Hong, S. B. (2007). Neural correlates of musicogenic epilepsy: SISCOM and SDG-PET. *Epilepsy Research* 77: 169–173.
40. Mórocz, I. Á., Karni, A., Haut, S., Lantos, G. and Liu, O. (2003). fMRI of triggible aurae in musicogenic epilepsy. *Neurology* **60**: 705–709.
41. Shibata, N., Kubota, F. and Kikuchi, S. (2006). The origin of the focal spike in musicogenic epilepsy. *Epileptic Disorders* **8**: 131–135.

Chapter 12

Musical Hallucinations

Stefan Evers

Musical hallucinations have been described in different neurological and psychiatric patients, but the pathophysiological background is not understood. Analysing the published cases, five sub-groups can be separated according to aetiology (hypacusis; psychiatric disorders; focal brain lesions; epilepsy; intoxication). There is a female preponderance of about 70%. Musical hallucinations occur over the age of 60, but those with focal brain lesions are significantly younger than other aetiological groups. Hemispheric dominance seems to play no major role in the pathogenesis of musical hallucinations but hypacusis is present in the majority of patients. Anticonvulsant and antidepressive substances have been effective in the treatment of musical hallucinations in some cases. The discussion on the pathophysiology of musical hallucinations comprises theories of deafferentiation (including auditory Charles-Bonnet-syndrome), sensory auditory deprivation, parasitic memory, or spontaneous activity in a cognitive network module.

Introduction

Musical hallucinations as a particular type of auditory hallucinations are a disorder of complex sound processing in which the perception is formed by instrumental music, songs, or sounds.[1,2] They are less common than unformed acoustic hallucinations such as tinnitus and must be differentiated from what is called an earworm or a sticky tune;[3] subjects are normally aware of the hallucinatory character of their perception. Probably the first reports on musical hallucinations were published by Baillarger in 1846 and by Coleman in 1849,[1] the first scientific descriptions were given by Petazzi in 1900 and by Bryant in 1907.

In the clinical setting, musical hallucinations are predominantly a problem of older subjects with hearing impairment. Despite their musical content, these hallucinations can be very disabling and impairing for sufferers. Thus in some cases, medical treatment is warranted. In this chapter, a review on the published case reports will be given followed by a systematic classification of the different aetiologies with a discussion on the possible pathophysiological mechanisms.

Previous Reviews

Reviews on case reports with musical hallucinations have been published in recent decades.[1,2,4,5] In one review,[1] 46 cases were analysed with a female preponderance of 80% and an average age of 60 +/− 19 years. 67% of the patients were (nearly) deaf, and in 40% musical hallucinations were the only symptoms. There was a psychiatric history in 26%, mostly depression. In 39% of the whole sample, there was evidence of a brain disease such as tumor, epileptic focus, or stroke. The authors concluded that musical hallucinations are more frequent in elderly women affected by deafness or brain disease and without a history of psychiatric illness.

Keshavan *et al.*[5] reviewed 59 cases from the literature and separated them into subgroups. They described a group of musical hallucinations associated with hearing loss, a second with distinct brain disease, and a third with a psychiatric disorder. The authors suggested that deafness is the most important factor and that non-specific brain damage is less frequently associated with musical hallucinations. They also found a female preponderance with a tendency for middle or advanced age across all groups. Furthermore, they noted that psychiatric disorders can make an additional contribution and that psychosis and depression can be associated with musical hallucinations in the absence of either brain disease or deafness.

In the review published by Klostermann *et al.*,[6] in patients with hearing loss, it was concluded that musical hallucinations caused by acquired deafness are a phenomenon predominantly found in elderly women. The majority of the cases did not show any pathology of the central nervous system, EEG and CT being normal in most cases. Mostly, the melodies perceived were repetitions of music well known from former times, in particular, religious or folk songs, and pop songs known from the radio and heard during childhood or adolescence. For treatment, drugs such as neuroleptics or anticonvulsants were unsatisfying. But in some cases, the perceptions disappeared after the underlying hearing loss had been treated.

Pasquini and Cole[7] analysed 32 cases aged 65 years of age or older with absence of major focal neurological disease (e.g. tumor, epilepsy, stroke) and drug toxicity. They concluded that musical hallucinations in the elderly occurred predominantly in females who had hearing impairments, one third of whom were depressed. The average age of the cases was 78 years, 84% being female and 16% male. Seventy nine per cent had a long-standing history of hearing impairment and 36% had a psychiatric disorder, most often a major depressive episode. In 18 patients, treatment and response to treatment could be analysed. Despite the limited data, the authors had the impression that, if the patient was depressed, antidepressive treatment was effective and more so than antipsychotic treatment. In non-depressed patients, treatment was only moderately effective in controlling the musical hallucinations.

Saba and Keshavan[8] were the first to analyse the prevalence of musical hallucinations in 100 patients with schizophrenia. They found 16 patients with musical hallucinations, 56% of whom were male and 44% female. The mean age was 34 years, with 69% of patients being between 26 and 50 years of age, but the figures are not representative, since the authors only examined schizophrenic patients with acute auditory hallucinations.

Gordon[9] concluded that hearing impairment predisposes to drug-induced musical hallucinations but that there is no consistent evidence of brain damage. In his opinion, antihypertensive drugs are particularly likely to cause musical hallucinations. It is supposed that the pathophysiological mechanisms leading to musical hallucinations are the same for predisposed patients taking non-specific drugs as for healthy subjects taking hallucinogenic drugs.

In the most recent review, 132 cases were analysed after a literature search in several medical reference systems.[2] Only musical hallucinations were included; musical illusions, complex psychosis (i.e. hallucinations of more than one sensory system), or hallucinations of simple sounds were not considered. In this review, five different pathologies associated with musical hallucinations were detected, and gives the basis for the following epidemioloigcal data. It is based on the year 2004 with three additional case reports published more recently.[10,11,12] Although it was decided to divide the case reports into groups with respect to the most probable aetiology, several patients showed additional conditions making it sometimes difficult to decide which of the pathologies could be the main factor. For example, a patient suffering from mild hearing impairment for several years suddenly had musical hallucinations after having started taking a certain drug. The medication could then be responsible as being the main cause of the aberrant perceptions, with the hearing impairment regarded as a cofactor. In focal brain disease, epilepsy, and psychiatric disease, hearing impairment as a concomitant factor was found in several, but not all, cases. In some patients, general brain atrophy maintained the musical hallucinations, but by itself was never found to be a sufficient trigger. In patients with coarse definite disease, epileptic activity was sometimes found in the juxtaposed brain areas but in several cases it was difficult to decide whether it was the lesion itself or the epileptic activity caused by the lesion that provoked the musical hallucinations.

Demography of Musical Hallucinations

The age of all published cases with musical hallucinations (n = 135) ranged from 20 to 90 years with a mean of 62 +/− 19 years. 70% were female. In 27% of the cases, a localised brain lesion or a localised epileptic focus was described, but the occurrence of musical hallucinations was not dependent on the side of lesion; all patients with full data were right-handed. Generalised brain atrophy was present in another 26%. Examining all cases more closely, it was obvious that one cannot assume a single aetiology; in a more detailed analysis, the total group of musical hallucinations could be subdivided into five different sub-groups, the most important clinical and demographic data presented in Table 1.

The prevalence of musical hallucinations is unknown, but have been observed in 0.16% of a large sample in a general hospital setting[13]; in a sample of elderly subjects with audiological complaints, its prevalence has been estimated at 2.5%.[14] In a lifetime prevalence study in psychiatric patients, more than one fifth of all patients sometimes experienced musical hallucinations. The highest prevalence with 41% occurred in the subgroup of patients with obsessive-compulsive disorder.[15]

Table 1. Demographic and clinical data of 135 patients with musical hallucinations for five different subgroups.

Diagnosis	Number	Side of lesion (if applicable)	Age in years (range)	Sex	Diffuse brain atrophy
All cases	135	20 right 16 left	62 +/− 19 (20–90)	40 male 95 female	34
Hypacusis	52	4 right 3 left	71 +/− 15 (35–90)	12 male 40 female	18
Psychiatric disorder	32	1 left	51 +/− 21 (20–88)	10 male 22 female	10
Focal brain lesion	21	13 right 8 left	50 +/− 16 (23–84)	8 male 13 female	2
Epilepsy	15	5 right 5 left	60 +/− 21 (22–85)	5 male 10 female	1
Intoxication	15	—	68 +/− 13 (40–88)	5 male 10 female	4

Aetiological Factors

Hypacusis

In about 50% of all cases, hypacusis was the only detected aetiological factor. The average age of these cases was 71 +/– 15 years (range 35 to 90). There were 22% male and 77% female patients in this group. In 35%, general brain pathology was found. Furthermore, several patients were suffering from additional conditions such as psychiatric disorders, focal brain lesion, epileptic activity, and toxic metabolic influences.

Psychiatric disorders

The average age of the more than 30 published patients in whom musical hallucinations could be attributed to psychiatric disorders was 51 +/– 21 years (range 20 to 88 years). Thirty two per cent of them were male, 68% female. In 45%, depression was diagnosed, in 35% schizophrenia, 10% were suffering from obsessive-compulsive disorder, and 5% showed neurotic symptoms. Obsessive-compulsive disorders might be much more relevant for musical hallucinations then previously assumed.[16]

Focal brain lesion

Out of those patients with focal brain lesions as the underlying cause, 62% had their lesion in the right hemisphere and 38% in the left hemisphere. The average age of this group was 50 +/– 16 years (range 23 to 84 years). 38% were male, 62% were female. In this sub-group, the following co-diagnoses were found: hearing-impairment, epileptic activity, and intoxication. The brain lesion was vascular in nature or a tumor in most cases, in two cases an infectious lesion was found, in the remaining cases the nature of lesion was not clearly identified. In addition, 34 patients with diffuse brain atrophy were reported without giving the cause of atrophy in most cases. The mean age of these patients was 74 +/– 13 years (range 29 to 90). 29% were male, 71% were female.

Epilepsy

In 15 published patients, 33% male and 67% female, epileptic brain activity was supposed to provoke musical hallucinations. In a third of the patients, epileptic activity was found in the right hemisphere, in a third the left hemisphere, or was diffuse or not specified in the other third. In some cases, the epileptic brain activity was either not localised or appeared in a generalised pattern; the patients in this subgroup had an average age of 60 +/− 21 years (range 22 to 88 years).

Intoxications

Intoxication or inflammatory encephalopathy was found to be the main cause of musical hallucinations in 13 patients, 31% male and 69% female. Their age was 68 +/− 13 years (range 40 to 88 years). In two cases, propranolol was thought to be the trigger; in one patient, imipramine seemed to maintain the musical hallucinations. Tramadol hydrochloride, morphine,[11] salicylate, and voriconazole[10] might have initiated musical hallucinations, each in one patient. In one case, severe hypocalcaemia in a patient suffering from post-thyroidectomy hypoparathyroidism was considered to cause the musical perceptions. The combination of pentoxifylline and amitriptyline resulted in musical hallucinations in one patient, alcoholism in another. In one patient, triazolam abuse had initiated musical hallucinations, and clomipramine obviously provoked the perceptions in another case. One patient had Hashimoto's encephalopathy and two others showed Lyme disease. In seven additional cases, it was assumed that toxic-metabolic influences provoked the hallucinatory state. In this subgroup, misuse of ergot alkaloids and alcohol abuse, the combination of lorazepam and tenazepam, benzodiazepines in general, oxycodone, amphetamines, and a quaternary combination of oxazepam, indomethacin, digoxin, and methylprednisolone were found.

It should be mentioned that some more drugs or drug-like substances have been reported to be potentially associated with musical hallucinations. In total, 18 drugs and substances were described which induced musical hallucinations: salicylates, benzodiazepines, triazolam, pentoxifylline, propranolol, clomipramine, amphetamine, quinine, triazolam, imipramine, a phenothiazine, carbamazepine, marijuana, paracetamol, phenytoin, procaine, alcohol, and general anaesthesia.[9]

Content of Musical Hallucinations

Most of the patients with musical hallucinations perceive familiar tunes. The variety of content in these patients ranges from religious, childhood or popular songs to any tunes from the radio; in older patients, religious songs seem to be most frequent.[16] Classical music was perceived in about 10% of the cases and folk music in 12%. The pieces of music were presented purely vocally in 26%, purely instrumental in 10%, and both vocal and instrumental in 47%. Unilateral perception was reported only in a minority, most of the patients having bilateral musical hallucinations. Emotions provoked by the perceptions were more frightening (about 80% of all cases) than pleasant (about 20% of all cases).

Keshavan *et al.*[5] pointed out that the relatively consistent feature of musical hallucinations, sometimes even associated with very personal remembrance, suggests that they are derived from memory traces and entitled this possible mechanism the 'concept of parasitic memory'. Hallucinatory perception may be constructed as a re-experience of stored perceptual experiences by stimulation of the relevant neuronal circuit. Voices, often in combination with instrumental music, are more frequently observed as musical hallucinations than purely instrumental music. Much more often, the musical hallucinations were considered as frightening and negative rather than pleasant or just positive or neutral. In schizophrenic patients, psychotic music perception often occurs as pseudohallucinations originating in memory representations.[17]

Laterality

Regarding the influence of laterality of a focal brain lesion associated with musical hallucinations, no commonly accepted concept has been shown so far. In his reviews, Berrios[1,4] suggested that the right or non-dominant hemisphere plays a major role in generating musical hallucinations. Interestingly, auditory hallucinations in the elderly are associated with asymmetric hearing impairment and incomplete suppression of sensory input from the left ear.[14] This is considered to be consistent with the right hemisphere being dominant in music perception,[18] at least in non-musicians.[19,20] On the other hand, Keshavan *et al.*[5] pointed out that right-sided lesions are only marginally more often associated with musical hallucinations among the cases with evident brain disease. In the most recent review, the lesions were found to be right-sided in 20 and left-sided in 16 patients; this difference was not significant and questions the hypothesis that right-sided lesions play a more important role in the aetiology of musical hallucinations than left-sided lesions.

In addition, two right-handed patients with musical hallucinations and a brain tumor of the right hemisphere were recently reported.[21] Hemispheric language dominance was evaluated with the Wada test and by EEG; the left and the right hemisphere were found to be dominant in the two patients, respectively, but hemispheric dominance has not been evaluated in most cases of musical hallucinations. Although all reported patients were right-handed, they should not necessarily be regarded as dominant for the left hemisphere, and vice versa, since the relationship between handedness and language dominance is not exact.[22] Hemispheric dominance in music perception depends on such individual musical experience as duration of musical education and degree of analytical listening to music.[20] In only ten of the case reports was information on musicals skills given. Among these, objective evaluation of musical ability was performed only in two cases by the Seashore test.[21] Most of the patients with musical hallucinations were reported to have no extraordinary musical skills.

In summary, the importance of hemispheric dominance for the development of musical hallucinations remains unclear and it does not appear to be relevant. To further clarify its role, it is suggested that in future studies on patients with musical hallucinations at least hemispheric dominance and musical talent should be evaluated.

Treatment

In most cases, no detailed information on treatment was given. There are only single descriptions of successful therapy from which no general recommendations can be made. One part of the therapy should be the treatment of hypacusis if applicable. In single cases, neuroleptic, antidepressive, and in particular anticonvulsive drugs may be successful. Collins *et al.*[23] supposed that increased external auditory stimulation reduces the severity of persistent auditory hallucinations. They based their hypothesis on a single case report of a 53-year-old psychotic woman in whom medical treatment had failed to make the voices heard disappear, although listening to tapes with music, or music and speech, succeeded.

No causal treatment of musical hallucination is known. When treatment was reported, most cases resolved by tackling the underlying cause, for example stopping a medication, treating Hashimoto's encephalopathy or Lyme disease, or ameliorating hearing. Roberts *et al.*[24] described the case of a 61-year-old woman in whom musical hallucinations disappeared after clipping two small aneurysms. In depressed people or patients suffering from schizophrenia, antidepressive was better than neuroleptic medication which sometimes improved the hallucinations; even electroconvulsive therapy seemed to help in one case.[25] Most cases responded successfully to carbamazepine. In a few cases, treatment with fluvoxamine, clomipramine, olanzapine, quetiapine, or valproate (the latter drug not given for epilepsy or mood modulation) was successful.

Pathophysiological Considerations

When analysing the case reports, no clear evidence could be detected as to the underlying aetiology of musical hallucinations. In fact, they seemed to be a phenomenon caused by heterogeneous mechanisms and do not present as a single medical entity.

Hearing loss or deafness is often reported and was probably the most important factor associated with musical hallucinations. They are even more intense when the surrounding noise is low. In this context, musical hallucinations can be interpreted as a deafferentiation phenomenon. Gordon[26] even suggested that musical hallucinations must exclusively be associated with an inner ear disease leading to 'a hyperactive state of the ear'. Patients with a combination of tinnitus and auditory hallucinations overwhelmingly report musical rather than vocal hallucinations whereas patients with schizophrenia exclusively report voice hallucinations.[27] However, there are clear case reports without any symptoms of sensory-neural hearing loss or middle or inner ear disease but with typical musical hallucinations. Therefore, otological problems represent a frequent but not essential condition for musical hallucinations. It has been suggested to compare musical hallucinations with visual hallucinations caused by focal lesions in the occipital lobe leading to visual deafferentiation resulting in visual deprivation (the Charles Bonnet syndrome).

Brain-imaging studies of patients with musical hallucinations showed a dysfunction of the temporal cortex, left more predominant than right lobe, which was confirmed by dysrhythmic EEG activity as measured by sphenoidal electrodes. There are older reports on provocation of auditory, in particular musical, hallucinations by direct stimulation of the superior temporal gyrus. It might therefore be that a few of the cases of musical hallucinations, in particular those with focal brain damage or with associated clinical epilepsy, represent focal epileptic seizures of the temporal cortex. In a case with unruptured aneurysms leading to focal epileptic seizures with musical hallucinations, this has even been shown by electrocorticography.[24] Recent functional brain imaging supports the very important role of the temporal lobes in generating auditory, including musical hallucinations,[28] but these efforts show that other brain areas are also involved, in particular distinct parts of the frontal lobe. Recently, it could be demonstrated by SPECT that the pattern of brain activations are different between musical and verbal hallucinations in one patient.[29] On the other hand, functional brain imaging in a patient with musical hallucinations showed activation of the left superior temporal lobe and of the left angular gyrus, regions relevant for language processing.[12]

Another type of deafferentiation might be due to a lesion of the acoustic radiation. When analysing case reports with musical hallucinations associated with the occurrence of a single brain lesion, several foci are located within the ascending and descending acoustic fibres or within the connecting fibres between the primary acoustic cortex and the association cortices. This includes, for example, a haemorrhage between the left geniculate body and the left acoustic cortex, brainstem lesions of the lateral or paramedian pontine tegmentum suggesting a supranuclear impairment of the auditory pathways, or the basal ganglia. Even small microangiopathic lesions have been proposed as the reason for a disconnection of the auditory pathways.

It remains, however, completely unclear how psychiatric conditions or intoxication can result in musical hallucinations. Even if one excludes purely psychotic disorders with complex hallucinations or with musical pseudohallucinations, there are patients with a co-occurrence of psychiatric disorders and musical hallucinations, some aspects of musical hallucinations might be explained by the concept of obsessive-compulsive disorders.[15,30] Recent neuroimaging studies showed that auditory hallucinations result from a complex neuronal network including different parts of the temporal and frontal cortex.[31] Possibly, general brain atrophy, intoxication, or psychiatric disturbances may result in a liability of this network to be activated by external triggers.

Griffiths[31] suggested a neuropsychological approach to explain musical hallucinations with spontaneous, and not processing, activity in a cognitive network module. Interestingly, no activity could be demonstrated by PET in the primary acoustic cortex, whereas in verbal auditory hallucinations, activation of Heschl's gyrus were noted.[32] Areas involved in the module for musical hallucinations were the posterior temporal lobes, the right basal ganglia, the cerebellum and the inferior frontal lobes, but this model does not explain how specific spontaneous activity is generated.

Memory or learning disturbances have also been considered in the discussion on the pathophysiology of musical hallucinations, which has lead to the concept of 'parasitic memory' by Keshavan *et al.*[5] This means that musical perception is never unlearned but represents a so-called autonomic, that is unchangeable, memory feature which can be experienced by chance or by external stimuli. Interestingly, nearly all musical hallucinations are very familiar musical sequences and not rarely heard or unknown tone sequences.

In summary, there are no biological findings explaining the pathophysiology of musical hallucinations completely, although there are theories describing its different natures. Most of the cases are in concordance with a deafferentiation phenomenon known from visual (and also from pain) processing. In a few cases, focal epileptic or similar abnormal brain activity is associated but the concept of a memory disturbance, called 'parasitic memory', triggered by unidentified stimuli, can explain musical hallucinations from a neuropsychological point of view.

Conclusions

Musical hallucinations are a heterogeneous and rare, but not uncommon, phenomenon in neurology and psychiatry. Moderate or severe acquired loss of hearing ability or deafness is probably the main aetiological factor, which occurs in about 60% of all cases. This supports the hypothesis that a possible mechanism for musical hallucinations is perceptual release. According to this theory, a sustained level of sensory input is usually necessary to inhibit the unwanted emergence of memory traces to consciousness. However, the phenomenon of musical hallucinations remains a heterogeneous group with respect to clinical characteristics and aetiology since psychiatric disorders, epilepsy, focal brain lesions, and intoxication are also suspected to induce musical hallucinations. In future studies, such patients should also be evaluated with respect to their hemispheric dominance and musical ability. Furthermore, brain imaging studies should be performed during attacks in order to detect differences between these hallucinations and real music perception with differences between the aetiological subgroups in the brain network activated by musical hallucinations. At least, it should be kept in mind that the majority of patients with this condition suffer rather than feeling happy or neutral and deserve adequate treatment. Medical treatment trials should comprise carbamazepine in the first instance, followed by other anticonvulsants, and then antidepressants.

References

1. Berrios, G. E. (1990). Musical hallucinations: A historical and clinical study. *Br J Psychiatry* **156**: 188–194.
2. Evers, S. and Ellger, T. (2004). The clinical spectrum of musical hallucinations. *J Neurol Sci* **227**: 55–65.
3. Fischer, C. E. *et al.* (2004). Musical and auditory hallucinations: A spectrum. *Psychiatry Clin Neurosci* **58**: 96–98.
4. Berrios, G. E. (1991). Musical hallucinosis: A statistical analysis of 46 cases. *Psychopathology* **24**: 356–360.
5. Keshavan, M. S. *et al.* (1992). Musical hallucinosis: A review and synthesis. *Neuropsychiatry Neuropsychol Behav Neurol* **5**: 211–223.
6. Klostermann, W. *et al.* (1992). Musik-Pseudohalluzinose bei erworbener Schwerhörigkeit. *Fortschr Neurol Psychiat* **60**: 262–273.
7. Pasquini, F. and Cole, M. G. (1997). Idiopathic musical hallucinations in the elderly. *J Geriatr Psychiatry Neurol* **10**: 11–14.
8. Saba, P. R. and Keshavan, M. S. (1997). Musical hallucinations and musical imagery: Prevalence and phenomenology in schizophrenic patients. *Psychopathology* **30**: 185–190.
9. Gordon, A. G. (1998). Drug-induced musical hallucinations. *J Nerv Ment Dis* **186**: 652–653.
10. Agrawal, A. K. and Sherman, L. K. (2004). Voriconazole-induced musical hallucinations. *Infection* **32**: 293–295.
11. Davies, A. N. and Quinn, T. (2005). Opioid-related musical hallucinations. *J Pain Symptom Manage* **29**: 327–328.
12. Mori, T. *et al.* (2006). Regional cerebral blood flow change in a case of Alzheimer's disease with musical hallucinations. *Eur Arch Psychiatry Clin Neurosci* **256**: 236–239.
13. Fukunishi, I. *et al.* (1998). Prevalence rate of musical hallucinations in a general hospital setting. *Psychosomatics* **39**: 175.
14. Cole, M. G. *et al.* (2002). The prevalence and phenomenology of auditory hallucinations among elderly subjects attending an audiology clinic. *Int J Geriatr Psychiatry* **17**: 444–452.
15. Hermesh, H. *et al.* (2004). Musical hallucinations: Prevalence in psychotic and nonpsychotic outpatients. *J Clin Psychiatry* **65**: 191–197.
16. Warner, N. and Aziz, V. (2005). Hymns and arias: Musical hallucinations in older people in Wales. *Int J Geriatr Psychiatry* **20**: 658–660.

17. Baba, A. *et al.* (2003). Musical hallucinations in schizophrenia. *Psychopathology* **36**: 104–110.
18. Zatorre, R. J. (1984). Musical perception and cerebral function: A critical review. *Music Perception* **2**: 196–221.
19. Bever, T. G. and Chiarello, R. J. (1974). Cerebral dominance in musicians and non-musicians. *Science* **185**: 537–539.
20. Evers, S. *et al.* (1999). The cerebral haemodynamics of music perception. A transcranial Doppler sonography study. *Brain* **122**: 75–85.
21. Evers, S. *et al.* (2002). Is hemispheric dominance relevant in musical hallucinations? Report of two cases. *Eur Arch Psychiatry Clin Neurosci* **252**: 299–302.
22. Knecht, S. *et al.* (2000). Handedness and hemispheric language dominance in healthy humans. *Brain* **123**: 2512–2518.
23. Collins, M. N. *et al.* (1989). Pilot study of treatment of persistent auditory hallucinations by modified auditory input. *Br Med J* **299**: 431–432.
24. Roberts, D. L. *et al.* (2001). Musical hallucinations associated with seizures originating from an intracranial aneurysm. *Mayo Clin Proc* **76**: 423–426.
25. Stephane, M. and Hsu, L. K. G. (1996). Musical hallucinations: Interplay of degenerative brain disease, psychosis, and culture in a Chinese woman. *J Nerv Ment Dis* **184**: 59–61.
26. Gordon, A. G. (1997). Do musical hallucinations always arise from the inner ear? *Med Hypotheses* **49**: 111–122.
27. Johns, L. C. *et al.* (2002). A comparison of auditory hallucinations in a psychiatric and non-psychiatric group. *Br J Clin Psychol* **41**: 81–86.
28. Engelien, A. *et al.* (2001). Functional neuroimaging of human central auditory processing in normal subjects and patients with neurological and neuropsychiatric disorders. *J Clin Exp Neuropsychol* **23**: 94–120.
29. Izumi, Y. *et al.* (2002). Differences in regional cerebral blood flow during musical and verbal hallucinations. *Psychiatry Res* **116**: 119–123.
30. Matsui, T. *et al.* (2003). Clinical features in two cases with musical obsessions who successfully responded to clomipramine. *Psychiatry Clin Neurosci* **57**: 47–51.
31. Griffiths, T. D. (2000). Musical hallucinosis in acquired deafness. Phenomenology and brain substrate. *Brain* **123**: 2065–2076.
32. Dierks, T. *et al.* (1999). Activation of Heschl's gyrus during auditory hallucinations. *Neuron* **22**: 615–621.

Chapter 13

Migraine Aura as Source of Artistic Inspiration in the German 'Dark Chanteuse' Alwa Glebe

Klaus Podoll

The German 'dark chanteuse' Alwa Glebe collaborated in an in-depth interview on the impact of migraine with aura on her art-making activities as a writer, painter, composer and singer. Her chronic headaches, recurring aura experiences and the mood associated therewith have all crept into her work, most often subconsciously. Moreover, her recurring migraine prodromes and postdromes include states of increased creativeness and clarity of analytical thinking, which she has used, sometimes deliberately, for the purposes of artistic creation.

Introduction

Previous studies on migraine- (or migraine aura-) inspired art have focused on the visual arts, mirroring the predominance of visual disturbances among migraine aura symptoms (Podoll, 2006). Although great composers such as Louis Hector Berlioz, Frédéric Chopin, Richard Wagner, Charles Gounod, Pyotr Ilyich Tchaikovsky, Gustav Mahler or Claude Debussy have left autobiographical testimonies of their migraines, none of them claimed that it had any influence on their work other than preventing them, at times, from composing and playing music (cf. Squires, 1941; Haan and Ferrari, 2000). In modern popular music, however, many songwriters and composers have explicitly expressed the experience of devastating migraine headaches (or, more rarely, the therapeutic effects of music on the latter; cf. Vertue, 1950) in their works; e.g., Aksak Maboul, David Fiuczynski and Rufus Cappadocia, DJ Signify, Hatepinks, Moonstar88, Nurse with Wound, Puddle of Mudd, The Coral, Troubled Hubble, Wise Guys, to quote just some of the more well-known artists and bands who used the word 'migraine' in the titles of their songs, instrumental pieces, LP or CD albums (Podoll, 2008a). As a matter of fact, despite the artists' strivings for originality, a number of bands have given themselves the names *Migraine* or *Migraines*, suggesting the attractiveness of metaphorical properties of migraine (Isler and Agosti, 2001) in contemporary popular culture. In contrast to this abundance of references to migraine headaches in modern popular music, there are only a few references to migraine aura. Edith Frost, an American artist who describes her music as 'pensive countrified psychedelia', recorded on her website: 'I'm a singer-songwriter — never had the occasion to mention SS [scintillating scotoma] in a song as far as I can remember. But then a lot of the songs are a bit disjointed and surreal, and I guess I do tend to mention lights and stars and things a lot, for no particular reason. Hmmm...' (Podoll, 2008a). The American band *Visual Aura* is named after the classic migraine of its youngest member, Ryan Gartner (Hodge, 2006). Similarly, the name of the alternative rock metal band *Migraine Eyes* from Finland refers to the visual migraine aura. Canadian singer-songwriter James Ruddy produced a song *Snowstorm in My Brain* that was based on his experience of persistent aura without infarction: 'I've got a snowstorm brewing in my brain/thunder and lightning giving me pain', (Podoll, 2008b). Except for Jeffrey 'Jeff' Scott Tweedy (2008), leader of the American band *Wilco*, none of the aforementioned artists have publicised the impact migraine had on their art in greater detail.

Following a call for entries on the Migraine Aura Foundation website, the German 'dark chanteuse' Alwa Glebe volunteered to contribute pieces of her migraine-inspired art to the website (Podoll, 2008a) and agreed to participate in a study addressing the impact of migraine on her art.

Artist's Biography

Alwa Glebe (see Glebe, 2008, for all biographical data) was born in 1960 in Hanover (Lower Saxony, Germany). During the 1980s, she spent some time in San Francisco, Glasgow, Hamburg, Kiel and Berlin before she moved to the South of Germany (Nuremberg and Fürth) and the Thuringia/Vogtland district. Alwa Glebe has been active on the music scene since the late 1970s. In 1979, she formed the German-British New Wave band *Index Sign*, released two albums with the legendary NO FUN Records and went on tour with the Scottish band *Simple Minds*. She was involved in several music and recording projects in and outside Germany before she studied design in Hamburg and philosophy and German literature in Hanover and Berlin. In 1993, she began to record her first solo album. In 1995, she published *Poems*, a now out-of-print collection of verse. From 1997 to 1998, she recorded her first CD album *Debüt*, released in 1999, and from 2000 to 2004 her second CD album *Will-o'-the-wisps* (in German, *Irrlichter*) which was released in 2005. In April 2006, Cuptose records re-released *Debüt*. She is currently working on her third CD album *Halcyon Days* to be released in 2010.

Artist's Medical History

There was no family history of migraine. Car sickness, fainting spells, abdominal pain, diarrhoea, nausea, vomiting and an increased sensitivity to meteorologic changes began to take hold in childhood from the age of around six.

Between the ages of 13 and 20, she had frequent recurring attacks of intense depersonalisation and derealisation lasting up to an hour, which became rarer in subsequent years.

'In my youth, I often had experiences in which I suddenly felt strange in my familiar surroundings. This is comparable with a sudden, strongly felt isolation, which is rather shattering as you no longer feel a part of the whole. This feeling has a component of asomatic or out-of-body experience. I'm at a bit of a loss as to how to describe it. I experienced such moments as immensely existential, oppressive, and unsolvable for me. Often they were so strong that the ground seemed to vanish beneath my feet (it really looked that way to me!). In these cases I quickly sat down anywhere.' She added: 'At that time, I never spoke with anybody about all these experiences. I weathered it on my own, trying to resolve these experiences for myself, while at the same time these events offered no explanation. Actually, I reacted helplessly, trying to cover them up through activity or coming to grips with them by writing them down in my 'diary', which I kept for some time. The feeling of being exposed to these situations, of losing control, and of feeling somehow detached from my usual self was especially unpleasant for me. I find it difficult to describe, especially after such a long time. I rather admit that I do not like the recollection of it. Still, I clearly remember the sensation and atmosphere of feeling detached and a stranger to those things I usually knew so well. This is a feeling I usually do not experience in every day life, but only in the creative process, especially while singing, painting and writing, where I also seem to step out of myself, and my environment starts to alter and becomes completely unfamiliar.'

Since the age of 14, she has suffered from recurring headache attacks lasting eight to 72 hours. These headaches were characterised by side-changing unilateral location, pulsating quality, severe intensity, aggravation by physical activity and association with nausea (often culminating in vomiting) and/or photophobia, phonophobia or osmophobia.

Prodromes and postdromes, i.e. symptoms before (Blau, 1980) and after headache attacks (Blau, 1991), included euphoric and depressive mood, increased creativeness and increased clarity of analytical thinking.

'Before the onset of a migraine, I have euphoric states (lasting one to three days) during which I am capable of a diversified, strongly distinctive, and occasionally extremely intensive and fast perception. Similarly, after a migraine, I feel like starting life anew. Thinking takes place at any time and cannot be cut off, but its form and intensity are varying, and sometimes also its quality. In this regard I see the close affinity to Nietzsche, to the alternating moods and corresponding thoughts as well as the constant 'renewal', which always demands 'self-conquest' — whatever is meant by this. I am also familiar with the phenomenon of a kind of 'sobriety' spreading after each ordeal survived — when in pain it is still absolutely possible to continue thinking — while at the same time an emotional vacuousness and coldness is emerging permitting sharper analysis.'

'Following the above described pre-migraine euphoric states', the artist added, 'immediately before or at the very onset of the first symptoms of the headache attacks, I get into melancholic states. Thus, I sort of lapse from euphoria into a 'black hole' and in this condition I'm unapproachable. This kind of mood is also prevalent immediately after an attack and is very pronounced after long lasting attacks, like the hormonally-triggered ones. I need some days to gather myself, during which time I am hardly active and tend to more or less let myself go — definitely a form of regeneration. The feeling of starting life anew emerges only after such melancholic states subside, when I feel able to start anew. Actually, such melancholic states are phases I have somehow to weather and conquer, mentally as well as physically, and I feel absolutely void then.'

Being strictly associated with the menstrual cycle in teenage years, her acute headache attacks became more frequent later on. Following an operation (septoplasty) for nasal septum deviation in 1998, there was a dramatic increase in headache frequency with up to 12–14 days of headache per month. The intensity and duration of the hormonally triggered migraine attacks significantly worsened, the severe headache then lasting unremitting for 6–7 days (i.e. status migrainosis).

Since the age of 14, the previously described headache attacks were sometimes heralded by subtle changes of brightness and perspective associated with *jamais vu* sensations, sparkles or lights akin to light reflexions, tiny stars, unilateral paraesthesias of the left extremities, unilateral but side-alternating tinnitus or amnestic aphasia with mild word-finding difficulties and phonemic paraphasias.

According to the artist, in the most impressive of her recurring migraine auras she experiences 'a well-known world which suddenly looks or feels completely different. Once, in late 2002, I experienced this aura very intensively, walking down the field path that I know inside out as I walk my dogs daily on it. The usual orderly perspective the crop land scenery offered suddenly became 'less orderly' or more distorted. I remember especially the mood I was in, this feeling of 'abstraction', in which I still perceived my environment, as I was still able to walk down the path as usual, while feeling things more 'on the whole' than in detail. Maybe this was due to the peculiar light and the quietness spreading, all things were more lightened and shining, if I am permitted to express it somewhat clumsily. The lines of the fields were more intensive, longer, and tended to become endless, which of course was not so in reality. I somehow felt at 'one' with all things and at the same time far away. This will sound funny, but I am unable to explain it otherwise. My dogs did not notice any changes or I failed to realise it, especially as they were hardly noticeable in this picture, being only a minor detail, while the whole was much more fascinating. If I were religious, I would presumably interpret this as a spiritual experience, but things being the way they are, these experiences are beyond my explanatory or interpretative possibilities.'

'I had a similar experience during a car ride with my husband in December 2003. In preparation for an important presentation we had driven this route several times, and the last ride — it was almost night and I was presumably already a bit 'off-colour', meaning the migraine was approaching because of all the preparatory stress — was a highly peculiar 'ride of lights', during which I perceived the usual street lights completely differently. Driving the route up and down had made the way that familiar to me that I realised the 'different' atmosphere apparently at once. I remember my husband talking to me, while I failed to react appropriately, giving 'twisted' answers instead because I was so much ensnared in my own world. There, I seem to have intensely perceived the mood of the perspective and the lighting conditions, amplified by the evening illumination. The street appeared to be an endless chain of lights leading nowhere and I had somehow lost time and space, like I had already been travelling for ages. When we arrived, this feeling was no longer present or had got lost in the disdainful office building, that I fail to recall. However, I still remember the movement of the drive itself, almost a weightless gliding. From my perspective, I was no longer on the street, but disconnected from it in the sense of being lost in reverie.'

Her ophthalmological and neurological examinations as well as sagittal and axial T1 and T2 weighted cranial MR imaging were normal except for non-specific neck-shoulder pain diagnosed as cervicobrachial syndrome. According to the second edition of the International Classification of Headache Disorders of the International Headache Society (2004), the diagnoses migraine without aura and migraine with aura could be made. There was no evidence for abuse of analgesics. While zolmitriptan proved moderately satisfactory for the acute treatment of her migraine attacks, no effective prophylactic pharmacotherapy was found. A frequency of up to 45% migraine headache days per year meant that she had to cope with a considerable burden in terms of suffering and temporary disability in social, leisure and vocational activities. Cognitive behavioural therapy, progressive muscle relaxation and biofeedback proved useful, at least to a certain degree, in terms of aiding and supporting her in this task. According to Alwa Glebe, however, it is her art that has been most helpful to her in coping with the chronic pain throughout most of her adult life.

Artist's Use of Migraine Prodromes and Postdromes with Increased Creativeness

Alwa Glebe is a multimedia artist who uses a wide range of media to communicate her art. Having started with creative writing and poetry, she became a painter, song text writer, composer and singer with 'the voice of an angel, a dark angel' (according to James 'Jim' Kerr from *Simple Minds*). According to Alwa Glebe, all facets of her art benefit from the 'intuitive insight' gained in some stages of her migraine attacks. Ever since her late teenage years, she has made use of her migraine prodromes and postdromes with increased creativeness. Similar examples of migraine postdromes featuring a transitory increase in creativity have been described in two cases by Stavia Blunt (this volume).

The artist reported: 'Due to the special disposition of the brain, when suffering a migraine I am capable of a diversified, strongly distinctive, and occasionally extremely intensive and fast perception. My artistic working method benefits much thereof, which is why I thoroughly appreciate and utilise these euphoric moments which mostly occur immediately before migraine onset and in which all things 'devolve' upon me. Of course I am aware that there is a price to pay, namely migraine itself, but on the other hand not to exploit all these fascinating possibilities would mean reduction to a mean, which in a creative respect is almost impossible for me. Happiness is irrelevant; it is neither an aim nor a way. If at all, there is only a form of 'happiness' when dark elements like pain and mourning are interacting mutually, thus contributing to 'elucidation' more than the so-called happiness resulting from comfort and satisfaction. Such happiness is stagnancy, meaning the artist's creative and individual death. While the fact that there are limits to 'elucidation' or insight causes melancholy, it also makes the world a wonderful mystery on the other hand, in which fascination and amazement are more important than any result, which will be — as so often is the case – 'outrun' by history.'

She continued: 'After each migraine attack, I start life anew, and this is meant in a thoroughly positive sense. For me, there has to be as much freedom as possible to centre and sort myself anew time and time again. Without this freedom, which is for me the only way to gain some peace and quiet, I would merely have to suffer or fight a life of migraines. However, over the years I've learned to accept migraine as a part of myself. I accept it without yielding to it, to which end the means of art are extremely helpful. This is no therapeutic development, but one full of insight. Migraine, and the associated heightened sensitivity to internal as well as external stimuli, opens room for myself undreamt of, and this is the case not only between attacks, but also during an attack. This may sound paradoxical, especially as migraineurs are said to retire to darkened and thus limited spaces. That's right, of course. Apart from these 'external' conditions enforced by migraine there are also 'internal' ones, which are so richly faceted, that they merit transformation. I have often deliberated to what extent migraine truly inspires me and to what extent it merely influences my work. Both seem to be the case. After having suffered 167 migraine days last year (2007), it has become impossible to consider this disease an isolated phenomenon occurring sporadically. Migraine has become the 'motor' of my life. Small wonder that the high-gear brain has to shift down every now and then. However, migraine is no punishment, and nor are migraineurs responsible for inducing such attacks. In principle, they are doomed by their vitality and accompanying efficiency. That is the tragic component, and handling it is a great challenge.' She added: 'Recently, it just occurred to me that this renewal of life after each migraine is quite congruent with Nietzsche's idea of enhancement of life instead of a mere sustainment of life.'

'As a rule, my ideas are emerging highly intuitively and quickly', she concluded, 'while elaborating them takes a long and intensive process. I have to admit I'm predisposed to be a little of a perfectionist. To strike the right balance and being able to filter critically is a very important criterion, in which less is more, which is also true for the music of Alwa Glebe.'

Artist's Use of Migraine Experiences as a Source of Inspiration in her Work as a Song Text Writer, Composer and Performer

According to Mick MacNeil (ex-*Simple Minds*), Alwa Glebe's self-written and self-composed songs are 'like pictures in a gallery' (Glebe, 2008), a metaphor employed by many critics of her first two CD album projects *Debüt* and *Will-o'-the-wisps*, respectively (Glebe, 2008). 'Basically, I apply my voice very pictorially', confirmed the artist in an interview with Andreas from *Amboss Mag*, 'trying to grasp with it the musical moods and textual intentions. This allows for multilayered facets — like with a paintbrush — among them a very individual, strongly expressed dramaturgy. Surrealistic sound paintings... I think this is a suitable image, just as comparing the texts with a knife. It is right that at times I can be very direct and dissecting, I do not write lovely pastoral lyrics. Kafka once compared a book with an axe, intended to split the frozen sea within us. I write texts that cut wounds, dissect and lay them open, that may hurt at times, while wrapping them in a mood of naturalness, as if there was only this world and no other, the world in pain. I have to admit that pain is an absolutely central element of my lyrics and not accidentally the supporting element of my voice' (Glebe, 2006c).

Pain is a *leitmotiv* in many of Alwa Glebe's songs, imbuing the programmatic track *In Pain* (in German, 'Im Schmerz') from her second CD album *Will-o'-the-wisps*. She wrote: 'The melancholic aspect of the pain is worked up in my music by giving voice to 'the beast', as I fondly call the pain. The song 'In pain' is definitely programmatic of the wish to express this feeling musically: its chronic manifestation prevailing within me. The constant repetition of the question 'why?' reflects the pointlessness of it all. Even the deepest insight into migraine does not change the fact that it will stay.' The '*Im Schmerz* track turned me into pieces!', one critic noted (Atomei, 2006), providing testimony of the song's effectiveness in conveying the *Stimmung* of the world in pain.

Im Schmerz	*In pain*
Briefe fluten blutend durch den Raum	letters flood bleeding through space
ein Husten	a cough
erkennend	cognisant
scheint mir wie ein Traum	appears like a dream to me
im Schmerz	in pain
Traum	dream
Du verdammter	you condemned one
lass mich nicht los	don't let me loose
fern	distant
Unbekannter	strange one
sterbend im Schoß	dying in the fold
im Schmerz	in pain
sag mir	tell me
warum	why
sag mir	tell me
warum	why

In her second CD album *Will-o'-the-wisps*, one critic wrote, 'Alwa Glebe possesses the voice of an apparition' (Wozny, 2006). Her vocal performance gives a proper reflection of another *leitmotiv* that runs through several songs of her creation, viz. the idea of instability of reality in a world of 'illusion', an idea obviously inspired (besides literary sources) by the artist's early migraine aura experiences of paroxysmal depersonalisation-derealisation, feeling 'distant' to herself and her environment, and her 'eyes' deceptions', i.e. her subjective visual sensations of migraine aura (Gowers, 1895). According to an interview statement from novelist and fellow migraineur Siri Hustvedt, the 'idea of visual instability is... philosophical, but certainly someone who suffers from this might be more inclined to have that kind of view of the world — that we don't always know what we're looking at. How do we read the world? How do we interpret it?' (Hodson, 2003)

Ich weiss	*I know*
Interpretation	Interpretation
neue Illusion	new illusion
trügerische Sicht	the eye deceiving you
am Abend	in the evening
in der Nacht	at night
im Licht	(in the light)
und ich laufe, ich laufe	And I keep going
ich laufe jeden Tag	keep going day by day
ich laufe auch wenn ich nicht mehr laufen mag	keep going even though I don't want to go on
und ich denke	and I think
ich denke	I think
ich denke viel zu viel	I think too much
ich denke auch wenn ich nicht mehr denken will	I think even though I don't want to go on thinking
Und ich rede	And I talk
ich rede	I talk
ich rede vor mich hin	I talk to myself
ich rede und rede	I talk and talk
und weiss nicht mehr mit wem	even though I don't know who I am talking to
und ich weiss gar nichts, so gar nichts	and I know nothing, I know absolutely nothing
am Abend	in the evening
in der Nacht (im Licht)	at night (in the light)
In Isolation	Isolated
ohne jeden Ton	without a single sound
mörderische Sicht	deadly sight
in der Nacht (im Licht)	at night (in the light)
Und ich laufe, ich laufe...	And I keep going, I keep going...

Mirroring her visual aura experiences of seeing sparkles or lights akin to light reflexions or tiny stars, the title track of Alwa Glebe's second album *Will-o'-the-wisps* has, according to the artist, 'also this 'character of lights', exemplified by the glittering, sparkling atmosphere and the title, which introduces into the mood.' On a side note, it is worth reporting that 'the studio recording of *Will-o'-the-wisps* was done while I was suffering from a migraine. I sung despite the migraine, because I really wanted to and I would not be detained. In the end it was rather successful; I put a lot of trouble into getting the song across well and pleasantly, as I especially love it, its glimmering, almost ethereal character, the pull of it and its 'seeming' infinity. At that moment I didn't realise that to many people it would be a sad song, on the contrary, I liked it very much.'

Irrlichter	*Will-o'-the-wisps*
Vielleicht ist es nur	Perhaps it's merely
der Traum vom Traum	a dream dreaming
der uns zusammenhält	that keeps us together
in einer Welt der Illusion	in a world of illusion
die mehr verspricht und hält	that holds and promises more
Vielleicht ist es nur	Perhaps it's merely
der Traum vom Traum	a dream dreaming
der uns so sehr gefällt	that pleases us so much
interpretierte Sicht der Welt	a rendering of the world
die mehr als anderes zählt	that counts for more than anything else
Wir sind wie Irrlichter	We are like Will-o'-the wisps
funkeln in der Nacht	aglow in the night
transluzente Gesichter	shimmering faces
um den Schlaf gebracht	unable to rest
Vielleicht zeigt sich nur im letzten Augenblick	Perhaps we only see at the very last moment
wohin die Reise geht	where the road is leading
und wo die Zeit niemals vergeht	and where time never goes by
was weiterhin besteht	but keeps on turning

vielleicht ist es so	it may well be so
wer weiß das schon	who can tell
wozu sich alles dreht	why everything keeps spinning
in dieser Welt der Illusion	in this world of illusion
die ewig kommt und geht	that forever comes and goes

In the song 'what would be if' (in German, 'was wäre wenn') from her *Will-o'-the-wisps* CD album, Alwa Glebe referred to her recurring auras experienced during the 'blue hour, my hour', walking with her dog on a familiar field path through 'crop land', 'nobody's land', when she had ineffable, 'unnamed' *jamais vu* sensations whereby this well-known environment appeared as 'unknown', and all memories of 'past things in new light', to her.

Was wäre wenn	*What would be if*
Ackerland	Crop land
Niemandsland	nobody's land
braunes Band der Lust	brown band of lust
unbekannt	unknown
unbenannt	unnamed
was hast Du gewusst?	what have you known?
Einfach nur ins Lot	Just perpendicular
was wäre wenn	what would be if
einfach nur ins Lot	just perpendicular
was wäre wenn	what would be if
Vergangenes im neuen Licht besehen	seeing past things in new light
Begegnungen, zu spät, um zu verstehen	encounters, too late to understand
zu spät, um zu verstehen	too late to understand
Blaue Stunde, meine Stunde	Blue hour, my hour
laß mich nie allein	never let me alone
einfach so	just so

'That's rather uncanny', the artist commented, 'the way this aura experience has apparently entered the text. I really do not know what to say, there is no end to this 'expedition of discoveries' and without your 'insight' into aura phenomena I would never have become conscious of all this. For me, this also reflects a longing — which corresponds with my feeling that I did not find this experience absolutely unattractive but was rather able to derive some pleasure out of it. Like the famous saying by Kafka: 'To draw sweetness out of disease'. She added: 'The line 'just perpendicular' (in German, 'einfach nur ins Lot') has for me a new meaning now. Maybe it is the idea to straighten up the image or the situation I am in, although at the same time I say 'what would be if', which is more suggestive of a 'wish' than reality, meaning that the image or situation cannot be straightened, at least not deliberately. Maybe this straightening of things is not my true wish, but rather what is interesting to me is the question of what things would otherwise or 'really really' look like. I can only imagine that I might have felt 'out of kilter', so that this arose because of the shifting perspectives. The straight lines of the field especially and the horizon before me become more distinct than they would have for example in a city, in the hustle and bustle — on an open field all things are much wider, more open, while at the same time being limited due to the plain perspective. Who would take notice of the horizon in a city? On the other hand I had not been afraid and had secretly enjoyed the feeling, which either contradicts the idea that I had been 'out of kilter', or I did not care, because it was quite natural. Would that be possible?'

Discussion

The self-reports of the German 'dark chanteuse' Alwa Glebe show that migraine can act as a source of inspiration not only in visual arts, but in writing, music and performance. The impact of her migraines on her own art only came to her conscious awareness after she had learnt of the Migraine Art 'movement' (Lambert, 2008) from the Migraine Aura Foundation website. In the course of her participation in the in-depth interview, she became consciously aware that the inspirational influence of her migraine had been much larger than previously thought. In addition to the pain, she had subconsciously transformed her visual aura experiences of seeing sparkles or lights akin to light reflexions or tiny stars into the lyrics and music of the title track of her second CD album *Will-o'-the-wisps* which has, according to the artist, 'also this 'character of lights', exemplified by the glittering, sparkling atmosphere and the title'. Furthermore, she had used a number of non-visual migraine aura symptoms (namely, depersonalisation, derealisation and *jamais vu* sensations) and the mood associated therewith as source of inspiration for some of her self-written and self-composed songs.

The subconscious action of the artistic inspiration by migraine on professional artists' art-making activities in multiple media requires expansions and modifications of the original concept of Migraine Art as defined by Podoll and Robinson (2008) in their monograph *Migraine Art: The Migraine Experience From Within*. 'Migraine Art is the idea that pictorial art techniques may provide an adequate medium, sometimes the best medium, to express and communicate the experiences that occur as signs and symptoms of migraines or sufferers' reactions to the disease. The term *art* is used here in its most inclusive sense, with no implication of the work being judged aesthetically... In the concept of Migraine Art, the relationship between a work of art and having a migraine is determined by the artist's spontaneous choice, or the request of someone else, for a personal migraine experience to be the subject matter.' As can be seen from Alwa Glebe's migraine-inspired work, migraine experiences can also be transformed into works of art without the artist's conscious choice to do so either on her own decision or acting on someone else's request. Moreover, her work shows that such migraine-inspired artwork can achieve other functions than expressing and communicating the experiences that occur as signs and symptoms of migraines or sufferers' reactions to the disease: as a matter of fact, all of the functions that have hitherto bee n attributed to good art by artists, philosophers of art and art scientists. That is, migraine-inspired art is *art*. The concept of *migraine-inspired art* can thus be distinguished from the original concept of *Migraine Art* by its application to art of all media, by its inclusion of subconscious processes of artistic inspiration and by its inclusion of aesthetical judgments.

References

1. Atomei, D. (2006). Review of Alwa Glebe's 'debüt'. *KOGAIONON Underground Magazine*, Romania, August 2006. KOGAIONON website. Accessed on May 10, 2008, at *http://www.kogaionon.com/english/archives_english/archives_reviewsenglish/kogaionon_archives_reviews_english_august2006.html.*
2. Blau, J. N. (1980). Migraine prodromes separated from the aura: Complete migraine. *Br Med J* **281**: 658–660.
3. Blau, J. N. (1991). Migraine postdromes: Symptoms after attacks. *Cephalalgia* **11**: 229–231.
4. Blunt. S. The creative brain: fundamental features, associated conditions and unifying neural mechanisms. In this volume.
5. Glebe, A. (1978). *Tagebuch.* Unpublished manuscript, Hanover.
6. Glebe, A. (2000). *Gedichte.* Nuremberg: Eigenverlag.
7. Glebe, A. (2006a). Interview with Kym Gnuch. *SONIC SEDUCER*, May 2006. Alwa Glebe's website. Accessed on May 10, 2008, at *http://www.alwaglebe.de/html/alwa_glebe_interview.html.*
8. Glebe, A. (2006b). Interview with Ecki Stieg, June 15, 2006. Grenzwellen website. Accessed on May 10, 2008, at *http://www.grenzwellen.com/gw_redaktion/modules/news/article.php?storyid=227.*
9. Glebe, A. (2006c). Interview with Andreas. *Amboss Mag*, December 2006. Accessed on May 10, 2008, at *http://www.amboss-mag.de/interviews.html.*
10. Glebe A. (2008). MySpace website. Accessed on May 10, 2008, at *http://www.myspace.com/alwaglebe.*
11. Gowers, W. R. (1895). Subjective visual sensations. *Trans. Ophthalmol. Soc. UK* **15**: 1–38.
12. Haan, J. and Ferrari, M. D. (2000). Mahler's migraine. *Cephalalgia* **20**: 254.
13. Hodge, B. (2006). Visual Aura. Heavy Frequency website. Accessed on May 10, 2008, at *http://www.heavyfrequency.com/archives/columns/fotm.php?id=154.*
14. Hodson, H. (2003). Darkness and light. *The Telegraph*, November 1, 2003. Accessed on May 10, 2008, at *http://www.telegraph.co.uk/arts/main.jhtml?xml=/arts/2003/01/11/basiri11.xml&page=2.*
15. International Headache Society (2004). The International Classification of Headache Disorders (2nd ed.). *Cephalalgia* **24**: 1–160.
16. Isler, H. and Agosti, R. (2001). Metaphorical properties of headache in Ancient Greek literature. *Cephalalgia* **21**: 522.
17. Lambert, H.-P. (2008). Littérature, arts visuels et neuroesthétique. Épistémocritique, vol. 2, January 15, 2008. Accessed on January 15, 2008, at http://www.epistemocritique.org/spip.php?article49.
18. Podoll, K. (2006). Migraine art in the internet: A study of 450 contemporary artists. *Int Rev Neurobiol* **74**: 89–107.

19. Podoll, K. (2008a). Migraine and music. Migraine Aura Foundation website. Accessed on May 10, 2008, at *http://www.migraine-aura.org/content/e24966/e24880/index_en.html.*
20. Podoll, K. (2008b). jruddy (song). Migraine Aura Foundation website. Accessed on May 10, 2008, at *http://www.migraine-aura.org/content/e27891/e27265/e42285/e43305/e62760/index_en.html.*
21. Podoll, K. and Robinson, D. (2008). Migraine Art: The Migraine Experience From Within. Berkeley, California: North Atlantic Books.
22. Squires, P. C. (1941). Peter Ilich Tschaikowsky. A psychological sketch. *Psychoanal Rev* **28**: 445–465.
23. Tweedy, J. S. (2008). Shaking It Off. *The New York Times*, March 5, 2008. Accessed on May 10, 2008, at *http://migraine.blogs.nytimes.com/2008/03/05/shaking-it-off/.*
24. Vertue, H. S. (1950). Music and migraine. *Guys Hosp Gaz* **64**: 350–354.
25. Wozny, M. (2006). Critique of Alwa Glebe's CD album 'Irrlichter'. *VIRUS! Magazine*, New York, January 29, 2006. Alwa Glebe's website. Accessed on May 10, 2008, at *http://www.alwaglebe.de/html/3.html.*

Chapter 14

Musical Palinacousis as an Aura Symptom in Persistent Aura without Infarction

Klaus Podoll

Palinacousis is an auditory illusion of persistence or perseveration of sound impressions after the cessation of the auditory stimulation. We analysed all cases published to date with respect to their demographic and clinical features and to the possible pathomechanism underlying the auditory illusion. In total, 32 cases (including four cases of musical palinacousis) were analysed and separated into five groups according to their aetiology (focal brain lesion; epilepsy; migraine; hypacusis; psychiatric disorder). Two additional cases reported musical palinacousis as an aura symptom in persistent aura without infarction, lasting >20 months in case one and ten and a half months in case two. In our cases, musical palinacousis could be distinguished from 'earworms' by its longer duration and by its association with verbal palinacousis.

Introduction

Palinacousis (or palinacusis) is an auditory illusion in which external acoustic stimuli such as speech, music or other environmental sounds persist or recur for variable periods of time after the initial acoustic stimulus has ended. Term and concept were introduced independently by Meyer (1962) in Germany, Bender and Diamond

(1965) in the US, and Bekény and Péter (1973) in Hungary. The most comprehensive study of this rare phenomenon is owed to Jacobs *et al.* (1971, 1972, 1973) who reported a series of seven cases, representing almost one quarter of the 32 cases reviewed in Tables 1–4. With respect to its probable aetiology, the previously published cases can be separated into five groups (in order of decreasing frequency): focal brain lesion; epilepsy; migraine; hypacusis; psychiatric disorder (cf. Evers and Ellger, 2004). Associated focal brain lesions of either hemisphere predominantly pertained to the temporal lobes, less frequently to the parietal, occipital or frontal lobes or to the medial geniculate body, the nuclear mass of the thalamic auditory relay nucleus. Patients with palinacousis due to focal brain lesion tend to perceive the illusory sound as on the side of auditory space that is contralateral to the lesion. Seizures were experienced by almost half of the cases, including aura, simple partial seizures or complex partial seizures. Whereas moderate or severe acquired loss of hearing ability is a main aetiological factor in musical hallucinations (Evers and Ellger, 2004), it is encountered only rarely in palinacousis. Similarly, psychiatric disorder plays a considerable role as aetiology of musical hallucinations, but only rarely so in palinacousis.

Likewise, migraine has heretofore only rarely been recorded as the probable or potential aetiology of palinacousis. Whereas migraine was registered as co-diagnosis in three of the aforementioned 32 cases, it was assigned as the cause of palinacousis in five cases. A 28-year-old patient of Bonhoeffer's (1940) had recurring attacks of migraine with aura terminating in the sensation of her own voice still re-echoing in a peculiarly strange way. Bille (1962) reported a 15-year-old boy who found that 'every sound had an echo (*echo-acousia*)', which lasted for about 30 minutes before a severe migraine attack. A 30-year-old female migraine sufferer reported by Barolin (1963) complained that all voices sounded as if they had been emitted in an echo chamber (cf. Barolin and Gloning, 1969). One of Sacks's (1970) patients found that the faintest sounds were followed by a protracted echoing or reverberation for some seconds after they had ceased. Podoll (2008) observed palinacousis (echoing sounds, especially electronic devices like phones, beeps, and buzzing) as a recurring migraine aura

Table 1. Published cases of palinacousis.

			Perseveration of		
Case	Sex	Age	voices	music	sounds
Bonhoeffer, 1940	F	28	yes	no	no
Bille, 1962	M	15	?	?	yes
Meyer, 1962	F	40	yes	no	no
Barolin, 1963	M	30	yes	no	no
Bender and Diamond, 1965	M	57	no	no	yes
Bender and Diamond, 1965	M	19	no	no	yes
Sacks, 1970	?	?	?	?	yes
Jacobs *et al.*, 1973, case 1	F	40	yes	?	yes
Jacobs *et al.*, 1973, case 2	F	16	yes	yes	yes
Jacobs *et al.*, 1973, case 3	?	?	yes	?	yes
Jacobs *et al.*, 1973, case 4	M	15	yes	?	no
Jacobs *et al.*, 1973, case 5	M	60	yes	?	no
Jacobs *et al.*, 1973, case 6	?	?	no	?	yes
Jacobs *et al.*, 1973, case 7	M	64	no	?	yes
Békény and Péter, 1973	F	39	yes	no	no
Malone and Leiman, 1983	F	75	yes	no	no
Patterson *et al.*, 1988	F	50	yes	no	yes
Shepard and Sandridge, 1992	?	?	no	no	yes
Masson *et al.*, 1993	M	60	yes	no	no
Auzou *et al.*, 1995	M	78	yes	no	no
Freudenreich and McEvoy, 1996	M	32	yes	no	no
Auzou *et al.*, 1997	F	51	yes	yes	yes
Fukutake and Hattori, 1998	M	49	no	no	yes
Terao and Matsunaga, 1999	F	75	yes	no	no
Griffiths, 2000, case 6	F	82	no	yes	no
Prueter *et al.*, 2002	F	20	yes	yes	no
Voll *et al.*, 2002	M	?	yes	yes	no
Di Dio *et al.*, 2007, case 1	M	49	yes	no	no
Di Dio *et al.*, 2007, case 2	F	72	yes	no	no
Wustmann and Gutmann, 2007	M	49	no	no	yes
Kim *et al.*, 2007	F	67	yes	no	no
Podoll, 2008	F	27	no	no	yes

Table 2. Clinical data of 32 published cases of palinacousis.

Clinical feature	Number of patients
Palinacousis with perseveration of:	
Voices	21
Sounds	15
Music	4
Side of lesion (if applicable):	
Left	13
Right	9
Site of lesion (if applicable):	
Temporal lobe	18
Parietal lobe	7
Occipital lobe	5
Frontal lobe	1
Medial geniculate body	1
Associated symptoms (frequency >1):	
Seizures	15
Verbal hallucinations	5
Confusion	5
Visual hallucinations	4
Palinopsia	4
Ears clogged	4
Musical hallucinations	4
Environmental sound hallucinations	4
Vertigo	4
Hemihypaesthesia	3
Visual illusions	3
Slurred speech	3
Global aphasia	3
Déjà vu	3
Loss of vision	2
Hearing loss	2
Hyperacusis	2
Motor perseveration	2
Reduced concentration	2

Table 3. Demographic data of all patients with palinacousis presented for the total group and for the different subgroups.

Diagnosis*	Number	Age in years (mean ± SD, range)	Sex
All cases	32	46.6 ± 21.0 (15–82)	14 male, 14 female
Focal brain lesions	17	51.7 ± 16.6 (16–78)	8 male, 7 female
Epilepsy	7	38.2 ± 24.8 (15–75)	3 male, 3 female
Migraine	5	25.0 ± 6.8 (15–30)	2 male, 2 female
Hypacusis	2	78.5 ± 4.9 (75–82)	2 female
Psychiatric disorder	1	49	1 male

* No reports on palinacousis induced by intoxication available.

Table 4. Clinical data of all patients with palinacousis presented for the total group and for the different subgroups.

Diagnosis	Number	Side of lesion (if applicable)	Co-diagnoses*
All cases	32	9 right, 13 left	2:2×; 3:1×; 4:9×; 5:3×
Focal brain lesions	17	7 right, 8 left	2:1×; 4:9×; 5:1×
Epilepsy	7	1 right, 4 left	2:1×; 5:2×
Migraine	5	1 left	
Hypacusis	2	1 right	3:1×
Psychiatric disorder	1		

* 1 = hypacusis; 2 = psychiatric disorder; 3 = focal brain disease; 4 = epilepsy; 5 = migraine; 6 = intoxication.

symptom in a 27-year-old female with a positive family history of migraine and a previous personal history of typical aura without headache since her teenage years and recurring attacks of migraine without aura and migraine with aura (Fig. 1) since the age of 26. 'As for sounds', the patient explained, 'I frequently get phantom cell phone rings, or mechanical buzzing or humming. I say phantom because my phone will ring, but then I will continue to hear rings

Figure 1. Illustration of attack of migraine with aura (Podoll, 2008).

intermittently over the next 15–20 minutes. Mostly sounds from electronic devices do this for me. I will get an echo from them.'

A similar account provided by Lippman and Lippman (1952) suggests that the migraine aura symptom of palinacousis may be underreported in the medical literature.

'All migrainoids occasionally hear sounds which don't exist. The most common occurrence is the ringing of a doorbell or a telephone bell. In a migrainoid's home the following scene is a familiar one:

You are having dinner. Your wife suddenly gets up from the table. You hear her go to the telephone, take down the receiver, and say, "Hello?" She repeats the words several times, hangs up, and returns to the table with a puzzled face. She murmurs something about having been *sure* that she heard the telephone ring.

Dinner is resumed. Presently she says, "Do *you* hear the telephone ringing?"

You frown and say, "No." You continue eating.

Now watch her for the next five minutes. You'll notice every now and then she appears to be listening intently. She probably continues to hear the phone at intervals all through the rest of the meal. But she won't mention it again. She has learned from experience to rely on someone else's testimony about the telephone bell' (Lippman and Lippman, 1952, p. 40).

In this chapter, we report on two migraineurs in whom persistent musical palinacousis was encountered as a symptom of persistent aura without infarction — formerly labelled prolonged migraine aura status (Haas, 1982) — a rare complication of migraine with aura with one or more aura symptoms persisting for > one week without radiographic evidence of infarction.

Case Histories

Case one

This 38-year-old male had a positive family history for migraine with his mother being affected. For as long as he could remember, he had seen worm-like, hollow 'floaters' in the whole visual field of both eyes. Since the age of 21, he suffered from recurring attacks of nausea along with dizziness and intense abdominal pain, right in the centre of his stomach, followed later by a fatigued feeling that would persist for the next day or two. Occasionally, he had attacks of depersonalisation with his vision becoming 'flat' and a feeling he described 'as though I am simply witnessing the scene, like I am not controlling my body'. Since the age of 27, he had recurring attacks of migraine without aura. From the age of 33, he experienced recurring attacks of unilateral side-changing tinnitus with a duration ranging from ten minutes to four hours. Since around spring 2005, he had recurring hallucinations of falling rain outside the window. In Autumn 2005, aged 36, he noted a significant worsening of the intensity of his 'floaters', which has lasted ever since, and an increased frequency of the aforementioned tinnitus attacks. Between March and mid-June 2006, he had three attacks of migraine with aura with a rapid onset of nausea, dizziness and confusion followed by headaches and then diarrhea. In late June 2006, he had the onset of muscle twitching of the eyelids and both upper and lower limbs, which caused him to seek medical care. In the absence of clinical and electromyographic findings of neurogenic disease, the diagnosis of benign fasciculation syndrome was made. In mid-July 2006, he had a severe unilateral headache followed by persistent grainy vision, binocular diplopia, musical palinacousis and bilateral somatosensory symptoms (a tingling in both hands and a weird feeling on his face as though he had walked into a spider web). Sometimes the volume of hearing fluctuated (oscillocusis) or music sounded 'wrong', like frequencies were missing (paracousis). By August 2006, he had developed intense somatic anxieties and a moderate depressive episode with somatic symptoms so that he was prescribed mirtazapine (30 mg per day). Ten days later, he experienced the onset of 'visual snow' involving the entire visual field, increased negative afterimages, trails, corona phenomenon ('shimmering auras around objects'), increased halos around lights, increased sensitivity to patterns inducing visual discomfort, hypacusia ('feeling of fullness in ears') and bilateral tinnitus; all of these persistent perceptual disturbances showed only a partial remission over a follow-up period of 20 months. From the same time onwards, he also experienced daily matutinal kaleidoscopic hallucinations depicted in Fig. 2. 'Upon awakening, I see a complex pattern of small geometric shapes that covers my entire visual field. It usually grows in intensity for the first ten to 20 seconds after I open my eyes, and then fades over the next few minutes... The colors that predominate in the patterns are blue and yellow...' This aura recurred daily over the period of follow-up.

Repeated medical, neurological, ophthalmological and ENT examinations between July 2006 and December 2006 yielded unremarkable results. VEP, BAEP, audiometry, EMG and nerve conduction study were normal. Sagittal and axial T1, FLAIR and T2 weighted cranial MR imaging were performed twice within two and a half weeks and were normal except for five small nonspecific white matter T2 hyperintensities within the right and left frontal lobes ('unknown bright objects') seen on both examinations. A third control half a year later showed a stable appearance of the MR imaging

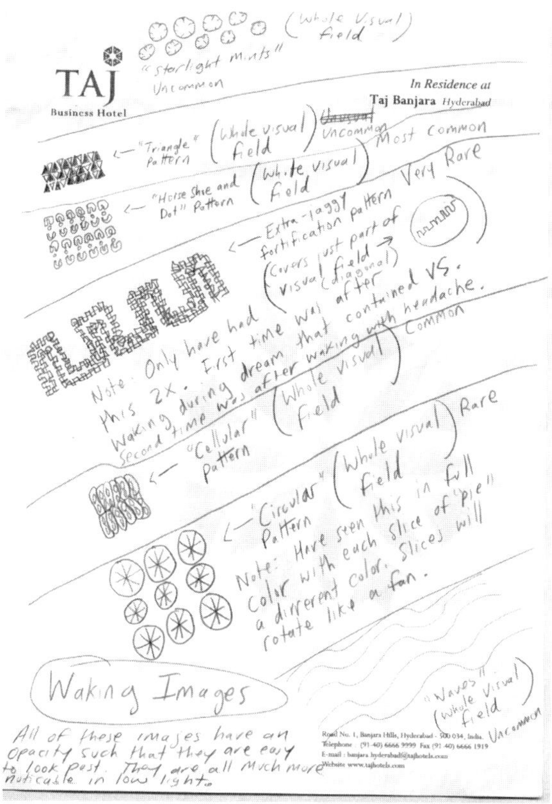

Figure 2. Daily matutinal kaleidoscopic hallucinations with 'starlight mints', 'triangle', 'horseshoe and dots', 'cellular' and 'circular' patterns simultaneously covering both temporal and nasal fields of both eyes.

findings. The patient's history and clinical findings fulfilled the diagnostic criteria of persistent aura without infarction, with the onset of its fourth episode being 'typical of previous attacks' (with tinnitus) 'except that one or more aura symptoms persists for >1 week' (International Headache Society, 2004). The fourth episode was also remarkable in that it combined the features of the repetitive (daily matutinal kaleidoscopic hallucinations) and continuous variety (persistent perceptual disturbances) of persistent aura, respectively.

The symptom of persistent musical palinacousis brought on by the third episode of persistent aura without infarction was described by the patient as follows: 'Well, since mid-July 2006 I have had continuous songs in my head. I do not perceive them as having come from the environment, and if I am speaking to someone or listening to someone else speak (or to TV or CD), then I don't hear the songs. Only when my mind is quiet do I hear them. They are usually songs I have listened to within the past 24 hours or so. They have remained stable since mid-July 2006 (over 20 months); they have not grown in intensity, etc. They are not disabling or disturbing.'

The patient's persistent musical palinacousis may feature only a part of a song rather than the full musical piece. 'I have a... condition', he related, 'in which music I listen to will continue to replay in my head for hours. Really, it never stops. I even wake up with a song in my head. It's always what I have listened to in the past day or so. I love music but it can get old at times. But, on the bright side, I'm never without tunes... For me, it's always just a part [of a full song] that keeps looping... For me, it very, very rarely will stop when I am very tired, but usually it goes non-stop.'

The patient added: 'I have occasionally had palinacousis with other sounds besides music, typically voices from conversations right before bed. But this is rare, it has perhaps happened only two or three times, and the voices are always just repeating verbatim what the person said when we were talking: there is never anything new, nor am I given commands, etc. It's usually just some phrases repeated over and over, not a full replay of the conversation. Once again, this is rare for me and not terribly distressing.'

Case two

This 27-year-old male, whose mother had a history of migraine, had seen 'visual noise' (in German, 'Augenrauschen') for as long as he could remember. 'I must have had this "visual noise" for my entire life, that's why I don't know what it's like to "see clear". Already as a child I saw the well-known noise: tiny white dots in the sky and during the night and on white (monochrome) surfaces.' In his fourteenth year (but not in later years), he suffered around six attacks of migraine without aura. Since his twenties, at the latest, he had experienced recurring attacks of tinnitus or musical palinacousis ('earworms'; in German, 'Ohrwürmer'). At the age of 26, following a period of psychosocial stress (separation from girlfriend, two moves and start of new relationship), he noted a significant worsening of the intensity of the persistent 'visual noise' commencing simultaneously with the onset of a tinnitus of the left ear persisting ever since. Also, from that time on, he got recurring attacks of vertigo or derealisation. Two months later, despite relief from the aforementioned stress by living in a happy new relationship, he fell into a moderate depressive episode with somatic symptoms and subsequently got stuck with musical palinacousis which persisted over the next ten and a half months. A month later, he experienced two attacks of typical scintillating scotoma separated by seven days. Three weeks after the full remission of musical palinacousis, he caroused all night long, and noted upon awakening next day that the persistent musical palinacousis had come back. On follow-up two months later, it had remitted only partially. His ophthalmological and neurological examinations as well as sagittal and axial T1 and T2 weighted cranial MR imaging were all normal. The patient's history and clinical findings matched the diagnostic criteria of persistent aura without infarction, with the onset of the second, third and fourth episode being 'typical of previous attacks' (with tinnitus or musical palinacousis, respectively) 'except that one or more aura symptoms persists for > one week' (International Headache Society, 2004).

The patient provided the following description of the persistent musical palinacousis brought on by the third and fourth episode of persistent aura without infarction: 'Since late August 2006 I have persistent "earworms". I could even tell you the exact day when it occurred for the first time: August 29, 2006. Songs, advertisements, etc. are rehearsed in my head for hours. (Everyone has this sometimes, but not daily!?) When I hear a song, it is rehearsed in my head. But it can also be that I awake in the morning and a current title is buzzing in my head. All pieces of music are familiar to me even if some are already somewhat older. Actually the earworms always come when I think about nothing and just relax, and this is quite exhausting.'

Discussion

According to the second edition of the *International Classification of Headache Disorders* of the International Headache Society (2004), persistent aura without infarction is defined as a complication of migraine that is characterised by aura symptoms lasting for > one week without neuroradiologic evidence of cerebral infarcts, which cannot be explained by other conditions. The said diagnostic criteria require that the present attack in a patient with migraine with aura is typical of previous attacks, except that one or more aura symptoms persist for over one week (i.e. the aura symptom that now persists for > one week must have occurred earlier in the patient's history as a transient migraine aura symptom). This criterion is met by both patients reported in this study who experienced recurring attacks of migraine aura without headache featuring tinnitus (cases one and two) or musical palinacousis (case two) before these phenomena recurred as persistent symptoms lasting for > one week. Other causes such as migrainous strokes could be excluded by cranial MR imaging that was normal except for five unspecific white matter lesions ('unknown bright objects') in case one (cf. Kruit *et al.*, 2004). Both patients had multiple episodes of persistent aura, a clinical feature previously recorded by Liu *et al.* (1995). Each of these was defined by the onset of at least one persistent aura symptom (as was the case in the first, third and fourth episode of case one and in all four episodes of case two) and/or a clinically significant worsening of at least one pre-existing persistent aura symptom (as in the second episode of both cases). The outcome of each episode of persistent aura ranged from full remission (third episode of case two), over partial remission (third and fourth episode of case one), to stable disease (first and second episode of case one and first, second and fourth episode of case two). It is worth noting that the 'floaters' in case one and the 'visual noise' in case two were reported to have been present for as long as the patient could remember, suggesting a very early childhood onset of the first episode of persistent aura before the age of two to four which is eventually obscured by childhood amnesia. Whereas such supposed very early childhood onset remains to be confirmed by direct neuropediatric observations of the given migraine complication in the said age group, Liu *et al.* (1995) have documented its occurrence in a nine-year-old girl who complained for two months 'that she saw constant blobs of white and gray, squiggles, clouds, comets, bubbles, "lines of ants" and "a million dots" (p. 665). As another interesting clinical feature of the two case histories reported in the present study, the documented persistent aura phenomena include five symptoms characteristic of basilar-type migraine in case one (i.e. tinnitus, hypacusia, diplopia, a variety of visual symptoms — viz.: grainy vision, 'visual snow', kaleidoscopic hallucinations, increased negative afterimages, trails, corona phenomenon, increased halos around lights, increased visual discomfort — simultaneously in both temporal and nasal fields of both eyes, simultaneously bilateral paraesthesias) and two symptoms of this migraine subform in case two (i.e. tinnitus and 'visual noise' in the entire visual fields of both eyes), so that it is suggested that the persistent aura here represents a *complication of basilar-type migraine.*

The two patients' descriptions of persistent musical palinacousis are similar to those reported by patients with musical palinacousis attributed to other aetiologies. Of the 32 patients with palinacousis reviewed in Table 2, only four had experienced musical palinacousis, which was, as in our case one, associated with perseveration of voices and other sounds as well. Case two from Jacobs *et al.* (1973), a 16-year-old girl with left parieto-temporo-occipital ependymoma, 'heard not only the voices and words, but also the "wind", "sirens", "music", and "orchestration" of the *Ulysses* theme' by Stanley Myers 'in her right ear for four days after she listened to it at the cinema' (p. 276) where she had seen Joseph Strick's film *Ulysses* shot in 1967 and based on James Joyce's novel of same name. 'On another occasions', the aforementioned authors continued, 'she heard an illusory "chorus" sing *Silent Night*' by Joseph Mohr and Franz Xaver Gruber, 'accompanied by a "piano", and "chimes", 24 hours after attending a Christmas party where she had actually heard the song' (p. 276). Similarly, Auzou *et al.* (1997) reported a 51-year-old female with left temporo-parietal glioblastoma multiforme who developed seizures and palinacousis several times per day with perseveration of voices, music (not otherwise specified) and sounds for an undetermined number of days. Case six from Griffiths (2000) was an 82-year-old female with progressive bilateral deafness and a posterior circulation stroke after which she experienced around three years of continual musical hallucinations. 'She also experienced palinacousis; she described several prolonged hallucinations in the form of hymns triggered by listening to the television programme *Songs of Praise*' (Griffiths, 2000, p. 2068), a long-running BBC television programme based around traditional Christian hymns. In the case from Prueter *et al.* (2002), a 20-year-old female schizophrenic patient who was not psychotic at that time, the perseverated sound consisted of her own and others' voices and pop songs the patient had heard on the radio during the preceding minutes, which she would hear for up to three hours repeatedly during the day over a period of two months. In musical palinacousis, Jacobs *et al.* (1973) observed that the persistent or recurrent illusory sensations often comprise only a portion of the original sensory event, such as 'musical fragments of different complexity' (p. 276). This feature, which is similarly encountered in palinopsia (Jacobs *et al.*, 1973) where the persistent visual image may likewise be only a part of an object, is well illustrated by case one of the present study where the patient noted that 'it's always just a part [of a full song] that keeps looping' in his experience of persistent musical palinacousis. In the four previously quoted cases of musical palinacousis from other aetiologies, its duration ranged from four days to two months, which is far exceeded by the duration of >20 months in the third episode of case one and ten and a half months in the third episode of case two, respectively, as observed in the present study. This very long duration, together with its co-occurrence with verbal palinacousis in case one, best discriminates musical palinacousis occurring as symptom of persistent aura without infarction from the common phenomenon known as 'earworms' (Levitin, 2006; Sacks, 2007), haunting melodies (Reik, 1953), tunes that get 'stuck in your head' (Kellaris, 2001) or musical imagery repetition (Bennett, 2002), respectively. In a discussion on Jacobs *et al.*'s (1971) pioneering study on palinacousis, Robert Cohn (1971) had asked: 'I would like to ask the authors of this very nice paper if their phenomenon isn't in some way similar to that which occurs when one goes to a concert and then for the next hours, hears the music again and again?' For its occurrence as a persistent migraine aura symptom, the answer can be given in the words of one of our patients (case two): 'Everyone has this sometimes, but not daily!?', and certainly not for over half a year, as was the case for both patients reported in this study.

References

1. Auzou, P. *et al.* (1995). Palinacousie avec hémianacousie par lésion temporale gauche. *Rev Neurol* **151**: 129–131.
2. Auzou, P. *et al.* (1997). Enregistrements EEG contemporains d'episodes de palinacousie et de palinopsie. *Rev Neurol* **153**: 687–689.
3. Barolin, G. S. (1963). Atypische Migränen. Klinik — Differentialdiagnose — EEG. *Wien Klin Wschr* **75**: 293–301.
4. Barolin, G. S. and Gloning, K. (1969). Neuropsychologische Störungen bei Migräne. *Wien Z Nervenheilk* **27**: 306–320.
5. Bekény, G. and Péter, A. (1973). Meningocerebral cysticercosis with auditory perseveration (palinacusis). *Confin Neurol* **35**: 236–247.
6. Bender, M. B. and Diamond, S. P. (1965). An analysis of auditory perceptual defects with observations on the localization of dysfunction. *Brain* **88**: 675–686.
7. Bennett, S. (2002). *Musical Imagery Repetition (MIR)* (Dissertation, Cambridge University, Cambridge). Accessed on April 12, 2008, at *http://www.seanbennett.net/music/Musical_Imagery_Repetition.doc.*
8. Bille, B. (1962). Migraine in school children. A study of the incidence and short-term prognosis, and a clinical, psychological and encephalographic comparison between children with migraine and matched controls. *Acta Paediat* **51**(Suppl. 136): 1–151.
9. Bonhoeffer, K. (1940). Dauerausfallserscheinungen bei Migräne. *Dtsch Med Wschr* **32**: 521–523.
10. Cohn, R. (1971). Discussion on Jacobs *et al.* Palinacousis or persistent auditory sensations. *Trans Amer Neurol Assoc* **96**: 126.
11. Di Dio, A. S., Fields, M. C. and Towan, A. J. (2007). Palinacousis — auditory perseveration: Two cases and a review of the literature. *Epilepsia* **48**: 1801–1806.
12. Evers, S. and Ellger, T. (2004). The clinical spectrum of musical hallucinations. *J Neurol Sci* **227**: 55–65.
13. Freudenreich, O. and McEvoy, J. P. (1996). Palinacousis after closed-head injury in a patient with schizophrenia. *J Clin Psychopharmacol* **16**: 94.

14. Fukutake, T. and Hattori, T. (1998). Auditory illusions caused by a small lesion in the right medial geniculate body. *Neurology* **51**: 1469–1471.
15. Griffiths, T. D. (2000). Musical hallucinosis in acquired deafness. Phenomenology and brain substrate. *Brain* **123**: 2065–2076.
16. Haas, D. C. (1982). Prolonged migraine aura status. *Ann Neurol* **11**: 197–199.
17. International Headache Society (2004). The International Classification of Headache Disorders (2nd ed.). *Cephalalgia* **24** (Suppl. 1): 1–160.
18. Jacobs, L., Feldman, M. and Bender, M. B. (1971). Palinacousis or persistent auditory sensations. *Trans Amer Neurol Assoc* **96**: 123–126.
19. Jacobs, L., Feldman, M. and Bender, M. B. (1972). The persistence of visual or auditory percepts as symptoms of irritative lesions of the cerebrum of man. *Z Neurol* **203**: 211–218.
20. Jacobs, L., Feldman, M., Diamond, S. P. and Bender, M. B. (1973). Palinacousis: Persistent or recurring auditory sensations. *Cortex* **9**: 275–287.
21. Kellaris, J. J. (2001). *Identifying Properties of Tunes that Get 'Stuck in Your Head': Toward a Theory of Cognitive Itch.* (Symposium conducted at the meeting of the *Society of Consumer Psychology* Winter Conference 2001.)
22. Kim, J. S., Kwon, M. and Jung, J. M. (2007). Palinacousis in temporal lobe intracerebral hemorrhage. *Neurology* **68**: 1321–1322.
23. Kruit, M. C. *et al.* (2004). Migraine as a risk factor for subclinical brain lesions. *JAMA* **291**: 427–434.
24. Levitin, D. J. (2006). *This is Your Brain on Music. The Science of a Human Obsession.* New York: Dutton-Adult.
25. Lippman, C. W. and Lippman, M. (1952). *Understanding Your Migraine Headache.* New York: Greenberg Publisher.
26. Liu, G. T. *et al.* (1995). Persistent positive visual phenomena in migraine. *Neurology* **45**: 664–668.
27. Malone, G. L. and Leiman, H. I. (1983). Differential diagnosis of palinacousis in a psychiatric patient. *Am J Psychiatry* **140**: 1067–1068.
28. Masson, C., Sztern, A., Cambier, J. and Masson, M. (1993). Palinacousie en relation avec une hémorragie temporo-pariétale droite. *Presse Méd* **22**: 596.
29. Meyer, H. -H. (1962). *Zur Klinik der Wahrnehmungsstörungen (Psychopathologische Phänomene bei temporaler Epilepsie).* In Kranz, H. (Ed.), *Psychopathologie heute*, pp. 193–204. Stuttgart: Thieme.
30. Patterson, M. C., Tomlinson, F. H. and Stuart, G. G. (1988). Palinacousis: A case report. *Neurosurgery* **22**: 1088–1090.
31. Podoll, K. (2008). *xjetgirlx.* Accessed on May 17, 2008, at *http://www.migraine-aura.org/content/e24966/e22874/e63421/index_en.html.*
32. Prueter, C., Waberski, T. D., Norra, C. and Podoll, K. (2002). Palinacousis leading to the diagnosis of temporal lobe seizures in a patient with schizophrenia. *Seizure* **11**: 198–200.

33. Reik, T. (1953). *The Haunting Melody: Psychoanalytic Experiences in Life and Music.* New York: Farrar, Straus & Cudahy, Inc.
34. Sacks, O. W. (1970). *Migraine: The Evolution of a Common Disorder.* Berkeley; Los Angeles: University of California Press.
35. Sacks, O. (2007). *Musicophilia. Tales of Music and the Brain.* New York: Knopf Publishing Group.
36. Shepard, A. and Sandridge, B. (1992). Palinacousis: A case report. *Neurology* **42**: 164.
37. Terao, T. and Matsunaga K. (1999). Musical hallucinations and palinacousis. *Psychopathology* **32**: 57–59.
38. Voll, C. L., Hickie, M. and Kelly, L. M. (2002). Palinacousis in association with a right parietal glioblastoma multiforme. *Can J Neurol Sci* **29** (No. 2 Suppl. 1): S60.
39. Wustmann, T. and Gutmann, P. (2007). Palinakusis bei Alkoholhalluzinose. *Psychiat Prax* **34**: 302–304.

Chapter 15

Coloured-Hearing Synaesthesia in Nineteenth-Century Italy

Lorenzo Lorusso and Alessandro Porro

In nineteenth-century Italy, several scientists (e.g. Botta, Verga, Berti and Lussana) became interested in the coloured-hearing phenomena. They published their observations on a number of coloured-hearing cases and started a debate on the neurological features of the phenomenon. These were the first hypotheses on colour-sound localisation, influenced by Gall's theories (Phrenology). The observations were the result of important discoveries on the mechanism of hearing sound performed in the eighteenth and nineteenth centuries by Domenico Cutugno and Alfonso Corti. Like French, German and English contributions to early research on synaesthesia, these Italian scientists provided original ideas on the colour-hearing process.

Historical Overview

Synaesthesia, which occurs when an individual receives a stimulus in one sense modality and experiences a sensation in another, has a long history. We can trace the interest in these kinds of cross-sensory relationships back to about the sixth century BCE and Pythagoras' quest to assign a particular colour to each musical note. In the fourth century BCE, Aristotle codified his idea that humans have five distinct senses, which was at odds with the concept of cross-modality (Tornitore, 1988; Manzoni, 2007). Later discussions in both philosophy and science were largely informed by the Aristotelian distinctions and subsequently historical documents show that synaesthesia was treated as a neurological abnormality.

A later figure generally cited in chronologies on synaesthesia is John Locke (1632–1704). Although it is often said that his work affirms that the knowledge of cross-sensory correspondence existed in earlier centuries (Locke, 1690; Mazzeo, 2005), this conclusion is misleading. Briefly, the Locke citation that is generally given in support of cross-sensory correspondence refers indirectly to his reformulation of the 'Molyneux Problem' conundrum: if a man born blind is to gain sight at a later age would he be able to identify the things around him? In his discourse, this philosopher speaks of the case of a blind man who associated the colour scarlet with the sound of a trumpet. Yet, while the cross-sensory connection inference is evident, the case is somewhat problematic in terms of validating the synaesthete experience since Locke's argument implicitly accepts the prevailing idea that the senses are separate (Ione, 2005; Mazzeo, 2005).

In 1644, the Englishman Kenelm Digby (1603–1665) wrote two treatises: the *Nature of bodies* and the *Immortality of reasonable souls*, where he described the importance of the phonetic language for deaf-mutes. Demonstrating the process of ocular hearing — 'to hear with eyes' or to hear the sound of words with eyes — Digby spoke of the art of observing the motion of sounds as if they are visible. In 1648, this aspect was worked up by philosopher John Bulwer (1606–1656) (Digby, 1644; Bulwer, 1648; Tornitore, 1999). Later, in 1677, Robert Boyle (1627–1691) discussed the ability of blind people to perceive colour by touching a wrinkled colored cloth (Boyle, 1677). These kinds of cases stimulated debate on Aristole's theory of the senses (Tornitore, 1999).

Newton's research on the nature of light and colours led to other important developments. In 1664, he used a prism to break up the sunlight. Two years later, in 1666, he affirmed that white light could be decomposed into several coloured rays, each of which is refracted with a different grade respect in relation to the others. These refractions generated seven primary colours in ordinary succession from minor refrangible red to more refrangible violet. Although Newton's research was influenced by Boyle's spectral pentachromatic light, Newton's correlation of light with seven notes was his own: what he concluded is that there is a relationship between amplitude of the band of colours and the wavelength of a musical cord. This conclusion led Newton to incorrectly suggest that sound/colour relationships create different vibrations, which in term stimulate the eyes and the ears and spoke and propagate to the brain. Newton's work was important for three reasons. Firstly, he approached the subjectivity of synaesthesia objectively. Secondly, what had previously been seen as a philosophical topic became an area for scientific study. Finally, and perhaps of most importance, is that when Newton extended the relationship between sound/colour to audition/vision, he introduced a sensorial analogy (Newton, 1671–1672; Tornitore, 1999). It is important to note that Newton's view of how sensations are connected offers a sharp contrast to those of W. Goethe. Whereas Newton viewed colour as a physical problem, involving light striking objects and entering our eyes, Goethe realised that the sensation of colour reaching our brain is also shaped by our perception. Goethe's work also demonstrated that Newton's colours hypotheses were unsatisfactory (Goethe, 1810; Chance, 1933).

Significant shifts relating to cross-modularity began to appear in the nineteenth century, which is a period of great importance to the history of synaesthesia. Scientific research was stimulated by the introduction of experimental psychology and the development of neurophysiology. Romanticism also flourished at this time. Indeed, many of the Romantics of the era had an interest in synaesthesia and, as discussed below, some of the literary figures used language that deliberately expressed the kinds of sensual relationships associated with synaesthesia. There was also a great interest in the relationship between humans and nature at this time. It was based on the idea that our sensorial messages had their origins in nature.

Key research shows how the scientific trajectory took form. Through his treatise, *The law of specific nerve energies* (1826), Johannes Müller, following the earlier insight of Charles Bell (1811), indicated that each sense modality has its own characteristic sensory quality, regardless of the physical means by which the peripheral nerve is stimulated (Bell, 1811; Müller, 1826). Müller's concept is deeply embedded in the analysis of brain function, and seems to negate the possibility of cross-modal activation in the cortex. Whether Bell and Müller's treatises were influenced by the romantic expression associated with music and a poetic form is open to debate.

The first medical reference to synaesthesia (1710) is attributed to Thomas Woolhouse (1650–1734), an English ophthalmologist. He discussed the situation of a blind man who claimed to experience coloured visions upon hearing certain sounds (O'Malley, 1957). The first historical description of synaesthesia seems to have been a medical treatise written in Latin published by GTL Sachs, in 1812: *Historia naturalis duorum leucaethiopum auctoris ipsius et sororis eius* (which could be translated into English as: 'The natural history of two albinos, the author himself and his sister') (Sachs, 1812). The author was known as 'Mister Sachs whitish' because of his albinism. The work, which was translated into German by JHG Schlegel, attracted interest among the medical community owing to its description of specific and invariant colour sensations evoked by vowels, consonants, musical notes, the sound of instruments, numbers, dates, days of the week, city names, periods of history and the stages of human life. Johannes Purkinje also investigated subjective visual sensations-hallucinations derived from the eye (Purkinje, 1819). His investigations covered numerous aspects of human biology. In congruence with this work, he published a series of papers devoted to subjective visual sensations, hallucinations, after-images and a wide variety of visual phenomena derived from the eye. In fact, his investigations represent the fullest treatment of the topic to this day. His classification system of 28 categories of entoptic and related phenomena seem closely related to synaesthesia, although Purkinje did not include the word in this classification.

Gustav Theodor Fechner (1801–1887) also conducted research in this area, seeking to resolve the mind-brain dichotomy through the science of psychophysics, or the direct measurement of sensation. His scientific approach to coloured sound led him to classify the colour-tone associations of 347 individuals, but he did not emphasise sensory quality as in synaesthesia (Fechner, 1876). In 1881, the psychiatrist Eugene Bleuler (1857–1939), in collaboration with his student Lehmann who had synaesthesia, found that about 12% of a sample of 600 people reported sensation of colour-vowel association. Bleuler sought to integrate Freudian psychoanalytic theory with newly emerging experimental psychology to describe the fractured mental state of patients affected by schizophrenia (Bleuler, 1881).

In Italy, the first publications on synaesthesia, better known as coloured hearing, appeared in the first half of the nineteenth century, with papers of Carlo Botta (1801), Andrea Verga (1864), Antonio Berti (1864) and Filippo Lussana (1865). All of these scientists contributed to the neurological debate surrounding synaesthesia.

Origin of the Word

Synaesthesia's Greek etymology (*syn* = together + *aesthesis* = perception) conveys the essence of coupling two or more senses such that a voice or music, for example, is not simply heard but also seen, felt, or tasted (Cytowic, 2002). The word 'synaesthesia' was disseminated for the first time during the Conference of Physiological Psychology in 1890. The year before, the term made its first appearance in the *Century Dictionary*, including the word 'coloured hearing', and began to have a common meaning (Simpson and Weiner, 1989; CNRS, 1992; Mazzeo, 2005). The term synaesthesia first entered the scientific vernacular in 1892 through Jules Millet's thesis on coloured hearing (*audition colorée*) (Dann, 1998). In 1865, the French physiologist Alfred Vulpian wrote the term 'synaesthesia' (CNRS, 1992; Mazzeo, 2005).

Although the term 'synaesthesia' came into use slowly, it is still possible to identify work in this area, as the previous section shows. Earlier we noted that Sachs offered his thoughts on a phenomenon that sounds like synaesthesia. He attributed his coloured hearing to the Daltonian phenomenon, which he called chromatodispsia because those affected were seeing too much colour. The ophthalmologist Cornaz considered the phenomenon to be a common, hereditary condition affecting only the male gender and thought of as an alteration of the eye perception, which he called 'hyperchromatopsie' or '*alteration du sens de la vue*' (Cornaz , 1848). In 1863, Perroud argued that the alteration was not only a peripheral sense, but also a defect of the central system. With the spread of experimental psychology, Perroud considered it a psychological defect because the problem laid not so much in a modification of perception as in the collection of ideas (Perroud, 1863). One year later, another French author, Chabalier, changed the term hyperchromatopsie (hyperchromatopsia) into pseudochromesthesia, confirming the psychological theory as '*un trouble des idées*' or mind disorders (a disorder of ideas or a false association of ideas) (Chabalier, 1864). At the same time, an Italian psychiatrist, Antonio Berti, read the French publication and concluded that the phenomena were a disturbance in the recall of ideas. Berti renamed it 'dyschromesthesia' (Berti, 1865). Another Italian, Filippo Lussana, described the relationship between sound and colour, calling this phenomena coloured voice or coloured hearing (Lussana, 1873; Tornitore, 1986). As the term synaesthesia spread among the nineteenth-century scientific community, we find that their studies were based in particular on sound and colour rather than on examining all the senses (Tornitore, 1986; Mazzeo, 2005). Further research on coloured hearing appeared in many countries: in English colour-hearing or colour-hearer, in French *audition colorée*, in Italian *audizione colorata* (O'Malley, 1957). In 1873, Nüssbaumer published his experience and that of his brother in Medizin Wochenschrift (Nussbaumer, 1873; Vescovi, 1895). In 1881, the London Medical Report and Lancet Cincinnati were the first Anglo-American publications to describe the condition (Lussana, 1884; O'Malley, 1957; Tornitore, 1986; Mazzeo, 2005) contributing to the start of a new neurological debate on synaesthesia.

Cultural Influence

As noted above, some of the interest in synaesthesia within the scientific community was due to the development of the literature, art and music of the Romanticism era. A rapid socio-economic transformation during this period was accompanied by a strong desire to retain a relationship with pre-industrial society and nature. This nurtured Romanticism and gave rise to studies of symbolism. Analogies between humanity and nature, based on different sensorial messages drawn from the natural world, were also formulated at this time. Much of this found expression in writing, we associate with synaesthesia in nineteenth-century literature. For these figures, synaesthesia was the key of universe and at the root of the creation of every work of art. Poets could correlate elements of nature with all the inner senses and poetry could reveal the mysteries underlying physical appearance. The interior world could be interrelated with the exterior world. Decadent poets such as Charles Baudelaire, Arthur Rimbaud, Julie Pierre Théophile Gautier, Joris-Karl Huysmans decided that synaesthesia was their *raison d'être* (O'Malley, 1957). Baudelaire's book, *Writings on Art* is a wealth of observations on profound connections among colour, sound and smell (Baudelaire, 1863). Twentieth-century artists and musicians such as Paul Klee, Vassily Kandinski, Johannes Itten, and Alexander Skrjabin deemed it fundamental to hear their interiority, since the soul is the root from which the five senses — and consequentially all arts — arise (Ione, 2006).

The Italian Contribution to Coloured-Hearing Synaesthesia

Influenced by the early French descriptions, the Italian neuroscientists of the nineteenth century became interested in coloured hearing after various cases of the phenomena were reported. They tried to explain the coloured-hearing process relying on the neurological knowledge of time. Through their research, these Italian scientists also developed research on the physiology of hearing.

Discoveries on the mechanism of hearing sound

Early in the eighteenth century, the physician Domenico Cotugno (1736–1822), who had a cultural interest in music (Gradenigo, 1918), improved the description of the sound mechanism formulated in 1704 by Antonio Maria Valsalva (1666–1723) (Valsalva,

1704) and in 1683 by Guichard Joseph Duverney (1648–1730) (Duverney, 1683). In 1761, Cotugno described inner hearing in his publication, *De aquaeductibus auris humanae internae* (Bianchi, 1923), where he formulated a theory on the physiology of hearing, postulating that shrieking caused vibration at the first part of the cochlea and grave sound at the summit of the cochlea (Mongardi, 1911). In conjunction with his discovery of the aural aqueducts, he found the labyrinth to be filled with fluid, and not air (Cotugno, 1761). Valsalva first found the presence of fluid, although he did not identify its role in the acoustic apparatus (Valsalva, 1704; Gradenigo, 1917). Cotugno recognised that the fluid was essential for sound transmission; in particular, he postulated that the vestibule perceived sound and the cochlea discerned tones. He formulated the early theory of resonance of the single sound (Cotugno, 1761; Levinson, 1936). The details of this theory were later improved on with the advent of new knowledge on acoustics and anatomy produced by Hermann von Helmholtz (1821–1894) (Helmholtz, 1863). In 1789, Antonio Scarpa (1752–1832) concluded his research on the macroscopic structure of the ear and discovered that the acoustic nerve crossed from internal hearing to the brain (Scarpa, 1789). The histological approach permitted Alfonso Corti (1822–1876) to describe the spiral organ that bears his name (Corti, 1851). Corti recognised that this spiral organ originated from the cochlea epithelium and represented the peripheral apparatus that transmits auditory excitement by the cochlea nerve. Ernst Reissner (1824–1878) completed the first study in 1851 (Reissner, 1851). Von Helmholtz formulated the final theory of resonance in which a single simple sound causes vibration in the specific sector of the resonate system through the radial fibres of the basilar membrane. These basilar membrane fibres, which differ in length and tension, vibrate as a result of resonance, according to the frequency of the sound perceived (Helmholtz, 1863). Over two centuries later, great advances were made in hearing apparatus, the neurological basis of sound transmission and its relation to the brain. These results explained the underlying mechanism of coloured-hearing synaesthesia.

Association of ideas phenomenon

In 1801, the physician, musician and politician, Carlo Botta (1766–1837) published a brief essay on the relationship between individual sound and different colours, according to Newton's theory on spectral light (Botta, 1801). He affirmed that only experts in music and musicians could posses this peculiar characteristic. He did not try to explain the phenomena, but he was interested in music therapy and its medical application in different disorders (Botta, 1860). Later, in 1864, renewed interest in relationship between sound and colour was demonstrated by the translation of a French paper by the psychiatrist, Andrea Verga (1811–1895) (Verga, 1864). Verga cited Chabalier's article in which 'pseudochromesthésie' was considered an alteration of the eye or a possible mental disturbance based on false association of ideas (Chabalier, 1864; Tornitore, 1986). One year later, in 1865, another psychiatrist, Antonio Berti (1812–1879) (Zago, 2006), published a brief comment on 'pseudochromesthesia', calling it a false sensation of colours, expressing skepticism about this 'new disorder'. He concluded that it was a psychic phenomenon related to an association of ideas based on mnemonics or memory technique. Berti also postulated that it was a subjective sensation that had no visual involvement but concerned memory (Berti, 1865). Berti's contribution terminated the pre-synaesthetic era based on visual disturbance (Tornitore, 1986). In the same period, the physiologist Filippo Lussana (1820–1897), under the pseudonym of Filinto, published another paper entitled *Lettere di Fisiologia morale dei colori* (*Letters on the moral physiology of colours*) (Lussana, 1865). In these letters, he provided a new name: coloured voice or language of colours and sought to advance a new hypothesis and mechanism (Fig. 1).

Lussana's scientific theory on coloured-hearing synaesthesia

Lussana adopted Newton's physic theory of sound/colour to explain coloured-hearing synaesthesia. He distinguished primary and composed colours, the latter made by the combination of the first.

Figure 1. Filippo Lussana.

According to Lussana, light rays are conducted by the optic nerves to quadrigeminal bodies and the optic thalamus where bright sensations are transformed into specific sensations. These sensations became ideas upon arriving at the centre of chromatic talent localised at the third frontal cerebral convolution connected with languages centres based on the phrenology theory of cerebral localisation: above the orbital area, where they are the centre for colour and melody of sound. Both centres are connected with a language area that confers a moral aspect to colours and sounds. The proportion between colour and sound is based on the timing of different musical notes and those of chromatic musical scale. Lussana elaborated a comparative scale based on colour and note vibrations (Fig. 2). He tried to explain how these vibrations could be transformed into emotions by different specific cerebral centres of chromatic and musical talents in relationship to the centre of language (Lussana, 1873).

dei toni musicali al minuto secondo	1° *do*	*re*	*mi*	*fa*	*sol*	*la*	*si*	2° *do*
	512	$512 \times 9/8$	$512 \times 10/8$	$512 \times 4/3$	$512 \times 12/8$	$512 \times 5/3$	$512 \times 15/8$	512×2
	512	576	640	683	765	853	960	1024
	$8/8$	$9/8$	$10/8$	$11/8$	$12/8$	$13/8$	$15/8$	$16/8$
dei raggi colorati trilioni al minuto secondo	*rosso*	*ranciato*	*giallo*	*verde*	*azzurro*	*endaco*	*violetto*	....
	500	532	563	607	630	690	735	....
	500	$500 \times 8{,}5/8$	$500 \times 9/8$	$500 \times 9{,}7/8$	$500 \times 10/8$	$500 \times 11/8$	$500 \times 11{,}8/8$	....
	$8/8$	$8{,}5/8$	$9/8$	$9{,}7/8$	$10/8$	$11/8$	$11{,}8/8$	....

Figure 2. Comparative scale between colour and sound based on number of vibrations.

Lussana and the centre of creativity

When the Italian physiologist Lussana stressed that there was a moral physiology of colours because sounds and colours have an 'emotional language', he also advocated an innate correspondence between cerebral centres and colours. Other authors believed that this phenomenon was occasional (Berti, 1864). Lussana's hypothesis had two primary merits: firstly, he hypothesises a neurological substrate for coloured hearing due to the nearness of colour and language centres, based on Gall's theory of phrenology. Secondly he correlated 'phonic language' and colour. Lussana cites Isaac Newton (1642–1727), who had incorrectly shown a correspondence between chromatic radiation and sound, separating the solar spectrum into seven intervals corresponding to seven notes of the musical scale. He had also suggested a parallel between 'the harmony of colours' and 'the melody of sound' (Newton, 1671–1672; Lussana, 1865; Finger, 1994).

In 1873, Lussana reported on two brothers he had met in 1865, one of whom was medical student with coloured-hearing synaesthesia (Lussana, 1873). This case provided Lussana with the opportunity to correlate the theory of colour, physiology and neurology and to fuse these three fields. In one theory of nerve specificity that he attributed to François Achille Longet (1811–1871) (Longet, 1850), Lussana formulated the correspondence between colours and sounds with the cerebral centre of colour being contiguous to the centre of language. He affirmed that the coloured-hearing mechanism was based on sound and hearing, but not on vision. The centres of sound, vision and language are adjacent and in mutual communication. Lussana, citing Nussbaumer, maintained the perception of sound caused colour to be seen because excitation of the auditory apparatus stimulated the cerebral centre of vision. Lussana's new idea was to demonstrate that these different centres were interconnected (Lussana, 1873; Tornitore, 1986). In accordance with the father of phrenology Franz Joseph Gall (1757–1828) (Gall, 1822–1826), Lussana located these centre in the frontal lobe, especially in the upper orbital area of the third cerebral convolution. He considered these areas to be at the centre of chromatic talent or the centre of creativity and correlated them with new studies on the cerebral map (Lussana, 1865; Lussana, 1873; Lussana, 1884) (Fig. 3). Lussana was the first to formulate a neurobiological theory of coloured-hearing synaesthesia, related to the newly emerging theory of language (Tornitore, 1986).

Overview on Contemporary Findings

A large number of publications on synaesthesia were produced at the end of the nineteenth century, which confirms that great interest had been roused in coloured-hearing synaesthesia. Between 1881 and 1931, 74 papers appeared in the literature (Marks, 1975; Harrison, 1995; Mazzeo, 2005). Individuals with the condition were the primary source for these accounts, which were quite subjective. The introspective type of description, unfortunately, proved to be a most unreliable method of disclosing the nature of human cognition since subjects tended to give information about their mental processes that was inconsistent with data that are more objective. Behaviourism emerged as the dominant psychological method in response to this

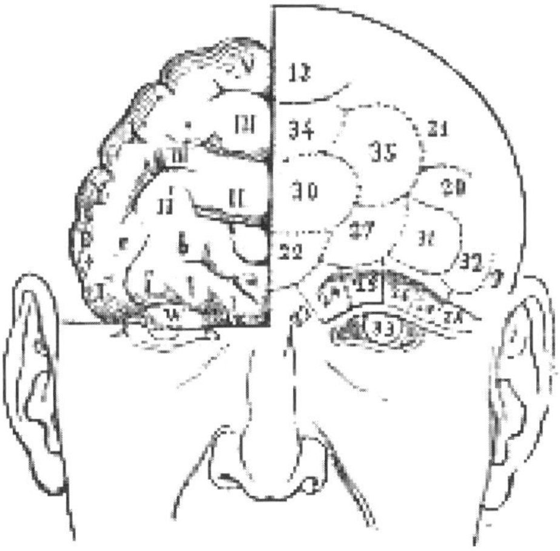

Figure 3. Localisation of centre of chromatic talent (I' 26), melody (c 32) and language (W 33) formulated by Lussana on the phrenology theory.

phenomenological data and in a bid to bring psychology into the field of natural science. As the name implies, this psychological approach concentrates only on observable behaviour whereas research on synaesthesia was based solely on self-reports and inner mental condition. Consequently, these accounts merely demonstrated the fortuity of behaviorist thinking, as evidenced by the fact that only 16 papers appeared on this topic between 1932 and 1974 (Marks, 1975; Harrison, 1995; Mazzeo, 2005). Behaviourism had failed to use scientific language to study mental states. The cognitivist model subsequently became the legitimate method of scientific investigation. With this, research on synaesthesia had a second renaissance, with renewed interest based a 'neurological reality approach'. Recent synaesthesia studies show that it is now seen as a way of illustrating how neuroscience might seek to reconcile subjective experience with objective analysis (Harrison, 1995). The advent of neuroimaging techniques provides a new and more objective method of studying the brain *in vivo*. This approach has provided objective information on differences in brain activity in subjects with synaesthesia for the first time. The neurological structure involved in synaesthesia is the limbic system (the cingulate gyrus, hippocampus), which has ascendancy over the cortex (posterior infero-temporal cortex and the junction of the occipital and parietal cortices) (Paulesu, 1995). In this context, synaesthesia becomes rational rather than emotional, although recent scientific studies have been unable to solve this rational/emotional dichotomy. Another important aspect is percept: visual, auditory, olfactory and others. This sensation structure allows information to be discretely identified as being specific to a sensory system. In the model which considers the senses individually, more importance is laid on the auditory than on the visual component. Most imaging studies have used spoken words to induce stimuli. Auditory information (in some cases exclusively linguistic) is carried beyond those areas of the brain concerned with auditory information and on to those dealing with visual information (Paulesu, 1995; Baron-Cohen, 1997; Mulvenna, 2006; Ward, 2006). Finally, research shows a relationship between synaesthesia and creativity. Cross-modality seems to be experienced disproportionately by artists, writers and musicians. This persistent connection with creativity is now being re-examined. According to the recent results, it seems that people with synaesthesia may have a greater aptitude for creative/abstract thinking (Mulvenna, 2006; Mulvenna, 2007).

Conclusion

In 1883, coloured-hearing synaesthesia was considered a rare phenomenon, affecting an estimated one in 20 people (Galton, 1883), fluctuating to one in 250,000 by the 1980s. In 1996, one person in 2,000 was thought to have synaesthesia. Subsequent estimates have continually risen and current large-scale screening of the population indicate that as many as one in 30 people at least have one subtype of synaesthesia. The attribute seems to run in families and recent genetic investigations suggested an X-linked trait (Baron-Cohen, 1996; Mulvenna, 2006). Although the modern era of synaesthesia research is still in its infancy, there are a number of ways to understand the cognitive processes underlying synaesthesia, which now appear to be widespread among the general population. In summary, since the nineteenth century studies on coloured-hearing synaesthesia have yielded important insights for the neurosciences: intra- and cross-modal perception, perceptual consciousness, brain development and plasticity, interaction with language and memory, and individual differences in cognition which are essential for creativity.

Acknowledgements

We are grateful to the Director of Biblioteca Civica 'Angelo Mai' — Bergamo, Giulio Orazio Bravi, and his personnel for their kind support of our research on Lussana's archive.

We are also grateful to Tonino Tornitore, because he is the first pioneer on the origin of the history of the coloured hearing synaesthesia.

Many thanks to Amy Ione, for her comments and suggestions on our manuscript.

References

1. Baratoux, J. (1883). De l'audition colorée. *Journal d'Hygiéne* **378**: 623–625.
2. Baron-Cohen, S. *et al.* (1996). Synaesthesia: Prevalence and similarity. *Perception* **25**: 1073–1780.
3. Baron-Cohen, S. and Harrison, J. (1997). *Synaesthesia: Classic and Contemporary Readings.* Oxford: Blackwell.
4. Baudelaire, C. (1863). *Ecrits sur l'Art.* Paris: Flammarion.
5. Bell, C. (1811). *Idea of a New Anatomy of the Brain: Submitted for the Observation of his Friends.* London: Staham & Preston. Reprinted in London: Medical Classics, (1936), pp. 105–120.
6. Berti, A. (1865). Della pseudocromestesia. *Archivio Italiano per le malattie nervose* **2**: 22–28.
7. Bianchi, L. (1923). Domenico Cotugno. *La Riforma Medica* **39**: 1–4.
8. Bleuler, E. and Lehmann, K. (1881). *Zwangsmässige Lichtempfindungen durch Schall und verwandte Ersheinungen auf dem Gebeite der andern Sinnes-empfindungen.* Leipzig: Fues's Verlag.
9. Botta, C. (1801). Memoires de la nature des sons et des tons. *Memoires de L'Acadèmie des science, littèrature et beaux arts* **12**: 191–214.
10. Botta, C. (1860). *Scritti minori.* Biella: Tipografia di Giuseppe Amosso.
11. Boyle, R. (1677). *Experimenta et considerationes de coloribus.* Genova: Amstelodami.

12. Bulwer, J. (1648). *Philocophus, or deaf and dumbe man friend.* London: Humphrey and Moseley.
13. Chabalier, D. (1864). De la pseudochromesthésie. *J de Médecine de Lyon*: 102.
14. Chance, B. (1933). Goethe and his theory of colors. *Annals of Medical History* **5**: 360–375.
15. CNRS (1992). *Trésor de la langue française. Dictionnaire de la langue du XIX et du XX siecle. Centre National de la Recherche Scientific.* Paris: Gallimard.
16. Cornaz, C. A. (1848). *Des abnormités congénitale des yeux et de leurs annéxes.* Lausanne.
17. Corti, A. (1851). Recherches sur l'organe de l'ouïe des mammifères. *Zeischrift für wissenschaftliche Zoologie von Siebold und Kölliker* **3**: 1–64.
18. Cotugno, D. (1761). *De aquaeductibus auris humanae internae.* Neapoli: Simoniana Tip.
19. Cytowic, R. E. (2002). *Synesthesia: A Union of the Senses.* Cambridge, MA: MIT Press.
20. Dann, K. T. (1998). *Bright Colors Falsely Seen: Synaesthesia and the Search for Transcendental Knowledge.* New Haven: Yale University Press.
21. De Vescovi, P. (1895). *Audizione colorata e visione cromatica dei suoni.* Roma: Tipografia di Giovanni Balbi.
22. Digby, K. (1644). *The Nature of Bodies.* Paris: Gilles Blaizot.
23. Digby, K. (1644). *Immortality of Reasonable Soules.* Paris: Gilles Blaizot.
24. Durveney, G. J. (1684). *Tractatus de organo auditus, continens structuram, usum et morbos omnium auris partium.* Nuremberg: Impensis Johannis Ziegeri.
25. Fechner, G. T. (1876). *Vorschule der Aesthetik.* Leipzig: Breitkop & Hartel.
26. Finger, S. (1994). *Origins of Neuroscience.* Oxford: Oxford University Press.
27. Flournoy, T. (1893). *Des phénomès de synopsie (audition colorée).* Paris: Felix Alcan.
28. Gall, J. F. (1822–1826). *Sur les fonctiones du cerveau.* Paris: JB Ballière.
29. Galton, F. (1883). *Inquires into Human Faculty and its Development.* London: London Press.
30. Goethe, W. (1791–1792). *Beiträge zur Optik.* Weimar: Zur Farbenlehere.
31. Gradenigo, G. (1916). Fu Helmholtz veramente l'ideatore della teoria sulla audizione che porta il suo nome? I precursori: Duverney (1683), Valsalva (1704). L'ideatore: Cotugno (1760). *Gior R Accad Med Torino* **22**: 453–463.
32. Gradenigo, G. (1917). La teoria sulla audizione Cotugno-Helmholtz. Nota II. I precursori: Antonio Maria Valsalva (1704). *Gio R Accad Med Torino* **23**: 217–229.
33. Gradenigo, G. (1918). Domenico Cotugno e la teoria sulla audizione. *La Riforma Medica* **24**: 671–673.
34. Grazzi, V. (1883). L'udizione colorata *L'imparziale* **10**: 317–322.
35. Harrison, J. (2002). *Synaesthesia. The Strangest Thing.* Oxford: Oxford University Press.

36. Harrison, J. and Baron-Cohen, S. (1995). Synaesthesia: Reconciling the subjective with the objective. *Endeavour* **19**: 157–160.
37. Helmholtz, H. (1862). *Die Lehre von dem Tonempfindungen.* Braunschweig: Vieweg.
38. Ione, A. (2005). Synesthesia's multi-colored history meets the twenty-first century. *Actas Primer Congreso Internacional Arte y Sinestesia, Cuevas del Almanzora* (Almería — España), 25–28 Julio de 2005.
39. Ione, A. (2006). Neurology, synaesthesia, and painting. *Inter Rev Neurobiol* **74**: 69–78.
40. Ione, A. and Tyler C. (2004). Synesthesia: Is F-sharpe colored violet? *J Hist Neurosci* **13**: 58–65.
41. Levinson, A. (1936). Domenico Cotugno. *Annals of Medical History* **8**: 1–9.
42. Locke, J. (1690). *An Essay Concerning Human Understanding.* London: Basset.
43. Longet, F. A. (1850). *Traité de physiologie.* Paris: Victor Nasson.
44. Lussana, F. (1865). Lettera seconda. Fisiologia morale dei colori. *Archivio Italiano per le malattie nervose* **2**: 141–148.
45. Lussana, F. (1873). Fisiologia dei colori. In Sacchetto, F. (Ed.), *Piccola Biblioteca Medica, Volume V.* Padova: Premiata Tipografia.
46. Lussana, F. (1884). Sull'udizione colorata. *Archivio Italiano per le malattie nervose* **21**: 371–377.
47. Manzoni, T. (2007). *Aristotele e il cervello. Le teorie del più grande biologo dell'antichità nella storia del pensiero scientifico.* Roma: Carocci Editore.
48. Marks, L. (1975). On colored-hearing synesthesia: Cross-modal translation of sensory dimensions. *Psychol Bull* **82**: 303–331.
49. Mazzeo, M. (2005). *Storia naturale della sinestesia. Dalla questione Molyneux a Jakobson.* Macerata: Quodlibet Studio.
50. Mongardi, R. (1911). Da Cotugno a Helmholtz. *Arch Ital Otol* **22**: 393–400.
51. Müller J. (1862). *Zur vergleichenden Physiologie des Gesichtssinnes des Menchen und der Thiere.* Leipzig: C. Cnobloch.
52. Mulvenna, C. and Walsh, V. (2006). Synaesthesia. *Curr Biol* **15**: 399–400.
53. Mulvenna, C. M. (2007). Synaesthesia, the art and creativity: A neurological connection. In Bogousslavsky J. and Hennerici M. G. (Eds.), *Neurological Disorders in Famous Artists.* Basel: Karger.
54. Newton, I. (1671–1672). New theory about light and colours. *Phil Trans R S London* **80**: 3075–3087.
55. Nussbaumer, F. A. (1873). Ueber subjective Farbenempfindungen die durch objetive Gehörempfindungen erzuegt werden. *Wiener medic. Wochenschrift* **1–2**: 4–7; 28–31; 52–54.
56. O'Malley, G. (1957). Literature synaesthesia. *The Journal of Aesthetics and Art Criticism* **15**: 391–411.
57. Paulesu, E., Harrison, J., Baron-Cohen, S., Watson, J. D. G., Goldestein, L., Heather, J., Frackowiak, R. S. J. and Frith, C. D. (1995). The physiology of

colored hearing: A PET activation study of word–word synaestehsia. *Brain* **118**: 661–676.

58. Perroud, M. (1863). *De l'hyperchromatopsie.* Lyon: Memoires et Comptes Rendus de la Societé médicale de Lyon.
59. Purkinje, V. J. (1819). *Oper Omnia.* Prague: Calve.
60. Reissner, E. (1851). *De auris internae formatione.* Dorpat: Universitat Dorpat.
61. Sachs, G. T. L. (1812). *Historia naturalis duorum leucaetiopum auctoris ipsius et sororis eius.* Solisbaci: Sumptibus Bibliopolii Seideliani.
62. Scarpa, A. (1789). *Anatomicae disquisitiones de auditu et olfactu.* Ticini: Tipographeo Petri Galeati.
63. Simpson, J. A. and Weiner, E. J. C. (1989). *The Oxford English Dictionary.* Oxford: Clarendon Press.
64. Tornitore, T. (1986). *Storia delle sinestesie. Le origini dell'audizione colorata.* Genova: Brigati-Carucci.
65. Tornitore, T. (1988). *Scambi di sensi. Preistorie delle sinestesie.* Torino: Centro Scientifico Torinese.
66. Tornitore, T. (1999). Vedere i suoni, udire i colori. Le prime testimonianze sulle sinestesie soggettive e oggettive. *Atti dell'Accademia Ligure di Scienze e Lettere* **2**: 403–426.
67. Valsalva, A. M. (1704). *De Aure Humana.* Bologna: Costantino Pisario.
68. Verga, A. (1864). La pseudocromestesia. *Gazzetta Medica Italiana-Lombardia* **49**: 426–427.
69. Ward, J. and Mattingley, J. B. (2006). Synaesthesia: An overview of contemporary finding and controversies. *Cortex* **42**: 129–136.
70. Zago, S. and Randazzo, C. (2006). Antonio Berti and early history of aphasia in Italy. *Neurol Sci* **27**: 449–452.

Chapter 16

Crossed Wires: Synaesthetic Responses to Music

Ivan Moseley

'I have had a most rare vision... The eye of man hath not heard, *the ear of man hath not seen*, man's hand is not able to taste, his tongue to conceive, nor his heart to report what my dream was'.

Shakespeare *A Midsummer Night's Dream* Act 4, Scene 1

Synaesthesia: Definitions and Characteristics

The subjective phenomenon of synaesthesia, from *syn* together *aisthesis* perception, has been defined as 'an involuntary physical experience of a cross-modal association', or more simply, a 'crossing of the senses'. It may take a number of forms, such as *coloured-olfaction* in which a given smell excites a sensation of colour, *coloured gustation*, in which the primary stimulus is taste, and a variety of visual and auditory phenomena, such as *coloured numbers* (including dates) or *shaped numbers*, *coloured letters* (particularly vowels) and *coloured graphemes*, in which words generate a colour sensation, often depending on their first letter. Visual phenomena may not be involved at all, as in *tactile-gustation*, where touching a particular surface evokes an associated taste. However, the form with which I shall deal principally is *chromaesthesia*, in which sound evokes subjective visual phenomena, often in the form of a perception of colour.

The sounds may not be musical. Indeed, in one study of 778 individuals who claimed to experience synaesthetic phenomena, music was the cause of colour sensations in only 75 (9.6%); while one (0.1%) described taste phenomena stimulated by music.[1] Colour sensations are the sensory phenomena most ubiquitously reported by synaesthetic individuals,[2] but non-visual responses are not uncommon.

Joris-Karl Huysmans[3] envisaged a (fictional) character who made music-taste connections: 'dry curaçao matched the clarinet whose note is penetrating and velvety; kummel, the oboe with its sonorous, nasal resonance; crème de menthe and anisette the flute, at once honeyed and pungent, whining and sweet'. Beeli, *et al.*[4] reported a 27-year-old professional musician who had colour-responses to individual notes (e.g. red to C, violet to F sharp), but who also experienced tastes in response to hearing different intervals: 'sour' with a minor second or major seventh (its inversion); 'bitter' with a major second or minor seventh (again, its inversion); 'salty' and 'sweet' with minor and major thirds, respectively; 'cream' and 'low-fat cream' with minor and major sixths (inversions of the thirds); 'mown grass' — possibly a smell rather than a taste — with a fourth; 'pure water' with a fifth; and 'disgust' with a tritone. (No reference was made to visual 'combination tones' when the two notes forming the interval were sounded.)

Francis Galton, writing in 1883, noted that 'the existence of the colour associations with sound is... fully remarkable... the vowel sounds chiefly evoke them. The seers are invariably most minute in their description of the precise tint and hue of the colours. They are never satisfied, for instance, with saying "blue", but will take a great deal of trouble to express or to match the particular blue... no two people agree, or hardly ever do so, as to the colour they associate with the same sound. Lastly... the tendency is very hereditary'.[5]

The American neurologist Richard Cytowic has proposed a list of features which characterise what we may term true synaesthesia. The evoked sensations are *involuntary* and cannot be suppressed and are perceived as *projected into the environment*, not just existing within the head. They are *durable* and *generic*; thus, in one individual they will always be the same given the same stimulus. Finally, they are *memorable*, and may indeed be the best-remembered aspect of the situation in which they are evoked, and are *emotional*: 'having this experience causes ecstasy and is viewed as an accomplishment'.[6]

Synaesthesia may be idiopathic ('developmental') or the result of a known event or disorder. Acquired synaesthesia (particularly chromaesthesia) may be due to hallucinogens such as LSD, mescaline or peyote (*Lophaphora williamsii*, a small cactus which grows mainly underground, found, for example, in central northern Mexico), but is rare even when these substances are ingested. Théophile Gautier,[7] experimenting with hashish, found that '*J'entendais le bruit des couleurs. Des sons verts, rouges, bleus, jaunes, m'arrivaient par ondes parfaitements distinctes*', — 'I was hearing the sound of colours. Green, red, blue and yellow sounds came to me in perfectly distinct waves'.

It may also be due to pathology, such as epilepsy, often arising in the temporal lobe, but is seen in less than 7% of people with this condition. A notable musical example was George Gershwin, whose temporal lobe glioma was heralded by olfactory hallucinations in the form of a 'burning rubber' smell. Other types of damage to the central nervous system, which can result in inappropriate subjective sensory phenomena include concussion; sudden noise or bright lights can cause pain in the body or limbs for a time after a head injury. Visual phenomena, bizarrely, also occur in about 12% of individuals with spinal cord injuries, and with neuronal degeneration: blind individuals may experience flashes of light.

Brain damage causing synaesthesia usually involves the optic nerve; about one third of neurones in ventral premotor cortex can respond to more than one type of sensory stimulus. One of the most famous examples of chromaesthesia in a blind man was reported by John Locke: '... a studious blind man [who] mightily beat his head about visible objects, and made use of the explications of his books and friends, to understand those names of light and colour, which often came in his way, bragged one day that he now understood what *scarlet* signified. Upon which, his friend demanded what scarlet was? The blind man answered, 'it was like the sound of a trumpet'.[8]

At least one person in 25,000 has idiopathic synaesthesia, although some estimates are as high as 12% of the population; the level accepted almost certainly depends on diagnostic criteria. It is significantly more common in females; estimates vary from three to eight times as common. It probably has autosomal dominant inheritance, via the X-chromosome — but monozygotic twins, both male and female, of whom only one of each pair had synaesthesia have been reported,[9,10] in keeping with incomplete penetrance; generations may be skipped.[11] It is said that synaesthesia occurs more commonly in the left-handed, and that synaesthetes have a higher incidence of allochiria (right-left hand confusion). However, McManus, in an exhaustive study of left-handedness,[12] makes no mention of any association with synaesthesia, and points out that in some studies less than half of unselected individuals *never* experience right-left confusion, while observing that left-handedness is slightly more common in males, and debunking the 'vulgar error' that left-handedness is associated with 'greater creativity overall'. Other claims include that 15% of synaesthetes are first-degree relatives of someone with dyslexia, autism or attention deficit disorder; that they are more prone to *déjà vu*, clairvoyance, 'precognitive dreams' and the feeling of a presence; and that their mathematical abilities and spatial navigation are below average.

Mechanisms of Synaesthesia

A number of theories have been advanced to explain synaesthesia, the earliest of which spring from the idea of both light and sound as waves. Aristotle[13] suggested in 350BC that 'colours may stand in relation to each other in the same manner as concords in music, for the colours which are (to each other) in proportions corresponding with the musical concords, are those which appear to be the most agreeable'. Almost 2000 years later, Isaac Newton, in his *Optics* of 1704 indicated 'a relationship between the colours of the rainbow and the notes of the musical scale', such that an octave would be violet, or a perfect fifth green. Johann Wolfgang von Goethe first embraced, then rejected the idea of an analogy between musical tone and light: 'Colour and sound do not admit of being directly compared in any way, but both are referable to a higher formula, both are derivable, though each for itself, from this higher law. They are like two rivers which have their source in one and the same mountain, but subsequently pursue their way under totally different conditions, in two totally different regions, so that throughout the whole course of both no two points can be compared. Both are general, elementary effects acting in accordance to the general law of separation and tendency to union, of undulation and oscillation, yet acting thus in wholly different provinces, in different modes, on different elementary mediums, for different senses';[14] writers on synaesthesia frequently omit reference to Goethe's change of mind. The difference between electromagnetic light waves and mechanical sound waves became clear (although by no means uncontested at the time[15]), with the publication of Helmholtz's *On the Sensations of Tone*, the first edition of which appeared in 1863.[16]

Goethe did, however, point out that colour perception depended on the brain of the perceiver, as well as on the wave content of the incident light: 'We are never deceived; we deceive ourselves… that I am the only person in this century who has the right insight into the difficult science of colours, that is what I am rather proud of, and that is what gives me the feeling that I have outstripped many'.[14]

Cytowic postulates an *uninhibited natural state*: that all or most individuals are synaesthetes, but not consciously aware of the effects of the phenomenon. While most people, for example, would be unable to answer, or even consider a question such as 'what is the taste of 13 euros?', many could respond to 'what colour is Tuesday?', and a number of those who were unable to do so would not find the question ridiculous. The distinguished British psychologist Charles Myers FRS, who subsequently interviewed Skriabin while the latter was on a tour of Britain,[17] described a similar *basic unity of the special senses* (*Urempfindung*), which then becomes 'elaborated, differentiated and dissociated'.[18] It is a common observation that young children respond to stimuli in an undifferentiated way, and subsequently in a more modular fashion, which would be compatible with a state of *neonatal synaesthesia*, which many 'grow out of'. Thus, we can perhaps talk of a *unity of the senses* or *linkage*: a primal generalised sensory response becomes differentiated into hearing, colour, etc.

The answer may simply be that there is *crosstalk*; that auditory and visual pathways cross physically or that in synaesthetes their interconnections may be more numerous or used more easily. There is debate as to whether this would happen at the cortical level, or in the brain stem. If the latter, the region of the geniculate bodies would appear a good candidate: auditory responses are channelled to the cortex via the medial geniculate body, while the immediately adjacent lateral geniculate body is part of the visual pathways. At the cortical level, the connection could simply be *cognitive*; although the synaesthetic response is perceived as spontaneous by the individual, prior mental associations, such as red and trumpets, of which the subject is now unaware, may nevertheless be involved.[19] I discuss this further below.

The fact that some of these associations appear initially to be acquired may not exclude their being true synaesthesia. Darwin 'noticed that children have much more difficulty acquiring colour words than in picking up names for things, and was rightly surprised... he had expected that colours would be cardinal among our concepts, and easy to learn by abstraction. They are not, at first'.[20] One should also bear in mind that colour-blindness was not formally recognised until its description by Dalton (himself a deuteranope) around 1800. Charles Taylor has emphasised that awareness of the self as distinct from the world and God's creation occurred in the West only in relatively recent history.[21]

Music-Induced Synaesthesia: The Nature of the Stimulus

Whatever the central (or peripheral) nervous systems involved in music-generated colour experiences, what in the sensory stimulus provokes the synaesthetic response? The candidates include single notes, keys, timbre, 'sound complexes' and sequences of combinations of these, and each of these may vary, depending on the individual synaesthete.

Individual pitches

In the eighteenth century, Père Castel devised a *clavecin oculaire*, a keyboard instrument on which coloured lights would be seen by the listener, corresponding to the various degrees of the scale heard in the music being played.[22]

His selection of colours to match individual pitches (as seen in Fig. 1) is clearly related to the spectrum of visible light, as Skriabin's would be almost 200 years later.

W. S. Colman described a violinist who could strike a note on the piano, then, some time later, tune his instrument by seeking the same evoked colour.[23]

Brooks Kerr, a partially sighted jazz pianist, perceived individual pitches as having a colour: He fingered a C on the piano, explaining, "This note is red". He hit a D. "This one is dark blue". He hit an F. "This is yellow". His finger wandered to a G. "This one is light blue".[24] The composer György Ligeti (1923–2006) reported finding major chords 'red or pink', minor chords 'somewhere between green and brown'; however, 'when I say that C minor has a rusty red-brown colour and D minor is brown, this does not come from the pitch but from the letters C and D. I think it must go back to my childhood. I find, for instance, that numbers also have colours; one is steely grey, two is orange, five is green'.[25]

There are obvious problems: pitches which consistently evoke certain colour sensations must be identified as such, which is predicated on the subject recognising (albeit unconsciously) a given pitch. Only about 20% of musicians have 'perfect pitch', and such recognition may also be affected by changes in tuning and temperament. When musicians employ baroque tuning, approximately a semitone below the customary A = 440 Hz concert pitch, does a synaesthete sensitive

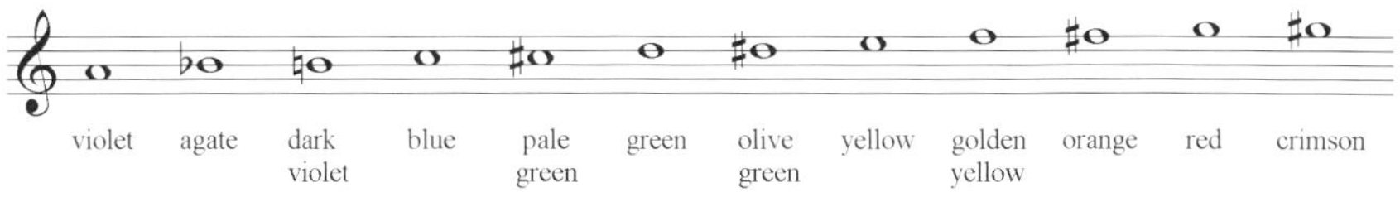

Figure 1. Castel's allocation of colours to pitches.

to individual pitches have the reactions to a sounded C that he or she would normally have to a concert-pitch B? Is that complicated by knowing the nominal key of the piece being played: would the notes of Bach's *Brandenburg Concerto No. 1* at baroque pitch have the same associations as a performance at concert pitch, because the subject regarded the piece as still being in F major rather than E major?

A different problem arises with enharmonic notation and relative pitches. On keyboard instruments, F sharp and G flat are the same; but do they elicit the same synaesthetic reactions when related differently to other notes? When a piece in C major begins to modulate towards the dominant key, G, an F sharp is perceived as what its name suggests: a sharpened fourth degree of the scale. If a G flat is introduced as part of a modulation towards D flat, are its synaesthetic consequences the same?

Scruton has identified ways in which we do not appreciate pitch and colour in the same way. There is 'no equivalent of colour-kinds — no areas of pitch continuum which belong together as do blues or greens'; and there is no equivalent of the octave in colour.[26]

The role of individual pitches in engendering synaesthetic responses is difficult to understand when they change very rapidly or form part of consonant or dissonant chords (C natural in C major and A flat major triads, or in a D flat major seventh, for example) and particularly in atonal music.

Keys

E. T. A. Hoffmann (1776–1822), the polymath writer, artist and composer of operas, orchestral and chamber works, created in *The life and opinions of the tomcat Murr, together with a fragmentary biography of Kapellmeister Johannes Kreisler on random sheets of waste paper* a fictional composer (with whom Brahms identified[27]), who had a 'coat the colour of C sharp minor, with an E major coloured collar'. Sacks[28] suggests that this seems 'too specific to be a metaphor… and that Hoffmann himself had colour-music synaesthesia or was… acquainted with the phenomenon'. Given that Hoffmann, a composer and wit, would have been aware that C sharp minor and E major have the same key signature, his reference to the coat may simply have been an ingenious musical joke. (Sacks seems to have trouble with key signatures: he reports Michael Torke, a synaesthetic composer, as saying that D major is blue. 'Context, however, is also important; thus Brahms's *Second Symphony* is in D major [blue], but one movement is in G minor [ochre]'; the symphony's second movement is in B major (as any horn player, faced with the notoriously difficult transposition — a diminished fifth down — knows only too well), the third in G major.[29])

Rimsky-Korsakov is said to have made the following connections: C major: white; G major: brownish gold, light; D major: daylight, yellowish, royal; and A major: clear, pink.[30] However, when Skriabin (1872–1915) was interviewed shortly before his death by Charles Myers, as well as making disparaging remarks about the older composer's claims, he said that 'whereas to him the key of F sharp minor appears violet, to Rimsky-Korsakov it appeared green; but this deviation [he] attributes to an accidental association with the colour of leaves and grass arising from the frequent use of this key for pastoral music'. Skriabin added that Beethoven's music was 'too intellectual' to evoke colour sensations, whereas the 'more psychological' modern music had strong effects: 'the colour underlines the tonality; it makes the tonality more evident'.[17]

Skriabin is often cited as a music-colour synaesthete, but it has been suggested that his aligning of colours with keys is related closely to the cycle of fifths (a series in which each note is the dominant of, i.e. a fifth above, the one which precedes it: C, G, D, A, E, B, F sharp, C sharp, G sharp, D sharp (E flat), B flat, F, C), suggesting that its basis was intellectual rather than purely sensory.[31]

The American composer Amy Beach (1867–1944) is said to have been a synaesthete who related colours to keys. 'For instance, Amy might ask her mother to play the "purple music" or the "green music".[32] 'Amy's mother encouraged her to relate melodies to the colours blue, pink, or purple, but before long Amy had a wider range of colours, which she associated with certain major keys... C was white, F sharp black, E yellow, G red, A green, A flat blue, D flat: violet or purple, and E flat pink.'[33] The second of these statements raises questions about whether this was true, idiopathic synaesthesia, or pseudo-synaesthesia, colour by association, particularly given the identification of C major with white (shared with Rimsky-Korsakov[30]), remembering that on most keyboard instruments the C major scale uses only the 'white' keys and its antipode, F sharp major as black.

Some of the problems mentioned above in connection with single pitches apply here: pitch perception, tuning, rapid changes and enharmonics (the last of these possibly more forcefully) and atonality.

Timbre

For the Swiss composer Joachim Raff (1822–1882) sounds of instruments produced colour impressions of various kinds. Thus, 'the sound of a flute produced the sensation of intense azure blue; of the oboe, yellow; cornet, green; trumpet, scarlet; the French horn, purple; and the flageolet, grey'.[34] The clearest and most distinct shades were those evoked by the high notes. Charles Myers,[35] a medically qualified psychologist and amateur violinist, himself associated colours with singing voices, but this may, of course, not be solely related to timbre; tessitura, attack and vibrato may also be involved.

The bandleader Duke Ellington noted that, 'I hear a note by one of the fellows in the band and it's one colour. I hear the same note played by someone else and it's a different colour',[36] while Leonard Bernstein (1918–1990), the American composer and conductor, spoke of experiencing 'timbre to colour' synaesthesia, in his *Young People's Concerts* series.

The contemporary composer Sean Day, who claims to be a genuine synaesthete, finds 'flutes... off-white, with shadings of blue and silvery flashes [and] cellos... dark cherry wood with green flecks', a description which again suggests colour-association.[1]

More complex stimuli

When Liszt first began as Kapellmeister in Weimar (1842), it astonished the orchestra that he said: "O please, gentlemen, a little bluer, if you please! This tone type requires it!" Or "That is a deep violet, please, depend on it! Not so rose!"[37] (It is of interest that in many texts on synaesthesia involving translation, 'rose' is often substituted for 'pink'. Many non-synaesthetes recognise some degree of connection between colours and their names in their native language — white, black, blue, brown — and these may be similar in a closely related language such as German: weiss, schwarz, blau, braun. However, the English word 'pink' is a striking exception: it does not have the deep-tone association of 'rosa' in German and Spanish or 'rose' in French. The name of the colour may even explain why it is one the English do not take seriously: 'shocking pink'!) This has been taken as clear evidence for synaesthesia, rather than that he may have been using a metaphor.

Donath[38] reported an individual who heard the flute as blue, the oboe as green and the violin as yellow, while the British neurologist Macdonald Critchley described a patient with a right frontal meningioma who associated the sound of a military band with 'nasty red', while a subdued, mellifluous dance band was 'misty blue'.[39]

Olivier Messiaen (1908–1992) was highly critical of explanations based on such simple phenomena as the foregoing, finding them naïve and childish, and said that colour resulted from 'sound complexes [chords] of greater intricacy than isolated notes or a conventionally tonal vocabulary'; he claimed to have synaesthetic responses even when reading a score. Their presence, however, was 'only fleeting; this extreme evanescence is complicated by the simultaneous appearance of several colours in dizzying juxtapositions which endure scarcely long enough to be glimpsed before they give way to a new array [so that] even to describe what is seen is not a simple task'![40]

It is well known that Messiaen made use of 'modes of limited transposition';[41] he also associated colours with these modes. Mode 2 (see Fig. 2) he associated with violet/purple; blue; rose/mauve, while Mode 3 (2) (see Fig. 3) 'the best of all my modes', evoked sensations of horizontally layered stripes: from bottom to top, dark grey, mauve, light grey, and white with mauve and pale yellow highlights — with flaming gold letters, of an unknown script, and a quantity of little red

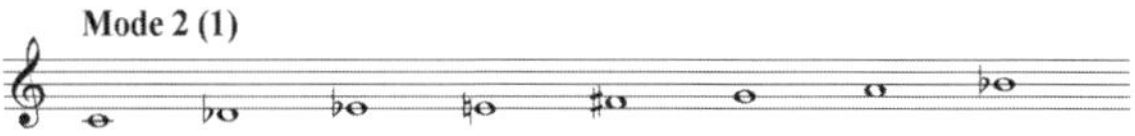

Figure 2. Messiaen's Mode 2, second version.

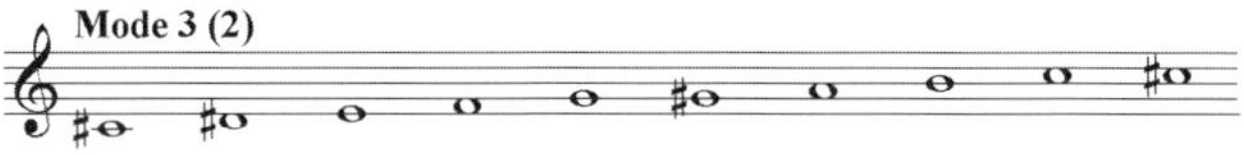

Figure 3. Messiaen's Mode 3, second version.

or blue arcs that are very thin, very fine, hardly visible... dominant are grey and mauve'.[40]

Messiaen was, however, strikingly inconsistent in his views; firstly, as to whether he was actually a synaesthete. 'Without suffering from physiological synaesthesia (as did my friend Blanc-Gatti, the painter, who had a disorder of the optic and aural nerves that allowed him actually to see colours and shapes when he heard music), when I hear a score or read it, hearing it in my mind, I visualise corresponding colours which turn, shift and combine, just as the sounds turn, shift, and combine, simultaneously'... 'One of the great dramas of my life... consists of my telling people that I see colours whenever I hear music, and they see nothing, nothing at all. That's terrible. And they don't even believe me... When I hear music — and it was already like that when I was a child — I see colours. Chords are expressed in terms of colours for me — for example, a yellowish orange with a reddish tinge. I'm convinced that one can convey this to the listening public'.[40]

Secondly, as to how personal sound-colour associations are: 'The influence of [Blanc-Gatti] and of Dukas's scene with the gemstones [in *Ariane et Barbe-Bleue*] caused me to think about my own sensations. I was my own physician, and I elaborated a theory that is doubtless valid only for me' but 'I do believe that perceptions would agree with each other in broad outlines. One couldn't see a yellow instead of a red, or a blue instead of a green. But there are small details — I, for example, see grey, mauve and gold tones in the second transposition of the third mode, of which I'm especially fond. Other people see it differently, but whatever happens, they'll see something grey and violet; it wouldn't be exactly mauve and gold, perhaps, but it would be a similar colour, most certainly not red'.[40]

Finally, in 1979 he claimed that 'the sound-colour relationship is important 'above all else', even rhythm', after having said in 1958: 'Let us not forget that the first, essential element in music is rhythm'.[40]

I know of no studies exposing individuals with synaesthetic responses to music to different recordings, with possible differences of pitch, timbre, etc. Jane Mackay addresses this question in her recently published *The Turn of the Screw: Visual Responses to Britten's opera*:[42] 'All the vocal and instrumental colours... are derived from Steuart Bedford's 1991 recording... Another recording or a live performance would probably not prompt major changes in synaesthetic perception but different voice colours and diverse instrumental balances would certainly stimulate other factors which might have been incorporated'.

Colour in Composition

Messiaen was not, of course, the first composer to indicate colour-correspondences in his music. Skriabin included a part for 'clavier à luce', to control a battery of coloured lights in his tone-poem *Prometheus* of 1910 (see Fig. 4). The lower of the two lines indicated a colour which would bathe the auditorium, and the upper a rapidly changing display, with a new colour for each key.

In 1922, Arthur Bliss gave the movements of his *A Colour Symphony* the names of colours, with explanations: *Purple*: 'The colour of amethysts, pageantry, royalty and death' — a slow processional march; *Red*: 'The colour of rubies, wine, revelry, furnaces, courage and magic' — a scherzo ending in 'a blaze of scarlet flame'; *Blue*: 'The colour of sapphires, deep water, skies, loyalty and melancholy' — like 'the lapping of water against a moored boat or stone pier'; and *Green*: 'The colour of emeralds, hope, youth, joy, spring and victory'.

Figure 4. A transcription of part of the 'clavier à luce' part from *Prometheus*.

Messiaen's approach was to score his music so that the colours it was intended to represent would be transmitted to his audience. He would, however, indicate these correspondences in the score; thus, in his *Couleurs de la Cité Célèste* he marks certain passages 'yellow topaz, light green chrysoprase and crystal'; 'here, the principal colour and the background of the music is the brass — the clarinets are in the background'; 'the ensemble must give the impression of the abundant colours of stained glass in the sun'; 'red with blue spots... the solo piano to keep the pedal depressed, and let the strings resonate while the gong is played, thus colouring the latter's harmonics'.

Painting in Response to Composition: Synaesthetic or Not?

Creation of graphic works in response to pieces of music as a manifestation of synaesthesia 'is a problematic issue... since many artists believed they possessed it but perhaps not to the extent that scientists would now equate with a true manifestation of that condition'.[43] This statement probably applies to a number of artists frequently cited as paradigmatic of synaesthesia, such as Kandinsky, Klee, Kupka and perhaps even Blanc-Gatti. A heightened sensitivity to the sounds of music which leads an individual to wish to express his or her reactions in another form — be it graphic, sculptural, poetic, balletic, etc. may often not fall within the sphere of synaesthesia.

In a study of word-colour (*not* music-colour) synaesthetes by Ward *et al.*,[44] drawings by these individuals in response to music were more often thought, by an unselected group of observers, to be 'more aesthetically pleasing' when combined with the music than drawings by a control group who did not report synaesthetic experiences. Considerations such as those in the preceding paragraph must modify any positive conclusions in terms of validation of the synaesthetic reaction on the basis of this experiment.

Neurological Investigation of Musical Synaesthesia

One of the problems in investigating synaesthetic experiences is that they are, by definition, subjective. It is also clear that the criteria for classifying subjective experiences as synaesthetic vary from near 'normality' — the recognition of some colour associations of words, numbers, etc., which is almost universal — to the demanding criteria postulated by Cytowic: that the sensations evoked are involuntary, and cannot voluntarily be suppressed.

We may also note, when considering whether synaesthesia is 'congenital' or 'developmental' (in the specific sense of acquired spontaneously during postnatal development) that a large proportion of the stimuli which elicit apparently synaesthetic responses are strictly artefactual. Thus, written numbers, letters and musical tones, are, like days of the week, human constructs; they do not exist in 'nature', in the same way as odours, tastes and different surfaces do, and which may elicit coloured olfaction, coloured gustation or tactile gustation. A small proportion, not more than about 7% of synaesthetes, have reactions such as smell-colour or pain-colour.[1]

Ramachandran and Hubbard[45] report that many people who do not undergo synaesthetic experiences doubt their true existence, suggesting that synaesthetes 'are just crazy'; that they are simply making cross-modality associations; that they are employing metaphor (e.g. 'bitter cold' or 'sharp cheese'); or that their experiences are drug-mediated.

Given the entirely subjective nature of the phenomena experienced, one may look for objective correlates. The experiment comparing the reactions of non-synaesthetes to drawings in response to words by, on the one hand, people who claim to have synaesthesia and, on the other, people who do not[44] points in this direction but it itself dependent on subjective feelings and criteria. An entirely plausible explanation is that people who are visually imaginative may render the 'emotional' content of a piece of music more comprehensibly to others than those with poor visual imagination. Studies of graphic arts students suggest that people who report synaesthetic experiences may score higher on tests of 'creativity'.[46]

Electrophysiological or other objective data showing a difference between synaesthetes, specifically those with *music-induced chromaesthesia*, and other subjects are, as far as I can ascertain, lacking. Sadly, there are few, if any, systematic neuroimaging studies of synaesthetic responses to music. Reasons for this may include the relative rarity of the condition, difficulties in designing an appropriate and reproducible experimental paradigm, and a recognition of the immense complexity of the cerebral response to music and the influences upon it of such elements as musical training, possession of perfect pitch, perceived dissonance and familiarity, which have been demonstrated in a number of studies, some described elsewhere in this volume.

Paulesu *et al.*,[47] comparing a group of women in whom colour sensations were aroused by words, but not by tones, found that on positron emission tomography both groups showed increased regional cerebral blood flow (rCBF) in areas classically associated with language, but that the synaesthetic women had increased rCBF in visual associative areas, such as posteroinferior cerebral cortex and the parieto-occipital junction, but *not* the primary visual cortex. Nunn *et al.*,[48] using functional magnetic resonance imaging (fMRI) had effectively similar results.

The significance of these findings is twofold. The first message, perhaps the more important in terms of validating the existence of synaesthesia, is that there appeared to be a genuine physiological difference (rCBF mirroring cerebral cortical activity) between the

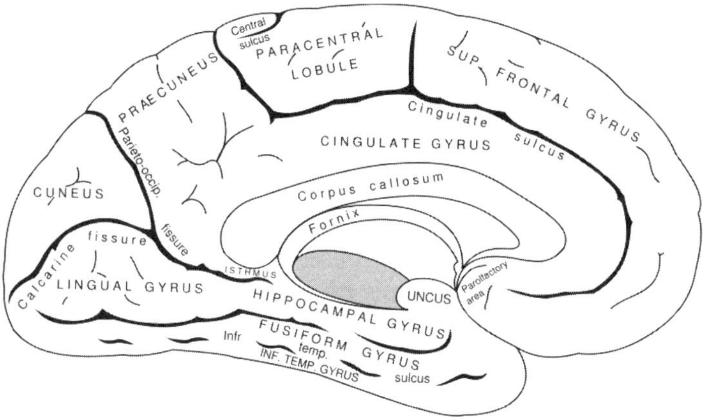

Figure 5. Diagram of the medial surface of the left hemisphere, showing the fusiform gyrus.

women who reported synaesthetic phenomena and a control group. The second, indicating which parts of the brain seemed to be involved, is clearly of interest for future research.

The Site of 'Cross-Talk'

Ramachandran and Hubbard, on the basis of studies too complex to be considered in detail here, but which hinge on identifying synaesthetes by means of test–retest consistency (particularly over relatively long periods of time), rapidity of response and performance in situations which tend to interfere with a synaesthetic reaction (placing a number which elicits red, say, against a red background)[49] suggest that synaesthetic experiences are indeed the result of 'cross-talk'.[45] Taking grapheme-colour synaesthesia as their paradigm, they postulate 'a mutation causing defective pruning and cross-activation between [cortical areas] V4 (or V8) and the number area, which lie right next to each other in the fusiform gyrus'.

It remains to be seen whether such studies can be plausibly extended to music-evoked phenomena.

References

1. Day, S. (2005). http://home.comcast.net/~sean.day/index.html.
2. Rich, A. N. and Mattingley, J. B. (2002). *Nature Rev Neurosci* **3**: 43–52.
3. Huysmans, J. K. (1884). *À Rebours*. Paris: Charpentier.
4. Beeli, G., Esslen, M. and Jäncke, L. (2005). *Nature* **434**: 38.
5. Galton, F. (1883). *Inquiries into Human Faculty and its Development*, p. 107. London: Dent.
6. Cytowic, R. F. (1993). *The Man who Tasted Shapes*, pp. 76–78. London: Abacus.
7. Gautier, T. (1966). In Solomon, D. (Ed.), *The Marijuana Papers*, pp. 121–166. New York: Signet Books.
8. Locke, J. (1823). *The Works of John Locke. A new edition, corrected*, Vol. 2, p. 191. London: Tegg.
9. Smilek, D. *et al.* (2002). *Neurocase* **8**: 338–342.
10. Smilek, D., Dixon, M. J. and Merikle, P. M. (2005). *Neurocase* **11**: 363–370.
11. Hubbard, E. M. and Ramachandran, V. S. (2003). *J Consciousness Stud* **10**: 77–84.
12. McManus, C. (2003). *Right Hand, Left Hand*, pp. 77, 86–89, 326. London: Phoenix.
13. Aristotle. *De sensu et sensili*. Quoted by Eastlake, C. L. (1840). In *Goethe's Theory of Colours*, p. 418. London: John Murray.
14. Eastlake, C. L., *ibid.*, pp. 298–299.
15. Anon. (1896). *Musical Times*, 1 August, pp. 481–484.
16. von Helmholtz, H. L. F. (1954). *On the Sensations of Tone as a Physiological Basis for the Theory of Music*. New York: Dover.
17. Myers, C. S. (1914). *Br J Psychol* 7: 112–117.
18. Myers, C. S. (1911). *Br J Psychol* **4**: 228–238.
19. Peillaube E. (1902). *Rev Philosophie* **2**: 701–718.
20. Hacking I. (2007) *London Review of Books* (1 November): 17–19.
21. Taylor C. (1989). *Sources of the Self. The Making of Modern Identity*. Cambridge: Cambridge University Press.
22. Castel, L. B. (1725) *Mercure de France* (November): 2552–2557.
23. Colman, W. S. (1894, 1898). *Lancet* **1**: 795–796; **1**: 22.
24. George D. (1981). *Sweet Man: The Real Duke Ellington*, pp. 225–226. New York: Putnam.
25. Ligeti, G. (1981). *Ligeti in Conversation*, p. 58. London: Eulenburg.
26. Scruton, R. (1997). *The Aesthetics of Music*, pp. 21–22. Oxford: Clarendon Press.
27. Kross, S. (1982). *19th-Century Music* **5**: 193–200.
28. Sacks, O. (1997). *Musicophilia*, p. 166. London: Picador.
29. Sacks, O. *ibid.*, p. 170.
30. Yastrebtsev, V. (1908). *Russ Muzyk Gaz* **39–40**: 842–845.
31. Galeyev, B. M. and Vanechkina, I. L. (2001). *Leonardo* **34**: 357–362.

32. Brown, J. W. (1994). *Amy Beach and her Chamber Music: Biography, Documents, Style*, p. 16. Metuchen, NJ: Scarecrow Press.
33. Jenkins, W. S. (1994). *The Remarkable Mrs Beach, American Composer*, pp. 5–6. Warren, MI: Harmonie Park Press.
34. Krohn, W. O. (1982). *Am J Psychol* **5**: 20–41.
35. Bartlett, F. C. (1948). *Obituary Notices of Fellows of the Royal Society* **5**: 767–777.
36. George D. *q.v.*, p. 226.
37. Anon. (1895). *Neuen Berliner Musikzeitung*, 29 August, translated by Sean Day.
38. Donath, J. (1923). *J Psychol u Neurol* **29**: 112.
39. Critchley, M. (1977). In Critchley, M. and Henson R. A. (Eds.), *Music and the Brain*, pp. 216–232. London: Heinemann, London.
40. Samuel, S. (1994). *Olivier Messiaen: Music and Color*. Portland: Amadeus.
41. Messiaen, O. (1944). *Technique de mon language musical*. Paris: Leduc.
42. Mackay, J. and Plant, A. (2007). *The Turn of the Screw: Visual Responses to Britten's Opera*, p. 20. London: Sounding Art Press.
43. Guy, F. In Guy, F., Shaw-Miller, S. and Tucker, M. (2007). *Eye Music: Kandinsky, Klee and all that Jazz*, p. 14. Chichester: Pallant House Gallery.
44. Ward, J. *et al.* (2006). *Perception*, http://www.ucl.ac.uk/library/media/sightandsound.
45. Ramachandran, V. S. and Hubbard, E. M. (2005). *Neuron* **48**: 509–520.
46. Domino, G. (1989). *Creativity Res J* **2**: 17–29.
47. Paulesu, E. *et al.* (1995). *Brain* **118**: 661–676.
48. Nunn, J. A. *et al.* (2002). *Nature Neuroscience* **5**: 371–375
49. Ramachandran, V. S. and Hubbard, E. M. (2001). *J Consciousness Stud* **8**: 3–34.

Chapter 17

The Recognition of Music in Frontotemporal Lobar Degeneration

Julene K. Johnson

The recognition of music was investigated in patients with frontotemporal dementia (FTD), semantic dementia (SD), Alzheimer's disease (AD), and matched controls. Patients were administered a battery of tasks that assessed pitch and melody discrimination, detection of pitch errors in familiar tunes, and naming of familiar tunes. There were no group differences on pitch and melody discrimination tasks. However, patients with SD had difficulty when asked to identify pitch errors in familiar tunes compared with controls, AD and FTD patients. All patient groups scored below controls on naming familiar songs, but the SD patients also named fewer songs than FTD or AD patients. The performance on naming familiar songs, but not other music tasks, was strongly related to measures of semantic knowledge. In summary, the current study suggests that patients with SD may have difficulty processing tonal knowledge of familiar tunes and also naming familiar tunes. Further studies addressing the recognition of familiar and novel music can contribute to theoretical studies of SD and brain-behavior relationships of music processing.

Introduction

The temporal and frontal lobes are preferentially involved in the recognition of music in humans.[1–3] Studies suggest that different music features, such as pitch, rhythm, or timbre, are processed by different networks of the brain and involve both hemispheres.[4,5] Brain damage can selectively damage the recognition of different music features. For example, damage to some parts of the brain can affect the ability to process melodic information, while damage to other areas can affect the perception of rhythm.[4]

The temporal and frontal lobes are commonly affected in neurodegenerative diseases such as frontotemporal lobar degeneration (FTLD) and Alzheimer's disease (AD). Semantic dementia (SD) and frontotemporal dementia (FTD) are two subgroups of FTLD.[6] SD is characterised by a progressive loss of semantic knowledge and atrophy of the temporal lobes, amygdala, and the anterior portions of parahippocampal gyrus and fusiform gyrus.[7–11] Patients with FTD have changes in interpersonal and personal conduct, emotional blunting, and early loss of insight that is associated with atrophy in the orbital frontal, insular, and anterior cingulate regions.[7]

Patients with SD have impairments in recognising both verbal and non-verbal information such as objects, environmental sounds, famous faces and voices, scents, and tastes.[12–14] There have been only a few studies that have investigated music abilities in patients with SD or FTD. For example, Miller and colleagues[15] discuss the emergence of a new interest in composing music in one patient with FTD. Two other studies describe a new, compulsive interest in music in two patients with FTD and one with SD[16,17]; this new compulsion is noteworthy because both patients previously disliked the type of music. Apart from these studies, there are none that evaluate recognition of music in patients with FTLD.

The purpose of the current study was to investigate music recognition in a group of patients with FTD and SD. For comparison, we included patients with AD because they are known to have semantic knowledge deficits[18] but to a lesser degree than SD.[19] Patients were administered a battery of tasks that assessed pitch and melody discrimination, detection of pitch errors in familiar tunes, and naming of familiar tunes.

Subjects

The patients were recruited from a university dementia clinic. Clinical diagnosis was determined after a detailed clinical history, neurological examination, a one-hour neuropsychological battery,[20] laboratory screening, and brain MRI. The FTD (n = 11) and SD (n = 20) patients met Neary criteria.[6] Patients with probable AD (n = 12) met NINCDS-ADRDA criteria.[21] We selected patients in the mild to moderate stages of dementia, operationally defined as a Mini-Mental State Examination (MMSE)[22] score greater than 15 or a Clinical Dementia Rating (CDR)[23] of less than two. Healthy controls (n = 17) were recruited from the community and underwent an evaluation identical to the patients. None of the controls showed evidence of impairment on neuropsychological testing or had a history of a neurological or psychiatric disorder.

Demographic information (i.e. age and education) and years of music lessons were compiled. Professional musicians were excluded from this analysis. None of the subjects had a history of hearing impairment or hearing aid use. All participants provided informed consent that was approved by the University of California, San Francisco committee on human research.

Neuropsychological Battery

We also compiled neuropsychological test results administered within three months of the music testing. Memory was evaluated using the ten-minute delayed recall trial of the California Verbal Learning Test — Mental Status (CVLT-MS)[24] and the Wechsler Memory Scale — Visual Reproductions.[25] Executive function was assessed using the Delis-Kaplan Executive Function System (D-KEFS),[26] Trailmaking (number-letter condition, scaled score) and Stroop (Interference condition, scaled score) tasks. Language was assessed using a 15-item Boston Naming Test,[27] 16 items from the Peabody Picture Vocabulary

Test — Revised,[28] and category fluency (number of animals in one minute). The copy trial of the modified Rey-Osterrieth figure[20] and the Number Location condition from the Visual Object Spatial Perception battery (VOSP)[29] were used to assess visuospatial abilities.

Music Battery

Pitch discrimination

The pitch discrimination task required subjects to determine whether or not two successive tones (separated by a one-second inter-stimulus interval) were the same or different. Following two practice examples, 20 pitch pairs were randomly presented. Ten pitch pairs differed by three to eight semitones, and ten were the different tones.

Melody discrimination

The melody discrimination task was taken from the Montreal Battery for the Evaluation of Amusia (MBEA) (task one)[30] and was used to assess the ability to discriminate two melodies that differed by one key-violating pitch. Subjects were asked to listen to 30 pairs of unfamiliar melodies and determine if they were the same or different (15 trials of each).

Familiar tune error detection

The pitch error detection task was designed to evaluate tonal knowledge about familiar tunes by asking subjects to detect a wrong note in excerpts of familiar melodies. This task was designed considering previous studies that used altered versions to assess tonal knowledge about familiar tunes.[31,32] Twelve highly familiar tunes were selected from American popular music songbooks[33,34] and were expected to be easily recognisable to individuals who are familiar with United States culture. All of the selected familiar songs were originally written with lyrics, but only the melody was reproduced (monophonically on an electric keyboard) with an effort to preserve the original tempo, rhythm, and style (including accents, phrasing, loudness of notes). The melodic excerpt included the portion of the melody in which the song title was a part of the song text, although only the melody was reproduced. The highly familiar tunes ranged in length from 13 to 21 seconds. Subjects were instructed that some of the melodies would have a wrong note and were asked to determine if the excerpt was 'correct' or 'incorrect'. Two-thirds of the tunes included an alteration of a single pitch. For the altered tunes, pitch errors either preserved (key-preserving) or violated (key-violating) the key (tonality) of the tune. All pitch errors preserved the contour of the melody and occurred on a prominent beat (non passing tones).

Familiar tune title recall and recognition

Subjects were also asked to verbalise the title of the familiar tune used in the pitch-error task. The song titles were scored as correct if all content words of the title were provided (e.g. ignoring accuracy of prepositions or articles). If the subject did not provide a correct title, four written titles were presented on a card. The multiple-choice responses included the correct title and three foils: 1) an invented but semantically related title, 2) a real and semantically unrelated song title, and 3) an invented and semantically unrelated title.

Statistical Methods

A one-way analysis of variance (ANOVA) was used to examine possible group differences in the demographic, neuropsychological and music data. *Post hoc* analyses were conducted using Tukey's method for multiple comparisons using SPSS (version 11.5 for Windows, SPSS Inc, Chicago IL). The non-parametric Kuskall-Wallis test was used with non-normally distributed data. The alpha level was set at 0.05. We examined the relationship between music tasks and neuropsychological measures using scatter plots and Pearson product moment partial correlations controlling for MMSE. Correlations were conducted on all subjects collapsed across diagnosis.

Results

Demographic and neuropsychological results

Table 1 summarises the demographic data. Age, years of education, and years of music lessons did not differ among groups (all $p > 0.05$). As expected, there were group differences on the Clinical Dementia Rating scale — sum of boxes ($p < 0.001$), and all patient groups scored higher than controls.

Table 2 summarises the neuropsychological test results. There were group differences on all neuropsychological tasks. On the MMSE, the patients with SD and AD scored significantly lower than controls (both $p < 0.001$), while FTD patients scored similarly to controls. All patient groups scored below controls on tests of verbal and visual memory (all $p < 0.02$). On tests of executive function, all patient groups scored below controls on Trailmaking (all $p < 0.002$), but only AD and SD patients scored below controls on Stroop Interference (both $p < 0.001$). On the visuospatial tasks, only AD patients scored below controls on the modified figure copy and VOSP number location (both $p < 0.0001$). Both SD and AD patients, but not FTD, scored below controls on the Boston Naming Test (both $p < 0.03$). In contrast, only SD patients scored below controls on the PPVT-R ($p < 0.0001$).

Pitch discrimination

Table 3 summaries the results from the music battery. All but one control scored 20/20 correct on the pitch discrimination task. Because of

Table 1. Demographic data.

	Controls	AD	FTD	SD
N	17	12	11	20
Age (years)	66.3 (7.0)	65.3 (9.4)	59.8 (6.5)	66.2 (9.5)
Sex (men:women)	7:10	8:3	7:2	6:3
Education (years)	17.5 (2.2)	15.6 (2.9)	16.2 (2.9)	16.7 (3.0)
Music lessons (years)	2.8 (3.9)	5.1 (5.2)	3.7 (5.3)	4.6 (5.8)
CDR — sum of box	0.0 (0.0)	5.3 (1.9)	6.1 (2.5)	4.2 (2.6)

Table 2. Neuropsychological test results.

	Controls	AD	FTD	SD
MMSE (30)	29.6 (0.8)	22.1 (5.1)	26.3 (3.4)	23.2 (5.7)
Memory				
CVLT-MS — 10 min. recall (9)	7.2 (1.4)	0.9 (1.2)	4.2 (3.3)	1.9 (2.4)
Visual Reproductions — Delayed (17)	13.6 (3.7)	4.8 (2.6)	8.6 (4.6)	6.6 (3.2)
Executive Function				
DKEFS Trailmaking (19)	11.9 (2.1)	2.7 (3.0)	5.5 (3.9)	6.9 (4.1)
DKEFS Stroop — Interference (19)	11.4 (1.5)	2.4 (3.1)	6.9 (5.3)	5.3 (4.6)
Visuospatial				
Rey-Osterrieth figure copy (17)	16.1 (1.3)	9.45 (6.8)	15.3 (1.8)	15.5 (1.3)
VOSP Number Location (10)	9.5 (0.6)	4.9 (2.7)	8.5 (1.0)	9.5 (1.0)
Language				
Boston Naming Test (15)	14.7 (0.6)	12.1 (3.4)	12.5 (2.0)	3.0 (2.6)
PPVT-R Comprehension (16)	15.8 (0.6)	14.3 (1.4)	14.6 (1.5)	5.7 (3.6)
Category fluency (animals in one minute)	24.1 (4.3)	9.4 (5.4)	12.0 (7.3)	5.5 (4.1)

Table 3. Music battery results.

	Controls	AD	FTD	SD
Pitch Discrimination (20)	19.9 (0.2)	19.6 (0.8)	18.5 (2.9)	19.7 (2.2)
Melody Discrimination (30)	26.1 (2.4)	25.1 (3.2)	24.2 (3.2)	23.7 (5.0)
Familiar Tune Error Detection (12)	10.8 (1.8)	9.6 (2.2)	8.7 (2.8)	7.6 (2.9)
Familiar Tune Title Recall (12)	9.8 (2.4)	6.9 (2.9)	6.8 (3.1)	1.2 (2.0)
Familiar Tune Title Recognition (12)	11.9 (0.2)	11.8 (0.6)	11.1 (1.0)	7.7 (2.6)

the skewed distribution of the data, we used a non-parametric Wilcoxin test that confirmed that there were no group differences ($p = 0.09$) on pitch discrimination. There was a trend for the FTD patients to score the lowest. After controlling for MMSE, there were no significant correlations between pitch discrimination and any neuropsychological tests (all $p > 0.05$).

Melody discrimination

There were also no group differences on melody discrimination ($p = 0.24$). The mean score obtained by the controls is similar to the normative sample of 160 healthy adults reported by Peretz and colleagues[30] (mean = 27, SD = 2.3). After controlling for MMSE, there were no significant correlations between melody recognition or any neuropsychological measures (all $p > 0.05$).

Familiar tune pitch error detection

There were group differences in the ability to detect pitch errors in familiar tunes ($p = 0.007$). *Post hoc* analyses showed that only SD patients had lower scores when compared with controls, but not AD or FTD. After controlling for MMSE, there were no significant correlations between pitch error detection and any neuropsychological measures (all $p > 0.05$).

Familiar title recall and recognition

Controls provided correct titles for 81% of the highly familiar tunes. There were group differences ($p < 0.001$) on title recall, and all patient groups recalled significantly fewer song titles than controls. Patients with SD also recalled fewer titles (10%) than patients with AD and FTD. After controlling for MMSE, the recall of song titles correlated with scores on the Boston Naming Test ($r = 0.78$, $p = 0.003$), Peabody Picture Vocabulary Test-R ($r = 0.87$, $p < 0.001$), and animal fluency ($r = 0.62$, $p = 0.03$) but none of the other neuropsychological tests. These results suggest that naming tunes is strongly related to verbal tests of semantic knowledge.

When given four written song titles in a multiple choice format, there were significant group differences on title recognition ($p < 0.0001$), and SD patients were the only group who scored significantly lower than controls, and they also scored lower than both FTD and AD. SD patients were only able to correctly identify 64% of the correct titles from a four-item multiple choice. When SD patients incorrectly selected a multiple choice title, 43% were other familiar song titles, 41% were semantically-related (but novel) titles, and only 16% were invented song titles.

Discussion

The main finding from this study is that patients with SD had considerable difficulty detecting pitch errors in familiar tunes, while AD and FTD patients performed similarly to controls. In contrast, SD patients performed similarly to controls on pitch and melody recognition tasks. A second finding was that patients with SD also had difficulty generating titles for familiar tunes, even when given multiple choice titles. These data suggest that SD patients have difficulty processing tonal knowledge in familiar, but not unfamiliar, tunes and also retrieving titles (verbal labels) of familiar songs.

The current study evaluated the ability to detect pitch errors in familiar tunes, which is one method for assessing tonal knowledge about familiar songs. Numerous studies suggest healthy adults represent precise knowledge about familiar tunes. For example, adults retain considerable knowledge about the intervals of familiar tunes and are able to detect pitch or rhythmic errors in familiar tunes with high accuracy.[35–40] The brain areas that are involved in representing knowledge about familiar tunes are not well understood. Using positron-emission tomography (PET) imaging with healthy adults, Satoh and colleagues[41] found activation of bilateral parietal lobules, precuneus, and lateral frontal cortex when asked to identify an altered familiar melody. The altered melodies involved changes in both pitch and rhythm, whereas the current study only involved alterations in pitch. Other studies suggest that the rostral medial prefrontal cortex is involved in tonal expectancy violations.[42] Familiar songs are unique in that they are highly over learned melodic sequences that carry meaning and expectations for specific music structures. Although semantic knowledge is commonly linked to the verbal or visual domains, there is considerable debate about whether or not music involves semantic elements.[43] In the current study, patients with SD had difficulty using tonal knowledge about familiar songs to identify pitch errors. This raises the possibility that tonal knowledge in familiar songs has a semantic component. There are also music-specific expectations that accompany familiar tunes. Patients with SD also have impairments in recognising objects, environmental sounds, famous faces and voices, scents, and tastes,[12–14] which is felt to reflect deficits in semantic knowledge. Further studies are needed to better understand how patients with SD process possible semantic elements in music and non-verbal sounds, such as environmental sounds.

The second finding from the study involves the ability to name familiar music. Again, patients with SD had difficulty generating titles for familiar tunes and also recognising the title from multiple choices. Difficulty in naming familiar songs is not surprising because a deficit in naming (usually tested with pictures of objects) is one of the core features of SD.[6] It appears that naming familiar songs is similarly difficult for SD patients. Deficits in naming familiar tunes have been associated with left-sided or bilateral brain damage. In an early study, Shankweiler[44] demonstrated that patients with either right or left temporal lobectomies had difficulty naming familiar songs compared with controls. Ayotte and colleagues[45] found that patients with left-sided and bilateral middle cerebral artery (MCA) ruptured aneurysms named fewer songs than unilateral right patients. Another study[46] documented a patient with a left anterior temporal and left parietal stroke who had difficulty naming familiar songs, but was able to improve performance to 90% with multiple choice titles. In our study, the SD patients were only able to improve their performance up to 64% with multiple choice titles. This suggests that even with cues, the SD patients have difficulty recognising the correct song title. Also in our study, the recall of song titles correlated with verbal tests of object naming and semantic memory and not other neuropsychological tests. Thus, naming familiar songs may involve similar cognitive processes and brain areas as naming other types of information.

In summary, the current study suggests that patients with SD may have difficulty processing tonal knowledge of familiar tunes and also naming familiar tunes. Further studies addressing the recognition of familiar and novel music can contribute to theoretical studies of frontotemporal lobar degeneration and brain-behavior relationships of music processing.

Acknowledgements

John Neuhaus, PhD for statistical advice; Karen Minzer, Vanya Green, and Chiung-Chih Chang for help with data collection.

References

1. Peretz, I. and Zatorre, R. J. (2005). Brain organization for music processing. *Annu Rev Psychol* **56**: 89–114.
2. Tramo, M. J., Shah, G. D. and Braida, L. D. (2002). Functional role of auditory cortex in frequency processing and pitch perception. *J Neurophysiol* **87**: 122–139.
3. Zatorre, R. J., Evans, A. C. and Meyer, E. (1994). Neural mechanisms underlying melodic perception and memory for pitch. *J Neurosci* **14**: 1908–1919.
4. Peretz, I. (1990). Processing of local and global musical information by unilateral brain-damaged patients. *Brain* **113**: 1185–1205.
5. Platel, H., Price, C., Baron, J. C., *et al.* (1997). The structural components of music perception. A functional anatomical study. *Brain* **120**: 229–243.
6. Neary, D., Snowden, J. S., Gustafson, L., *et al.* (1998). Frontotemporal lobar degeneration: A consensus on clinical diagnostic criteria. *Neurology* **51**: 1546–1554.
7. Rosen, H. J., Gorno-Tempini, M. L., Goldman, W. P., *et al.* (2002). Patterns of brain atrophy in frontotemporal dementia and semantic dementia. *Neurology* **58**: 198–208.
8. Galton, C. J., Patterson, K., Graham, K., *et al.* (2001). Differing patterns of temporal atrophy in Alzheimer's disease and semantic dementia. *Neurology* **57**: 216–225.
9. Mummery, C., Patterson, K., Price, C., Ashburner, J., Frackowiak, R. and Hodges, J. (2000). A voxel-based morphometry study of semantic dementia: Relationship between temporal lobe atrophy and semantic memory. *Ann Neurol* **47**: 36–45.

10. Gorno-Tempini, M. L., Dronkers, N. F., Rankin, K. P., *et al.* (2004). Cognition and anatomy in three variants of primary progressive aphasia. *Ann Neurol* **55**: 335–346.
11. Chan, D., Fox, N. C., Scahill, R. I., *et al.* (2001). Patterns of temporal lobe atrophy in semantic dementia and Alzheimer's disease. *Ann Neurol* **49**: 433–442.
12. Snowden, J. S., Neary, D. and Mann, D. M. A. (1996). *Frontotemporal Lobar Degeneration: Frontotemporal dementia, progressive aphasia, semantic dementia.* New York: Churchill Livingstone.
13. Snowden, J. S., Thompson, J. C. and Neary, D. (2004). Knowledge of famous faces and names in semantic dementia. *Brain* **127**: 860–872.
14. Simons, J. S., Graham, K. S., Galton, C. J., Patterson, K. and Hodges, J. R. Semantic knowledge and episodic memory for faces in semantic dementia. *Neuropsychology* **15**: 101–114.
15. Miller, B. L., Boone, K., Cummings, J. L., Read, S. L. and Mishkin, F. (2000). Functional correlates of musical and visual ability in frontotemporal dementia. *Br J Psychiatry* **176**: 458–463.
16. Boeve, B. F. and Geda, Y. E. (2001). Polka music and semantic dementia. *Neurology* **57**: 1485.
17. Geroldi, C., Metitieri, T., Binetti, G., Zanetti, O., Trabucchi, M. and Frisoni, G. B. (2000). Pop music and frontotemporal dementia. *Neurology* **55**: 1935–1936.
18. Lambon Ralph, M. A., Powell, J., Howard, D., Whitworth, A. B., Garrard, P. and Hodges, J. R. (2001). Semantic memory is impaired in both dementia with Lewy bodies and dementia of Alzheimer's type: A comparative neuropsychological study and literature review. *J Neurol Neurosurg Psychiatry* **70**: 149–156.
19. Rogers, T. T., Ivanoiu, A., Patterson, K. and Hodges, J. R. (2006). Semantic memory in Alzheimer's disease and the frontotemporal dementias: A longitudinal study of 236 patients. *Neuropsychology* **20**: 319–335.
20. Kramer, J. H., Jurik, J., Sha, S. J., *et al.* (2003). Distinctive neuropsychological patterns in frontotemporal dementia, semantic dementia, and Alzheimer disease. *Cogn Behav Neurol* **16**: 211–218.
21. McKhann, G., Drachman, D., Folstein, M., Katzman, R., Price, D. and Stadlan, E. M. (1984). Clinical diagnosis of Alzheimer's disease: Report of the NINCDS- ADRDA Work Group under the auspices of Department of Health and Human Services Task Force on Alzheimer's Disease. Single *Neurology* **34**: 939–944.
22. Folstein, M. F., Folstein, S. E. and McHugh, P. R. (1975). "Mini-mental state". A practical method for grading the mental state of patients for the clinician. *J Psychiat Res* **12**: 189–198.
23. Berg, L. (1988). Clinical Dementia Rating scale (CDR). *Psychopharmacol Bull* **24**: 637–639.
24. Delis, D. C., Kramer, J. H., Kaplan, E. and Ober, B. A. (2000). *California Verbal Learning Test* (2nd ed.). San Antonio, TX: The Psychological Corporation.

25. Wechsler, D. (1997). *Wechsler Memory Scale* (3rd ed.). San Antonio, TX: The Psychological Corporation.
26. Delis, D., Kaplan, E. B. and Kramer, J. (2001). *The Delis–Kaplan Executive Function System.* San Antonio, TX: The Psychological Corporation.
27. Kaplan, E., Goodglass, H. and Wintraub, S. (1983). *The Boston Naming Test.* Philadelphia: Lea and Febiger.
28. Dunn, L. M. D. (1981). *Peabody Picture Vocabulary Test — Revised.* Circle Pines, MN: American Guidance Service.
29. Warrington, E. K. and James, M. (1991). *The Visual Object and Space Perception Battery.* Bury St Edmunds: Thames Valley Test Company.
30. Peretz, I., Champod, A. S. and Hyde, K. (2003). Varieties of musical disorders. The Montréal Battery of Evaluation of Amusia. *Ann NY Acad Sci* **999**: 58–75.
31. Dowling, W. J. and Fujitani, D. S. (1971). Contour, interval, and pitch recognition in memory for melodies. *J Acoust Soc Am* **49**: 524–531.
32. Johnson, J. K. (1997). *Music Cognition in Dementia of the Alzheimer Type* [Doctoral dissertation]. Dallas, University of Texas at Dallas.
33. Birnie, W. A. H. (1969). *Reader's Digest Family Songbook.* Pleasantville, NY: Reader's Digest Association.
34. Ralph, T. (1986). *The American Song Treasury: 100 Favorites.* New York: Dover.
35. White, B. W. (1960). Recognition of distorted melodies. *American Journal of Pschology* **73**: 100–107.
36. Dowling, W. J. and Bartlett, J. C. (1981). The importance of interval information in long-term memory for melodies. *Psychomusicology* 30–49.
37. Bartlett, J. C. and Dowling, W. J. (1980). Recognition of transposed melodies: A key-distance effect in developmental perspective. *J Exp Psychol Hum Percept Perform* **6**: 501–515.
38. Attneave, F. and Olson, R. K. (1971). Pitch as a medium: A new approach to psychophysical scaling. *Am J Psychol* **84**: 147–166.
39. Kalmus H. and Fry, D. B. (1980). On tune deafness (dysmelodia): Frequency, development, genetics and musical background. *Ann Hum Genet* **43**: 369–382.
40. Dowling, W. J. (1978). Scale and contour: Two components of a theory of memory for melodies. *Psychol Rev* **85**: 341–354.
41. Satoh, M., Takeda, K., Nagata, K., Shimosegawa, E. and Kuzuhara, S. (2006). Positron-emission tomography of brain regions activated by recognition of familiar music. *AJNR Am J Neuroradiol* **27**: 1101–1106.
42. Janata, P. (2005). Brain networks that track musical structure. *Ann NY Acad Sci* **1060**: 111–124.
43. Koelsch, S., Kasper, E., Sammler, D., Schulze, K., Gunter, T. and Friederici, A. D. (2004). Music, language and meaning: Brain signatures of semantic processing. *Nat Neurosci* 7: 302–307.
44. Shankweiler, D. (1966). Effects of temporal-lobe damage of perception of dichotically presented melodies. *J Comp Physiol Psychol* **62**: 115–119.

45. Ayotte, J., Peretz, I., Rousseau, I., Bard, C. and Bojanowski, M. (2000). Patterns of music agnosia associated with middle cerebral artery infarcts. *Brain* **123**: 1926–1938.
46. Eustache, F., Lechevalier, B., Viader, F. and Lambert, J. (1990). Identification and discrimination disorders in auditory perception: A report on two cases. *Neuropsychologia* **28**: 257–270.

Chapter 18

Maurice Ravel and the Music of the Brain

Ola Selnes

The nature of Maurice Ravel's neurological disorder, which developed when the composer was at the height of his career at about age 52, has been the subject of much debate. Proposed explanations have included subdural hematoma (from his taxi-cab accident), brain tumour, Alzheimer's disease, Pick's disease or some other less common degenerative brain disorder. The earliest symptoms, which may have appeared as early as 1927, included errors in his writing and memory lapses. He developed significant word-finding difficulties, and became prone to angry outbursts when he was unable to retrieve a word. Symptoms of apraxia became apparent in 1933. His handwriting deteriorated further, and by 1934, he was unable to sign his name. It has been suggested that the style of some of Ravel's later compositions might be a reflection of his underlying brain disease, specifically, the repetitive nature of one of his most famous compositions, *Boléro*, has been interpreted as a manifestation of musical perseveration consistent with possible frontotemporal dementia, but there is little evidence to support this interpretation.

Introduction

There are several famous composers who have suffered from known or presumed neurological disease. Mozart may have had a chronic subdural hematoma[1] and Donizetti and Schubert had a history of neurosyphilis. Modest Mussorgsky is believed to have had a history of alcohol-related Wernicke's encephalopathy.[2] The question of whether the musical output of composers with central nervous system disease is actually influenced by the nature of their underlying brain disorder has been of considerable interest. Can neurological dysfunction in musicians teach us something about the neural substrates of music and artistic creativity? Although one might expect that brain disease should invariably be associated with a negative impact on artistic creativity, a paradoxical enhanced artistic expression associated with frontotemporal dementia has been described.[3] Perhaps one of the most enigmatic of all examples of neurological disease in a famous composer is that of Maurice Ravel. He developed symptoms of a progressive neuro-behavioral disorder at the age of 52, when he was still at the height of his career. He completed several works after the onset of his earliest neuro-cognitive symptoms and, although the exact nature of his underlying neurological disease still remains unknown, it has been suggested that his brain disease actually may be reflected in the style of one of his later compositions, *Boléro.* Although a somewhat intriguing hypothesis, the evidence in its support remains weak at best.

Background

Maurice Ravel (1875–1937) was the oldest child of a Swiss railroad engineer and a Basque mother. He was brought up and spent most of his adult life in Paris. He was a man of small stature, but of immaculate dress and style. During his youth, he was generally in reasonable, but perhaps not robust health. He was an avid walker and swimmer. As early as 1912 (age 37), he wrote to Ralph Vaughan-Williams that the completion of his monumental composition *Daphnis and Chloe* had left him in 'a pitiful' state, and that he had to spend time in the country to recover from an emerging 'neurasthenia'.[4] Although initially rejected for being underweight, in March of 1915, he enlisted as a truck driver during World War I. His war-time letters describe frequent difficulties with insomnia and exhaustion. During the fall of 1915, he was treated for dysentery and had a protracted period of convalescence. His mother died in January of 1917, and Ravel went through a period of significant grief and depression. Between 1916 and 1921, he only finished two major compositions, *Le Tombeau de Couperin* and *La Valse*. Following this, the rate of his output slowed significantly, and he only completed an average of one work per year for the next 12 years of his active life as a composer.

Emergence of Early Neurological Symptoms

Although the exact timing of his earliest neurological symptoms remains undetermined, his good friend Hélène Jourdan-Morhange was concerned enough about changes in his behaviour to arrange for him to see Dr. Valery Radot in 1927. It was recommended that Ravel should take a full year of complete rest. Instead, he embarked on a very successful tour of North America. He arrived in New York on January 4, 1928, and during his three-month tour visited Boston, New York, Philadelphia, Chicago and several other cities. *Time Magazine* (January 23, 1928) described him 'at the end of the evening far more impressive as a composer than as a pianist'. However, there were some examples of 'absent-mindedness' during this tour. His appearance with the Chicago Symphony was reportedly delayed by nearly an hour because his evening shoes were not immediately available. He returned to France at the end of April 1928, and shortly thereafter began work on his most famous composition, *Boléro*.

Boléro

Originally called *Fandango*, the one-movement orchestral composition later known as *Boléro*, was originally composed as a ballet in response to a commission from the dancer Ida Rubinstein. It is now considered one of Ravel's most well-known compositions. The work was finished on October 15, 1928[4] and the premiere took place at the Paris Opera on November 22, 1928. Ravel was reportedly quite surprised by the success of the piece, and had predicted that many orchestras would refuse to perform it. In an interview with the *Daily Telegraph* in 1931, he described *Boléro* as:

> Consisting of orchestral tissue without music — of one very long, gradual crescendo. There are no contrasts, and practically no invention except the plan and the manner of execution.

There are other examples of Ravel being his own most severe critic. On one occasion, he said to his friend Arthur Honegger: 'I've written only one masterpiece — *Boléro*. Unfortunately, there is no music in it'.[5] Despite Ravel's reservations, the piece has become a staple of the classical musical repertoire.

Late Neurological Symptoms

On October 9, 1932, Ravel was involved in a taxi-cab accident in Paris, and he suffered some broken teeth and chest injuries. It has been suggested that this accident may have played a primary rather than incidental role in Ravel's subsequent neurological decline.[6,7] There was, however, no history of loss of consciousness and no other apparent neurological sequelae.[8] He recovered well enough to be able to attend a concert in Switzerland in December of 1932. Throughout the following year, Ravel's neurological symptoms progressively worsened. His last public performance as a conductor was in November of 1933. He was examined by Dr. Théophile Alajouanine, who concluded that Ravel suffered from a, 'Wernicke's aphasia of moderate intensity'.[9] In addition, Alajouanine noted an apractic component. The diagnosis of Wernicke's aphasia is somewhat surprising in that Alajouanine notes that Ravel's 'understanding of language remains much better than oral and written abilities'. This is a clinical presentation that is more characteristic of an aphasia secondary to anterior (frontal) cerebral involvement, such as e.g. Broca's aphasia. During 1933, his handwriting continued to deteriorate, and he had difficulties signing his name. During the summer of 1933, his symptoms of apraxia became more prominent. He was previously a very good swimmer, but on one occasion, he had to be rescued from the water because of his inability to co-ordinate his movements. He was a heavy smoker, and was occasionally observed to put the cigarette in his mouth the wrong way around.[10] During the ensuing years, Ravel consulted many physicians, but no cause for his progressive neurological decline was identified. His friends eventually convinced the neurosurgeon Clovis Vincent to perform exploratory surgery. The pre-operative diagnosis was ventricular enlargement. The surgery was performed towards the end of December, 1937. Although he appeared to be somewhat improved immediately after the operation, he then became unresponsive and died a few days later, at the age of 62. No autopsy was performed, and a final pathology-based diagnosis is thus not available, but there are several features of his clinical presentation that are consistent with corticobasal degeneration, however.[11] Corticobasal degeneration (CBD), first described in 1968, is an akinetic rigid syndrome that often presents with early word-finding difficulties, effortful speech and handwriting difficulties. Motor symptoms and apraxia often develop somewhat later. One feature of the clinical presentation of CBD is asymmetric motor impairment. There is no description of Ravel having such an asymmetric presentation, however. The age of onset is typically late in the fifth decade of life. Others have suggested that Ravel's neurological symptoms and course could be compatible with the diagnosis of primary progressive aphasia.[8] Both CBD and primary progressive aphasia are now considered part of the spectrum of tauopathies, and with considerable overlap in their clinical presentation.[12,13]

Boléro as a Reflection of an Underlying Neurological Disorder

It has been suggested that the somewhat unconventional structural form of *Boléro* might actually be a reflection of Ravel's evolving brain disorder. It is not entirely clear who actually first proposed this idea. As early as 1975, however, Kerner suggested that the 'unnerving stereotypy' of *Boléro* was as a precursor of Ravel's neurological disease.[14] He was hypothesising that the etiology was Pick's disease, now considered part of the spectrum of the frontotemporal degenerative disorders. In 1997, Cybulska asks the rhetorical question: 'Could it be that *Boléro* represents a form of musical perseveration?', which she notes, is one of the most prominent features of frontotemporal degeneration.[15]

There are several reasons why it is unlikely that the compositional style of *Boléro* was the unintended side-effect of an emerging brain disorder. First, when Ravel composed *Boléro*, his neurological symptoms were at best minimal. The work was started after his American tour which, despite its exhausting nature, Ravel not only tolerated but also enjoyed. In a letter dated April 4, 1928 to Madame Fernand Dreyfus written shortly before concluding his North American tour, he writes: 'I have never felt better than during this crazy tour'.

Second, there is evidence that Ravel approached the composition of *Boléro* as he did most of his other works: as an exercise in form. 'I am going to try to repeat it a number of times without any development, gradually increasing the orchestra as best as I can'.[4] Ravel brings innovation, rather than relying on past techniques, by not varying primary aspects (melody, rhythm and harmony) but rather 'secondary' aspects of the music (dynamics, timbre and orchestral density). Third, and perhaps most important, since his neurological disorder was a progressive one, one would expect that compositions completed after *Boléro* would be even more influenced by his brain disorder. Anyone who has had the pleasure of listening to the second movement of his *Piano Concerto in G major*, completed more than three years after *Boléro* and premiered in January of 1932, might agree that there are no traces of obsessive or relentlessly repetitive features in this composition. There is thus no strong evidence to support the claim that Ravel's neurological disease influenced the compositional style of his last compositions. As noted by Dr. Cardoso, the 'sole effect of his disease was to silence him'.[10]

Acknowledgements

The author thanks Dr. Pamela Talalay for her helpful comments and Robinson Muńoz MD, for translating some of the material from French.

References

1. Drake, M. E., Jr. (1993). Mozart's chronic subdural hematoma. *Neurology* **43**: 2400–2403.
2. Bentivoglio, M. (2003). Musical skills and neural functions. The legacy of the brains of musicians. *Ann NY Acad Sci* **999**: 234–243.
3. Miller, B. L., Ponton, M., Benson, D. F., Cummings, J. L. and Mena, I. (1996). Enhanced artistic creativity with temporal lobe degeneration. *Lancet* **348**: 1744–1745.
4. Orenstein, A. (1990). *A Ravel Reader*. New York: Columbia University Press.
5. Nichols, R. (1987). *Ravel Remembered*. New York: W. W. Norton & Company.
6. Otte, A., Audenaert, K. and Otte, K. (2003). Did Maurice Ravel have a whiplash syndrome? *Med Sci Monit* **9**: LE9.
7. Otte, A., De, B. P., Van De, W. C., Audenaert, K. and Dierckx, R. (2003). The exceptional brain of Maurice Ravel. *Med Sci Monit* **9**: RA134–RA139.
8. Amaducci, L., Grassi, E. and Boller, F. (2002). Maurice Ravel and right-hemisphere musical creativity: Influence of disease on his last musical works? *Eur J Neurol* **9**: 75–82.
9. Alajouanine, T. (1948). Aphasia and artistic realization. *Brain* **71**: 229–241.
10. Cardoso, F. (2004). The movement disorder of Maurice Ravel. *Mov Disord* **19**: 755–757.
11. Baeck, E. (1996). Was Maurice Ravel's illness a corticobasal degeneration? *Clin Neurol Neurosurg* **98**: 57–61.
12. Boeve, B. F. (2007). Links between frontotemporal lobar degeneration, corticobasal degeneration, progressive supranuclear palsy, and amyotrophic lateral sclerosis. *Alzheimer Dis Assoc Disord* **21**: S31–S38.
13. Sha, S., Hou, C., Viskontas, I. V. and Miller, B. L. (2006). Are frontotemporal lobar degeneration, progressive supranuclear palsy and corticobasal degeneration distinct diseases? *Nat Clin Pract Neurol* **2**: 658–665.
14. Kerner, D. (1975). Ravels Tod. *MMW Munch Med Wochenschr* **117**: 591–596.
15. Cybulska, E. M. (1997). Bolero unravelled: A case of musical perseveration. *Psychiatric Bulletin* **21**: 576–577.

Chapter 19

Cerebrovascular Disorders of Baroque Composers

Tomislav Breitenfeld, Darko Breitenfeld and Vida Demarin

Investigating composers' maladies through history, we noticed that Baroque composers lived surprisingly longer than expected for that time, our research showing that many suffered, or died from stroke. Giants among them — Heinrich Schütz, Johann Sebastian Bach, Georg Friedrich Handel and Christoph Willibald Gluck, suffered strokes, as well as another 16 prominent composers of the Baroque era. This showed that stroke is not only a global burden of the twentieth and twenty-first centuries.

Cerebrovascular disorders of Baroque composers will be discussed in this chapter, together with their influence on creativity, with emphasis on four of the most famous — Heinrich Schutz, Johann Sebastian Bach, Georg Friedrich Handel and Christoph Willibald Gluck. Our observations are part of a larger investigation of neurological disorders in famous composers throughout history.

Introduction

The term 'Baroque' is of roman origin meaning 'bizarre'. It was introduced in the seventeenth and eighteenth centuries for the architecture of middle Europe characterised by extravagant decoration. By the twentieth century, it was used to describe specific and easily defined musical style that was played from before 1600 to the deaths of Johann Sebastian Bach in 1750 and Georg Friedrich Handel in 1759.

After the periods of Mannerism and Renaissance, there was the climax of royal despotism that terminated with Louis XIV of France (*L'état c'est moi*). Developing colonies supplied Europe with more treasure from the new world which made for an era of spectacular progress in knowledge, science and philosophy with such names as Isaac Newton, Galileo Gallilei, Johannes Kepler, Spinoza, John Locke, René Descartes and Benjamin Franklin. Art also flourished with the works of Shakespeare, Molière, Cervantes, Bernini, Caravaggio, Rubens, Rembrandt, Vermeer and Velasquez.

In medicine, there was progress in the sixteenth and seventeenth centuries by increased knowledge in anatomy and physiology. Iatrochemistry and iatrophysics were introduced with the era of Thomas Sydenham, The English Hippocrates, as well as Gjuro Baglivi, famous physician from Dubrovnik, Croatia, who practiced in Rome. Antony van Leeuwenhoek constructed the microscope and Sir Christopher Wren helped to start the practice of intravenous injection. With improvement of hygiene the number of infectious diseases decreased, although the black plague still occasionally ravaged Europe. Hermann Boerhaave founded clinical medicine, followed by Albrecht von Haller who reformed physiology. Giovanni Battista Morgagni helped pathological anatomy to establish urology and orthopaedics. Ophthalmology was rescued from charlatans and new successful and harmless methods of operation were introduced for cataract, albeit too late for Bach and Handel.

The Baroque era introduced polyphony, ornamentation and contrasting elements in music, which was highly honoured by the church and royal courts and was also a new form of entertainment with the growth of merchant and other civil classes. Harmonic complexity was imposed with emphasis on contrast, introducing basso continuo, and opera was developing from more recitative to complex arias. New instrumental forms such as the concerto and sonata were inaugurated and new instruments created, with the violin becoming the most important string instrument (the era of Stradivarius) and solo singing was introduced.

Although Baroque music originated in Italy in the seventeenth century, it was adopted in Germany at the beginning of the eighteenth century with its culmination in the works of Johann Sebastian Bach and Georg Friedrich Handel, which had an enormous influence on the 'new style' with such classical composers as Haydn and Mozart.

Pathographies

HEINRICH SCHUTZ (1585–1672)

Heinrich Schutz was born in 1585 in Kostritz but moved to Kassel where he had his first musical lessons and served as a choirboy. In 1609, Schütz left for the University of Marburg to study law, before going to Venice where he spent four years studying music with Giovanni Gabrieli. From Kassel where he worked as an organist, he soon moved to Dresden as a Kapellmeister to the Elector Johann Georg I. In Dresden, he was responsible for providing music to major court ceremonies, whether religious or political. In 1619, Schutz published his first collection of sacred music, the *Psalmen Davids*, dedicated to the Elector. In 1620, he married Magdalena Wildeck, who died in 1625 leaving Schutz with two daughters whom he placed in the care of their maternal grandmother; he never remarried. Even though settled in Dresden, he left several times. In 1627, he was in Torgau where his *Dafne*, the first German opera, was performed. In 1628, he again went to Venice where he stayed for one year (probably meeting Monteverdi), and again in 1633, when he left Dresden to stay in Copenhagen for two years.[1]

Heinrich Schutz is considered the greatest religious German composer of the seventeenth century and the first of international stature, influenced mostly by Giovanni Gabrieli and Claudio Monteverdi. He was one of the last composers to write in a modal style, but unfortunately little of his music has survived. His merit was bringing new musical ideas from Italy to Germany to influence such future German baroque giants as Handel and Bach.[2]

There are no relevant medical records regarding his health. He died at the age of 87, obviously living longer than expected for the seventeenth century. This probably means that he had a calm and settled life with a lot of luck avoiding fatal infectious diseases and surviving even the Thirty Years' War. Available portraits show an elderly but healthy man with no facial asymmetry or other signs of a possible prior cerebrovascular event, although he was reported to have died of a stroke in early November 1672.[3]

JOHANN SEBASTIAN BACH (1685–1750)

Bach's origin is from a large family of musicians in Thuringia, mid-Germany, as far as Hans Bach in the sixteenth century.[1] At the beginning of the seventeenth century, the Bach family reached its peak with the sixth generation of musicians, especially with Johann Sebastian.[4] After Bach's sons, the activity of the family's involvement in music decreased.[5] Johann Sebastian Bach was born on 21 March 1685 in Eisenach and was educated at school in basic musical and theological-humanistic science.[6] He learned mostly from his father and later his brother, having moved after his mother's early death. When grown up, he took the most outstanding musicians appointments (Weimar, Arnstadt, Mühlhausen) travelling through Germany to get acquainted with Baroque music trends and opera.[7] From 1708, he worked in Weimar as a court musician and later as a concert maestro composing celebratory music. Then he joined the Duke of Cöthens as a court musician, composing non-clerical and instrumental music. In 1723, he went to Leipzig as 'Director Musices' where he was composing as a cantor of St. Thomas' Church, mainly clerical music such as cantatas, masses, passions and oratorios. This was his most natural period, and he felt the most satisfying one.[8] He married twice, as a young man to his cousin Maria Barbara Bach, with whom he had seven children. Ten years after she died, he married again in 1721 with Anna Magdalena Wulcken, the Duke's singer at the Cöthens court, with whom he had thirteen children. Very close to each of his wives, he was a good father to his numerous children, all of whom were taught music. With each marriage, two well-known musician-composers were born: Wilhelm Friedmann and Carl Phillip Emanuel from the first and Johann Christoph Friedrich and Johann Christian from the second.

Many of his children from both marriages died in early childhood, and in their adolescence several of his children suffered from alcoholism and were mentally handicapped. Very few of his children married and had their own children, so that in the two next generations (almost in the first) all the male descendents died out, while female descendants still lived (although naturalised) in Poland.[9] Genetically, the Bachs used to live until their sixties which was quite a good average for that time, but a few lived longer. They were robust, often very hasty, especially Johann Sebastian, who defended his own point of view sometimes with stubbornness, frequently conflicting with his colleagues, especially his superiors. He was appreciated to a large extent by the majority of his country, but more as an organist and a specialist for arms manufacture than as a composer. His sons were more popular throughout the world as composers than he himself. It took a hundred years from his death until he was rediscovered by Felix Mendelssohn.[10]

Johann Sebastian Bach was a healthy, strong man who was shortsighted from childhood, but according to his temper, nature and stature he probably had high blood pressure and diabetes was later also suspected.[11] His vision was said to have been damaged because of writing and copying notes from his early days in the dark (frequently only by candle light). In one of his rare authentic portraits in 1746 (by Elias Gottlob Haussmann) oral asymmetry is obvious, suggesting minor to partial right central facial palsy, probably due to a prior (mild?) stroke; from the same portrait his obesity is obvious.[10,11] During the last two years of his life, vision deteriorated rapidly with pain in his eyes,[12] so that he was advised to consult an oculist. It seems that his general condition was bad even earlier because of the reduction in his activities in the last five years of his life with lessening in his creativity. By the end of 1749, he was not able to write any more, either due to another stroke, or deterioration of his vision. He was forced to dictate his notes. At that time he was touring Europe and went to Leipzig where he consulted a bombastic public broker, who was considered a charlatan but a wise oculist, operator 'chevalier and gentleman' — an Englishman named John Taylor.[13] At that time, Taylor was famous as an operator-inventor of a needle for grey dimness of the eyes (cataract). Treating many eye illnesses he used to cut them too often, causing a lot of damage and was nicknamed the 'Münchausen of medicine'.[14] He was very arrogant and described his residing in Leipzig in the following way: 'I saw all kinds of various animals, like camels, dromedaries etc, but in Leipzig I operated on a famous old music master, I saved his vision, he was educated together with Handel whom I operated on later'.[15] It is one of those historical ironies or coincidences, that both composers who became blind at the same age were both operated 'because of the cataract', both operations failed, done by same 'specialist', but at a distance of a 1,000 kilometres from each other. Handel probably suffered from brain stroke with central blindness. On the other hand, Bach probably had a hemorrhagic glaucoma, characterised by pain and sudden onset, which suggests that neither had a cataract.

In the year 1749, Bach had his first talk with Taylor who had operated on him twice in 1750, the first operation at the end of March and the second at the beginning of April 1750. His vision did not improve (which is quite opposite to Taylor's statement); in addition inflammation ensued. Further details about Bach's operation and death were found when Berlin newspapers wrote in two of their issues about a Leipzig report that Taylor's operation was fully successful with great satisfaction of his contemporaries, who greatly appreciated Bach. Nevertheless, a Rostock doctor, Eschenbach in May of the same year denied these two favourable reports about Taylor's operations and listed complications in both cases, including Bach's inflammation. Another newspaper on the 3 August 1750 reported that several days previously, J.S. Bach died from the unlucky consequences of Taylor's eye operation; the only consolation in Bach's case was that if he really had suffered from hemorrhagic glaucoma, no-one would be able to help him anyway. After those failed operations, Johann Sebastian spent his days in a dark room, depressed and only did some dictation of what he had composed. In mid-July 1750, he suffered a fatal stroke, complicated by fever (pneumonia).[10] A couple of hours before he died, he thought that he could see again but this was probably hallucinations.[11] Two famous local doctors tried to help him but without success, and he died in the evening of 28 July 1750. He was buried nearby in St Thomas' Church[16,17]; his grave was known only by oral tradition and was mentioned in just one local newspaper, only incidentally. The coffin was made of oak-wood, which helped when his graveyard was sought. They found three coffins of that type, but only one corresponding to Bach's description — the skull of an older man was found with strong bones and other details were in concordance with Bach's portraits. The anatomy professor employed at that time a famous sculptor, Seffner, to make a portrait-like bust over a plaster cast of the skull with features resembling those of the great composer.[16–18] The skull was further examined by the famous Professor Politzer, an authority on otology at that time, who discovered particularly pronounced temporal bones and *fenestra rotunda*; also the first coil of the cochlea was noted to be unusually large, indicating a unique development of the cochlear ganglion. Impressions of the fusiform and inferior temporal gyri on the skull suggested a marked development of cerebral function of opposing areas of the brain with the implication that this could be related to Bach's perfect pitch and extraordinary musical genius.[16] Obviously Bach's skull was not normal in its appearance — many of the aspects of Bach's brain that are connected with hearing and music were enlarged and have left abnormal markers in the inside of the skull (e.g. temporal lobe, cochlea...). This raises an interesting question in terms of understanding the neurophysiology of a musical genius — is it the case that Bach was born with a musical brain, or did practicing music alter his brain architecture?[19]

From today's scientific point of view, the question of the authenticity of Bach's skull is still doubtful. An instructive example is the historical saga on Mozart's skull from Mozarteum. Although many scientists had conflicting evidence or denied the authenticity of Mozart's skull for decades, the results of DNA analysis, announced in early 2006, on the 250th anniversary of Mozart's birth, did not solve that mystery.[20,21]

According to available sources, Johann Sebastian Bach did not have serious health problems until the age of about 60, when symptoms of cerebrovascular disease occurred, his risk profile including age, obesity and possible hypertension and diabetes, all of which led to his fatal stroke in 1750. For the last two years of his life, he had sight impairment finally becoming blind, probably from hemorrhagic glaucoma and two unsuccessful operations performed by John Taylor.

After Bach's death, his already displaced sons took away their part of his legacy, Bach's two wives and unmarried daughters stayed together and died soon after; their only income was from social support. Even for respected, honoured musicians and composers, the baroque era had still not reached a higher standard — for which Bach's family is a good example.

GEORG FRIEDRICH HANDEL (1685–1759)

This famous German Baroque composer was born in 1685 (the same year as Johann Sebastian Bach) in the small picturesque town of Halle on the Saal river. His father was a barber-surgeon who wanted his son to become a lawyer, but little Georg was from childhood very keen on music. Starting to learn to play various musical instruments at an early age, beginning from a clavichord to an organ, at the age of eight, he played so well that the Duke of Saxony suggested to his father musical schooling for his gifted son. Even though he 'had' to study law, he soon dedicated himself entirely to music and his first job was in Hamburg as a violinist and cellist.

In 1706, he went to Italy where he became famous as a composer but even more as an organ virtuoso, competing with the Italian master, Domenico Scarlatti.[1] After 1710, he returned to Germany, to Hannover, following the Elector who later became King George I of England and soon went to London to take up permanent residence. Although he had a very dynamic life with many crises, failures, falls and impoverishment, Handel was highly esteemed, popular and accepted as the greatest English musical name until then.[1,22] Handel was intelligent, well educated, generous, and with a fun-loving temperament. Tall with a strong stature, he never married but later became known for a large fondness of drink. He suffered changing moods, either due to overworking or the failure of his business affairs, but probably due to his pyknic constitution and cyclothymia giving a bipolar disorder with shorter or longer depressions that always resulted in decreased activities, not only in fighting spirit, social and working enthusiasm, but also creativeness in composing and extensive performing activity both as a conductor or virtuoso, as well as an organiser of musical life in London.[23]

Handel's frankness, extrovert nature, cheerfulness and hypomania was occasionally interrupted by deeper depressions, particularly in the years 1729, 1734, 1737, 1743, and 1745; after the age of 70, he was especially low, due to his blindness and the impossibility of leading his previous dynamic life as performer or composer. Obviously suffering an episodic mental disorder, his habits showed frequent and excessive alcohol consumption, especially of port. He was also a smoker but heavy abuse of nicotine is not certain.[24] In his later portraits, obesity is obvious as well as possible gout. His family history points to strokes in his mother and grandmother.

During his life, he had two major medical problems. One of them was recurrent weakness of his right arm and the other was blindness, first noted in 1751 at the age of 66. He was prone to rheumatism, occasionally visiting spas such as Tunbridge Wells and others, even before the age of 50. In 1737, he had his first episode of paralysis of the right arm with possible mental confusion. After prolonged rehabilitation in Aix-la-Chapelle he recovered completely. In 1742, during his visit to Ireland he had another transient palsy. In the following year, 1743 he had another episode of his paralytic disorder, at which time he first noticed impairment of speech.[24–26] In 1751, Handel lost the sight of his left eye, followed by progressive visual loss so that he was almost completely blind during the last few years of his life. Supposed to be due to cataract, he was operated on three times, but always unsuccessfully. Two operations were performed by William Bromfield, but the last by the already mentioned Chevalier John Taylor who was well known as a self-promoting doctor and probable charlatan.[14,15] As previously mentioned, he also operated unsuccessfully on Johann Sebastian Bach.

Few composers have earned any fame from their medical history investigated by several medical experts. Besides Ravel and Mozart, it is rare for composers to have several pathographies,[2] but Georg Friedrich Handel is one of them. The aetiology of his recurrent paralytic right hand disorder remains unclear, although his family history (strokes on his mother's side), smoking, alcoholism, obesity and probable elevated blood pressure, are all cerebrovascular risk factors. But he suffered from rheumatism as well. According to Evers, Bäzner and Hennerici, all explanations point to a vascular origin for recurrent palsies and visual impairment, due to probable stenosis of his left internal carotid artery and recurrent embolism to the left hemisphere. His eye disease being vascular in origin is indicated by its sudden onset, one (left) eye affected first, absence of pain and unsuccessful cataract operations.[24,25] Yet Frosch and some other authors pointed to a rheumatic-degenerative disorder because of the absence of other symptoms of central nervous system disease and surprising complete (though prolonged) recovery after every episode of paralysis.[26] Previously, Kloverkorn suggested syphilis, which is not likely since no typical signs of syphilis were described.[27] We find both cerebrovascular and rheumatic theories convincing, but want to point out some other possibilities. In the eighteenth century, port wines contained lead and Handel's love of this could have lead to chronic nerve poisoning or saturnine gout.[26] Another possibility is lover's arm — when the arm is held for a long period in a forced position, especially with local pressure on peripheral nerves (falling asleep when groggy, possibly with a beloved person sleeping on his upper arm or shoulder); such damage is by no means rare, and has been considered seriously.[26] There is also the possibility of alcoholic damage to the optic nerve but a definite diagnosis will probably never be known. Most of his symptoms can be attributed to cerebrovascular disease and strokes, and there is no convincing evidence that these medical problems seriously affected his artistic creativity or lifestyle.

CHRISTOPH WILLIBALD GLUCK (1714–1787)

Christoph Willibald Gluck (or Cluc or Kluk, as he was sometimes orthographically spelled) was born in 1714 in the village of Erasbach in Bavaria, at that time occupied by the Austrian Empire. His father Hans Adam was a hunting and forest master for the noble Lobkowitz family, who relocated several times during Gluck's earlier years.[1] In 1731, the young Christoph Gluck left home to escape to Prague, where he supported himself as an organist in different churches, studying music and philosophy at the university, but with music remaining essentially self-taught. After Prague, he went to Vienna in 1735 and then in 1736 to Milan, hired as chamber musician. There he played the violin in the prince's court orchestra and probably also studied counterpoint and composition with the city's leading musician, Giovanni Battista Sammartini.[28] In 1741, he composed his first opera *Artaserse* and in 1745 travelled to London, where he met Georg Friedrich Handel. Later he went to Dresden and Bohemia and in 1748 was back in Vienna. When he grew up, he had many relationships, including with the buffo-singer primadona Gaspera Bacheroni, who had affairs with other men at the same time and gave him the so-called 'gallant' disease — lues. In 1750, Gluck married Maria Anna Bergin, daughter of a merchant with close ties to the Imperial Court but, probably because of his lues and its consequent sterility, they did not have children.

From 1752 until the 1770, he settled in Vienna, but when Italian opera made a comeback to the Viennese stage around 1760, Gluck was introduced to librettist Calzabigi and from the beginning their collaboration turned out to be great success — *Don Juan* in 1761 and even more in the next year when they triumphed with *Orfeo ed Euridice* — first performed in 1762 and soon becoming one of the most rewarding eighteenth-century operas.[31] In 1767, they collaborated successfully again on *Alceste*. During the 1770s, Gluck won further acclaim with *Iphigénie en Aulide* (1774), *Armide* (1777), and *Iphigénie en Tauride* (1779).[29,31] Gluck was probably the most important opera composer of the eighteenth century and a precursor to the musical dramas of Richard Wagner, with Hector Berlioz as his devoted fan. Gluck's historical importance rests on his establishment of a new equilibrium between music and drama, and his greatness on the power and clarity with which he projected that vision; he dissolved the drama in music instead of merely illustrating it.

In Paris in 1779, he had a transient ischemic attack but fully recovered. In 1781, he suffered an apparent stroke with accompanying right-side paralysis and aphasia during preparations for his last opera, *Echo et Narcisse*.[32] He partially recovered and returned to Vienna, where he became easy tempered, sometimes indifferent and suffered pneumonia and after long-lasting fevers, recovered very slowly. The following years were characterised by weakness and fragile health, but he succeeded preparing an important German version of *Iphigénie en Tauride*. In 1784, he suffered another stroke which almost stopped all of his future plans and he died suddenly from the final stroke in 1787 and was buried in Matzleinsdorf Cemetery (near Vienna),[32] his body being later transferred to the city's main cemetery. As an introvert, Gluck lived not for his surroundings but for his goals, and he expected others to do the same. In his sixties, he unfortunately had a major stroke, after which he did almost no composing, being partially handicapped because of his palsy and speech impairment (aphasia).

Several Baroque composers have suffered or died from stroke including the following[33]:

ADRIANO BANCHIERI — (1568–1634) Italian composer, organist, music theoretician and writer.

PIERRE MONTAN BERTON — (1727–1780) French composer who from 1759 was the director of the Paris Grand Opera.

JOHANN JAKOB FROBERGER — (1616–1667) German composer, organist and keyboard virtuoso.

HANS LEO HASSLER — (1564–1612) German composer and organist.

GOTTFRIED AUGUST HOMILIUS — (1714–1785) German composer and organist.

NICCOLO JOMELLI — (1714–1774) Italian composer.

ADOLPH CARL KUNZEN — (1720–1781) German composer and organist.

VIRGILIO MAZZOCCHI — (1597–1646) Italian composer.

CHARLES LOUIS MION — (1698–1775) French composer and teacher.

JACOPO PERI — (1651–1633) Italian composer.

GIACOMO PREDIERI — (1611–1696) Italian organist and composer.

GIOVANNI BATISTA SAMMARTINI — (1700–1770) Italian composer, director and organ player.

DANIEL SPEER — (1636–1707) German orphan composer.

AGOSTINO STEFFANI — (1654–1728) Italian churchman, composer and diplomat.

JOZEF ANTONIN STEPAN — (1726–1797) — Czech composer, teacher and keyboard virtuoso.

GIUSEPPE TARTINI — (1692–1770) Italian famous violinist and composer.

Discussion

This chapter has shown that despite a high mortality in past centuries, Baroque composers lived markedly longer than expected, those who had a stroke lived over 70 years (average 72 years). It is no surprise that almost all these were men, nine being German, and seven of Italian origin. Because of the difficulties in travel and communication, to achieve acknowledgment it was probably necessary to live long enough, although Johann Sebastian Bach's case challenges even this theory.

Two Baroque giants, Bach and Handel, never met even though they knew each other and intended to meet (Handel being much more famous at that time). It might be thought that composers knew relatively little of each other, but this is far from the truth, almost all travelling a good deal by slow, long-distance stage-coach. Because of these travels, early Baroque styles were adopted by Germans from Italy, culminating in the works of Handel and Bach.

References

1. Schonberg, H. C. (1992). *Lives of the Great Composers.* London: Abacus.
2. Bukofzer, M. (1947). *Music of the Baroque era.* New York: W. W. Norton & Co.

3. Smallman, B. (2000). *Schutz* ('The Master Musicians' series). New York: Oxford University Press.
4. Lange-Eichbaum, W., Kurt, W. and Ritter, W. (1985). *Genie Insinn und Ruhm (Händel)*. Munchen-Basel: Ernst Reinhardt.
5. Slater, E. (1971). The problems of pathography. *Acta Psychiat Scand* **219**: 133–144.
6. Ostwald, P. (1985). *Schumann: The Inner Voices of a Musical Genius.* Boston: North Eastern University Press.
7. Cherbuliez, A. -E. (1957). *Johann Sebastian Bach.* Frankfurt; Hamburg: Fischer.
8. Geiringer, K. (1977). *Die Musikerfamilie Bach.* Munchen: 2. Aufl. C. H. Beck.
9. Breitenfeld, D., Thaller, V., Breitenfeld, T., Golik-Gruber, V., Pogorevc, T., Zoricic, Z. and Grubisic, F. (2000). The pathography of Bach's family. *Alcoholism* **36**: 161–164.
10. Breitenfeld, T., Vargek Solter, V., Breitenfeld, D., Zavoreo, I. and Demarin, V. (2006). Johann Sebastian Bach's Strokes. *Acta Clinica Croatica* **45**: 41–44.
11. Breitenfeld, T. (2006). The eyes and brain of Johann Sebastian Bach. *Arch Ophthalmol* **124**: 1510.
12. Vollhardt, M. (1935). Uber das Augenleiden Johann Sebastian Bachs, seinem Operateur und wie es diesem später in Dresden erging. *Medizinische Welt* **50**: 1825–1829.
13. Zeraschi, H. Bach und der Okulist Taylor (1956). *Bach-Jahrbuch* **43**: 52–64.
14. Ober, W. Bach, Händel, and 'Chevalier' John Taylor, M. D. (1969). *New York State Journal of Medicine* **69**: 1797–1807.
15. Taylor, J. (1761). The history of the travels and adventures of the 'Chevalier' John Taylor, Ophthalmiator, written by himself. London.
16. Peipert, J. F. and Roberts, C. S. (1986). Wilhelm His, Sr.'s Finding of Johann Sebastian Bach. *Am J Cardiol* **57**: 1002.
17. His, W. (1985). *Anatomische Forschung uber Johann Sebastian Bach's Gebeine und Anlitz.* Leipzig: F. C. W. Vogel.
18. Wustmann, G. (1895). Die Auffindung der Gebeine Johann Sebastian Bachs. *Grenzbote* **54**: 415–425.
19. Breitenfeld, T. (2007). The eyes, brain, bones and skull of Johann Sebastian Bach. *Arch Ophtalmol* **125**: 1717–1718.
20. Karhausen, L. R. (2001). The Mozarteum's skull: A historical saga. *J Med Biogr* **9**: 109–117.
21. Stadlbauer, C., Reiter, C., Patzak, B., Stingeder, G. and Prohaska, T. (2007). History of individuals of the 18th/19th centuries stored in bones, teeth, and hair analyzed by LA-ICP-MS-a step in attempts to confirm the authenticity of Mozart's skull. *Anal Bioanal Chem* **388**: 593–602.
22. Mainwaring, J. (1980). *Memoires of the Life of George Frederick Handel.* New York: Da Capo Press.

23. Breitenfeld, D., Thaller, V., Breitenfeld, T., Golik-Gruber, V., Pogorevc, T. Zoricic, Z. and Grubisic, F. (1996). Pathographies of famous musicians (G. F. Handel). *Alcoholism* **32**: 163–168.
24. Bazner, H. and Hennerici, M. (2004). Georg Friedrich Handel's strokes. *Cerebrovasc Dis* **17**: 326–331.
25. Evers, S. (1996). Georg Friedrich Handel's Strokes. *J Hist Neurosci* **5**: 274–281.
26. Frosch, W. A. (1989). The 'case' of George Frideric Handel. *N Engl J Med* **321**: 965–969.
27. Kloverkon, G. H. Georg Friedrich Handel in Aachen. *Sudhoffs Archiv* **41**: 104–112.
28. Cooper, M. (1935). *Gluck*. London.
29. Kinsky, G. (1925). *Glucks Briefe an Franz Kruthoffer*. Wien.
30. Reissman, A. W. (1882). *Gluck, sein Leben und seine Werke*. Leipzig.
31. Schmid, A. and Chr. W. (1854). Ritter von Gluck. Dessen Leben und tonkunstlerisches Wirken. Leipzig.
32. Breitenfeld, T., Demarin, V., Vargek Solter, V. and Trkanjec, Z. (1999). Famous composers and neurological disorders — Christoph Willibald Gluck (1714–1787) — a pathography. *Eur J Neurol* **6**: 71.
33. Sadie, S. (1980). *New Grove World Dictionary of Music and Musicians*. London: Macmillan.

Chapter 20

From Sensibility to Madness in Nineteenth-Century Romanticism — Neurosyphilis in German-Speaking Composers

Hansjörg Bäzner and Michael Hennerici

The nineteenth century has been the most important period of Lied composition in central Europe. In this century, Franz Schubert, Felix Mendelssohn-Bartholdy, Robert Schumann, Franz Liszt, Johannes Brahms, and Hugo Wolf were the protagonists of this musical form. As far as can be determined from contemporary sources, Schubert, Schumann and Wolf all died from the consequences of neurosyphilis. In this chapter, we will focus on Robert Schumann and Hugo Wolf. Both shared such a variety of symptoms that interfered in their artistic expression, especially in respect to Lied compositions, that it seems improbable that they should have suffered from different diagnoses, as has been argued previously.

Introduction

Neurosyphilis was endemic in the nineteenth century, seeing as an effective cure was not available. At that time, treatment was largely based on only moderately effective mercury preparations, which were associated with severe side-effects. Although neurology as a discipline made enormous progress in the nineteenth century, and the diagnosis of neurosyphilis could be made clinically with great certainty, the infectious agent of *Treponema pallidum* and Paul Ehrlich's Salvarsan treatment were only discovered in the first decade of the twentieth century. As a consequence, the disease remained a threat to those who were sexually active.[1]

Syphilis was perceived to be a consequence of immoral, improper, or promiscuous sexuality, the diagnosis resulting in severe social stigmatisation. This fact most probably has led to many historical medical analyses being incomplete or misleading. The diagnosis of syphilis in these geniuses is still today not unanimously accepted by experts in the field of musical and medical history. Naturally, syphilis in Robert Schumann and Hugo Wolf cannot be confirmed retrospectively using modern diagnostic tools. In the beginning of the twenty-first century, however, in large part made possible by the publication of Robert Schumann's medical records[2–4] and the striking similarities with Hugo Wolf's case,[2] we postulate that the diagnosis of syphilis in both composers can be made with reasonable certainty. Characteristically, in both Robert Schumann and Hugo Wolf euphoric surges of creative energy, associated with emotional highs, produced clusters of productivity resulting in quite phenomenal series of song composition. At the same time, Schumann and Wolf shared a life-long fear of 'becoming mad'.

The age of Romanticism, stressing at times banked emotion as a source of experience, was probably the perfect environment nurturing the aesthetics typical for poems and music of the time. It remains speculative if this elated 'sensitivity' compensated for often unfulfilled dreams and wishes.

Hugo Wolf (1860–1903) — Biographical Notes and Personality Traits

One of his closest friends, Edmund Hellmer, introduces Hugo Wolf as follows: 'He was of very short stature [154 cm] — broad shoulders, strong and short his neck — almost graceful his small hands and feet. His head slightly bent forward — face and hands coloured like old ivory… his eyes were dark like black-ink spots and seemed as burning from an inner fire. His short-cut hair was smooth and blond, his beard on lips and chin brownish… his clothes were modest but always perfectly clean… scrupulous order is the principle of his life'.[5] Wolf's most important biographer, Frank Walker, has documented some remarkable *Conversations with Hugo Wolf*,[6] which nicely characterise the composer. His judgement on women's role in society is all but legendary: 'However, the woman must naturally always preserve decorum'. On another occasion, he displayed rather rude behaviour. Having laid down on a sofa and resting in the house of friends: 'I'm quite indifferent. I do what I please. If people don't like it, they should say so. All responsibility for my conduct I take on myself.' According to Robert Hernried, 'Wolf was both tough and gentle, overbearing and humble, intellectually acute and naively child-like, condemnatory and forgiving, a Titan and a humble mortal. His fiery temperament went from one extreme to the other, and all too often it must have been incomprehensible to bourgeois stodginess and complacency'.[7]

Wolf was born in Windischgraz which then belonged to the Austrian Empire and is now a small town in Slovenia called Slovenj Gradec. He was a very bad student except for musical studies, and was dismissed from several secondary schools as well as from the Vienna Conservatory due to lack of discipline. For most of his life, Wolf stayed in Vienna. He changed apartments on numerous occasions, having complained about the noise of others.[8–13]

Wolf's activities as a critic for the Vienna 'Salonblatt' began to pick up in the 1880s. He was quite merciless in his criticism, but fervent in his support of the genius of Liszt, Schubert, and Chopin. His most famous victim was Johannes Brahms. The intensity and expressive strength of his convictions made him numerous enemies in musical Vienna.[8,9] His judgement on the late works by Robert Schumann was strict at best: '[Schumann's] 'Faust' is very weak, except for a few passages in the third part. All his later things get feebler and have almost no substance. I am no great lover of his symphonies either. Only in a small frame, in miniature style, there he is a master'.[6] If we agree with this statement, we have also to admit, that Wolf himself had great difficulty in composing works of the 'large frame'. The 'Rosé Quartet' would not even look at his only string quartet after having been criticised in one of his regular columns. On another occasion, at the rehearsal of novelties, his tone poem of *Penthesilea* was met by the Vienna Philharmonic Orchestra under conductor Hans Richter with nothing but derision for the man who had dared to criticise Brahms.[8]

Robert Schumann (1810–1856) — A Brief Note on his Personality

As Robert Schumann's biography is very well known,[14,15] we will focus only on some details of his personality. He was quite tall, walked slowly, dragging his feet, at times tip-toeing. Contemporaries mentioned that he often kept his eyes almost closed, leading to speculation that he may have had reduced vision. According to his first biographer, Joseph von Wasielewsky, Schumann 'lacked the ability to put himself in close rapport with others, and to make his meaning clear to them; this was because he either was silent, or spoke so low that he could not be understood'.[15] Especially in his younger years he was known to drink alcohol, at times excessively, smoking cigars for many years.

The Infection

Wolf's musical gift and his charm earned him attention and patronage as soon as he was introduced to Vienna society as a very young man. Decisively, in his early Vienna days he most probably was infected with syphilis. If we take the memoirs of Alma Mahler at face value, 'Hugo Wolf as a very young man [17 years old] was taken by Adalbert Goldschmidt into the so-called 'Lehmgrube' (a brothel) where Goldschmidt played dance music, for which he was often 'awarded' with a young woman without charge. He presented his 'honorarium' once to his friend Wolf, and Wolf took away with him 'the wound that never will heal'.[16]

With respect to the infection in Robert Schumann's case, an episode noted in his diaries is important.[17] Schumann had a sexual relationship with a woman named 'Christel'. Schumann also called her 'Charitas' and meticulously noted and counted their rendezvous. On May 12, 1831, he speaks of a 'bad wound' that caused 'biting and devouring pains'. His friend and medical student Glock 'made an embarrassed face'. Schumann expressed his hope that 'only guilt brings Nemesis' and speaks of his 'incautiousness'. On June 4, the remark of 'the frenulum bitten by daffodil water' may indicate medical treatment of the primary syphilitic lesion. Many years later one can find the following entry in his diary: 'Troubled year 1836: Sought out Charitas and consequences thereof in January 1837'. This can be interpreted as a document of Schumann's growing anxiousness with respect to the potential sequels of his earlier adventure.

Some authors have argued against the diagnosis of syphilis, referring to the fact that Robert and Clara Schumann had eight children. This argument must be rejected: the infection ('Charitas') dates back to May 1831 and the marriage of Robert and Clara was on September 12, 1840. Since sexual transmission of Treponema pallidum occurs only when mucocutaneous syphilitic lesions are present and such manifestations are uncommon after the first year of infection, Robert was certainly not infectious given the latency of nine years after the primary infection.

Productivity in Clusters

The years 1888 and 1889 were amazingly productive years for Hugo Wolf. He travelled to the holiday home of the Werner family in Perchtoldsdorf to compose in solitude. Here, he produced more than 50 songs after poems by Mörike at astonishing pace. Still in the same year, the *Eichendorff-Lieder* followed, then in 1889 the 51 *Goethe-Lieder*. The *Spanisches Liederbuch* was begun in October 1889, totalling the number of songs composed in these two years to more than 140. This productivity is all the more remarkable when considering other phases in Wolf's career (such as 1884–1886 and 1892–1895), when his creativity seemed completely stalled[8–13] (Fig. 1).

Quite interestingly, we find the same episodic frenzied working pace in Schumann's oeuvre. In 1840, he composed more than 150 songs, in

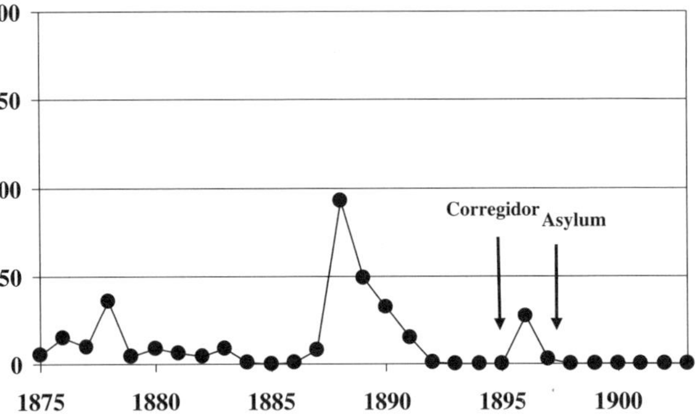

Figure 1. Creativity in clusters: Hugo Wolf's Lied compositions are clustered in several short periods. Shortly after the composition of his opera *The Corregidor* in 1895, he produced his last songs.

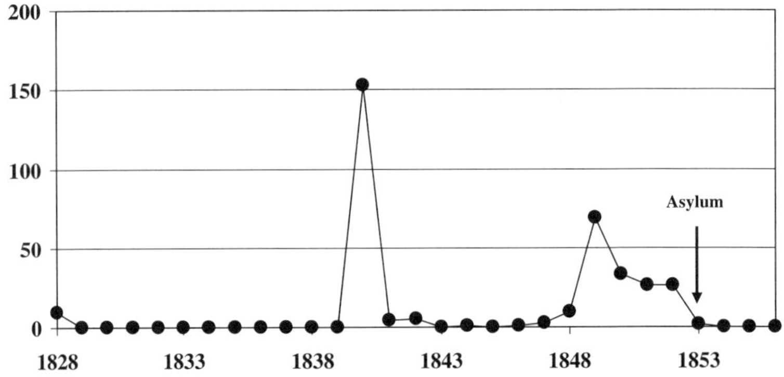

Figure 2. Creativity in clusters: Robert Schumann's Lied compositions are clustered in several short periods. After his admission to the asylum in Endenich he composed no further songs.

the years of 1849–1850 more than 100 songs were produced (Fig. 2). On October 19 1840, Clara Schumann notes in the joint diary of the Schumanns: 'Robert says he cannot compose just now, which depresses him. This grieves me very much. Doesn't he remember all he has composed during this last year? Must the mind never rest?

Surely it will burst forth later with all the greater power...' Except for Franz Schubert, who also died from syphilis, we were unable to discover a similar composing pattern in other song composers.

Neurasthenic Prodromal Phase

After a hopeful start in 1850, Schumann's position as a city music director in Düsseldorf became controversial. In Düsseldorf, Robert and Clara were warmly welcomed and had met a well drilled, though partly dilettante orchestra. Schumann wrote in a letter to his predecessor Hiller nine months before his arrival at Düsseldorf that he did not 'expect much culture in an orchestra, and [was] prepared to meet common musicians, but not rude or malicious ones'.[18]

The documents of the growing tensions between the authorities of the musical society and the Schumann family fall in a period that may well be called the prodromal phase of Schumann's final illness. As early as 1851, the executive committee of the General Music Society was alarmed by the fact that a large segment of the public had criticised Schumann's conducting, deploring that he often appeared to be dreaming and was clumsy and slow speaking. Further, discipline in the chorus and orchestra was steadily deteriorating. By the summer of 1852 Schumann's debilitating ill health forced him temporarily to relinquish his conducting duties until early December to Julius Tausch, his deputy.[18] By the end of 1852, deputy mayor Wortmann remarked that 'the orchestra misses, under Herr Schumann, the solid, sure indication of tempos, even beats, clear, definite, and understandable comments'.[19]

Joseph Wilhelm von Wasielewski (1822–96), the new concertmaster of the orchestra, reported that Schumann 'lacked the physical energy and endurance requisite for a director: he was always easily exhausted and was obliged to rest at intervals during a rehearsal. Nor did he exercise any sort of care or oversight'.[15]

Also in Hugo Wolf's case a neurasthenic prodromal phase seems likely. While he composed the score of his opera *The Corregidor* in 1895 he was extremely sensitive to any noise. It is reported that he shot and killed singing birds.[8] Although Wolf had hoped that his opera would have been performed in Vienna, Berlin, or Prague, he had to accept that with the promotion of his good friends in southern Germany *The Corregidor* was set on stage in Mannheim. During the rehearsals he was unpleasant and spoke of the musicians as 'idiots one cannot work with', and wrote to Melanie Köchert that 'the whole gang... was incapable'. Though the premiere was a success, only one further performance followed in Mannheim.[20]

Outbreak of General Paresis: Suicidal Attempts and Hospitalisation

Wolf's final illness stage began at the latest in summer 1897, when he intended to accompany his friend Haberlandt on a bicycle tour. For this purpose, he took bicycle riding lessons in June 1897 which had to be stopped due to severe co-ordination deficits. Wolf simply was unable to stay on the bike and suffered numerous falls. He wrote to his friend Faißt: 'Being a beginner, I have so far only learned to know the dark side of bicycle riding. My poor body is tattooed from bruises I have got in the driving lessons. However, I will continue facing the divine pleasure, which is waiting on me...'.[8,9] Earlier, namely in August 1896, further important evidence can be found in Wolf's medical history. After a train trip to Graz, Wolf was seen by an ophthalmologist for a conjunctival irritation which had been caused by a soot particle. Doctor Elschnigg diagnosed Argyll Robertson pupils and informed Wolf's friend Heinrich Potpeschnigg, a dentist, about his observation. Wolf, however, was not informed about this finding and its implications.[8,9]

September 1897 was the decisive point in Wolf's illness with the overt outbreak of general paresis. His friend Haberlandt was sent a sheet of music with Wolf's handwriting saying: 'Piping hot! Straight from the frying-pan! Am beside myself! Sell me up! Am blissful! Raving!' Haberlandt became suspicious of this strange letter and decided to see him at his apartment in the Schwindgasse. Wolf appeared quite deranged and spoke excitedly about his new opera *Manuel Venegas.* He sat down at the piano and played and sang from the fragmentary opera in a highly exalted mood, claiming that he had never composed anything so beautiful before. This performance, however, seemed to calm down his spirits and Haberlandt left his friends without anxiety; but Wolf continued composing, 'raging like a volcano', and invited all of his friends on the following Sunday to 'play the opera at Perchtoldsdorf to all the faithful'. This date was again preceded by strange letters such as the one to Melanie Köchert, where he urged Melanie to join the audience for his performance: 'If on the same day you don't put in an appearance I shall never again set a foot in your house.'[21] The following Saturday September 19, 1897 marked the onset of madness and megalomania. He met his friend Edmund Hellmer for lunch. While eating unusually fast, he picked the chop he had ordered and simply tore the meat away from the bone, gulping it down like an animal. Suddenly he whispered to Hellmer: 'Did you already know — I have been named director of the Court Opera?' Hellmer first believed Wolf was joking, but Wolf repeated the 'news' when Föll, another friend, arrived. The two friends finally were given a private concert in Wolf's apartment, then Wolf drank brotherhood with them and ended up talking about matters that under normal circumstances he would never have discussed openly — his family, his poverty and his relations with women.[8] The following day, Wolf ran through the streets of Vienna in order to find the singer Hermann Winkelmann in his duty as the new director of the opera. He aimed at engaging the singer for the same day's concert in Bokmayer's house. Everybody met there at 5 pm and Wolf planned to play recently composed material from his new opera *Manuel Venegas* on the piano. He started doing so, then tried to play Wagner's prelude to the *Meistersinger*, but had to interrupt because he could not recall the score by heart. Thereafter, he proclaimed to the audience his nomination as the new director of the Vienna Hofoper and that his first legal act would be dismissing Mahler from his post. Haberlandt had asked Dr. Gorhan from the local hospital to attend the concert as an act of precaution. Dr. Gorhan diagnosed the outbreak of madness and suggested bringing Wolf to an asylum immediately. The private asylum of Dr. Svetlin in Vienna offered to take Wolf as an inpatient the next morning. Hence, Wolf had to be accompanied back to his home by some friends, where he furiously attacked the housekeeper. His friends had great difficulties to bring this unpleasant scene to an end. On September 21, Wolf was admitted to Doctor Svetlin's asylum — a carriage was sent and Wolf was told that this carriage was bringing him to the opera for the signature of his new contract.[8–13]

A visit Wolf had paid just a short time before to his former roommate and friend from the days in the Vienna conservatory, Gustav Mahler, who had already begun working as the provisional director of the opera, preceded this catastrophe. Both composers had disputed the plans to produce Wolf's *The Corregidor* at the Vienna Court Opera, which Mahler had actually promised to produce previously. During the dispute, Mahler finally had expressed doubt as to whether the opera would be performed in Vienna at all. Wolf left Mahler raging bitterly and vowing revenge for this disappointment. In his extreme emotional state, Wolf produced the idea that removing the obstacle, Mahler would lead the way to his so long merited success, becoming director of the opera.

Hugo Wolf was discharged in late January of 1898 after four months in Dr. Svetlin's asylum. His symptoms had improved to a certain degree. For most of the subsequent months his friends would take him on several short holiday trips. This was also the case in October 1898 when the disease deteriorated again.

During a stay in Traunkirchen, Wolf left his apartment one morning tormented by ideas of persecution. For some time his friends had searched the vicinity, when he appeared on the edge of a nearby forest, dripping wet; he had tried to drown himself in the Traunsee, but had finally swum to the shore. Then he had wandered in the woods ashamed of what he had attempted to do. He himself asked to be put in a mental home, 'only for god's sake not Svetlin's'. Hence, he was brought to the Niederösterreichische Landesirrenanstalt in Vienna, where he would stay until his death in February 1903.[8–13]

Similarly in Schumann's case, according to several sources, the idea of entering a mental institution had been his own. Following a note in his diary on February 10, 1854 a series of auditory delusions occurred. They continued for the following weeks — usually in the form of music. His final piano composition, the *Ghost Variations*, has been composed as a result of these. Schumann believed the theme to these variations was sent to him by an angel. The hallucinations became more and more horrendous, at times Schumann felt himself attacked by wild beasts.

At the same time, he obviously was afraid of losing control of his actions, and feared harming either himself or someone else. Clara Schumann wrote in her diary that during the night of February 26, Schumann 'suddenly got up and wanted to have his clothes. It was necessary, he said, for him to go into the asylum because he was no longer in control of himself, and did not know what he might do... Robert laid out with plain deliberation everything that he wanted to take with him: watch, money, music paper, pens, cigars; and when I said to him: 'Robert, do you want to leave your wife and children?', he replied, 'it will not be for long. I will soon return cured'.[15]

On the following day, in the midst of the Rhenish carnival in Düsseldorf, he left the house and attempted to take his own life by leaping into the Rhine. In the days that followed, he said to his doctors that 'they should take him to an asylum, because only there could he return to health'. Dr. Richard Hasenclever, his treating physician decided to send him to an asylum in Endenich near Bonn, being acquainted with Dr. Franz Richarz who owned and led this private institution. Schumann arrived in Endenich on March 4 accompanied by Hasenclever and two attendants. At that time, it took eight hours to travel from Düsseldorf to Bonn-Endenich. The asylum in Endenich had a good reputation and by all accounts was well maintained.[22,23]

The recent publication of Robert Schumann's patient record from the Endenich asylum 150 years after his death is important for understanding his complex pathography.[2–4] Dr. Franz Richarz had never intended to publish the record owing to doctor-patient confidentiality; his daughter-in-law bequeathed the record to her godson, a psychiatrist, who was the uncle of Aribert Reimann, a well-known contemporary German composer. Irritated by the fierce and unfair discussion regarding Schumann's final disease, Reimann decided to allow publication of the record.[2–4] This has been a much welcome and most important contribution to a clearer picture of Schumann's final years. In the view of most authors who have since analysed the material, the diagnosis of neurosyphilis becomes more plausible than it had already been. However, curiously enough, the original sources published by the Schumann-Society in Düsseldorf, contain the detailed comments of a psychiatrist who makes every attempt to reject this diagnosis.[4] Thereby, he stands in a long tradition of authors who favour different disease theories, obviously trying not to tarnish the reputation of the composer and his family.[24–29] This given respect may have its origins in the attitude of Clara Schumann, their children and the very protective close friends Johannes Brahms and Joseph Joachim. However, the finally published sources give a very clear picture of neurosyphilis. The following excerpts aim at illustrating the most remarkable symptoms of Schumann's disease:

> **11 April 1854**: Said to the attendant, the authorities had ordered him being burned in hell: he had done too many bad things. During the night very restless, most time out of bed, not taking off the clothes... mourned as if in pain, was totally sleepless.
>
> **19 April 1854**: Restless at night, spoke loudly to himself until midnight, about the Veneris, being unhappy, becoming mad; got up later and aimed at leaving the room, became violent with the attendant.
>
> **27 September 1854**: Yesterday moved to the main building: very pleased and thankful thereabout, helping with the transfer... wrote a moderately long, thoughtful and calmly written letter to his wife, with omissions of some words, the date correct. Talked a lot to himself during the night, softly, the fingers moving on the blanket as if playing the piano. Asking the attendant if he had made bad things, this somebody had mentioned to him.
>
> **10 November 1854**: Almost continuously plays *Domino*.
>
> **31 December 1854**: Rejected sherry and part of the Rhine-wine, saying this tasted bitter, similar to the water...
>
> **8 January 1855**: ...talking of poison when taking his medication... is so much disorganised, that makes it impossible for him to have a short communication about a certain object...
>
> **12 January 1855**: Yesterday was visited by Mr. Brahms, was very pleased about this visit... talks during the round quite freely and understandable but very slowly and with a voice similar to a child (with omissions).
>
> **22 January 1855**: During the round, reports a spell that made him believe he would die. Said he had never had something similar before, these had

been cramps, namely in his right hand... during the rounds — half an hour after the spell... had convulsions in the fingers, and could not suppress them.

24 January 1855: ...Convulsions in the right hand.

9 February 1855: ...speaks today again about the silly things, that the silly voices were calling to him.

24 February 1855: Yesterday in a good mood, talking a lot. Was visited by Mr. Brahms, who found him much better than the last time.

7 March 1855: ...the deprivation of sleep being his the worst problem, the hearing of the silly voices, a bad demon menacing him, the same having played a lot of horrible animal faces seven months ago etc.

9 March 1855: His speech for most of the time totally incomprehensive.

12 March 1855: During the round sitting on the sofa, a spell of anxiousness with convulsions in his extremities. Complains of headache, pressure in the chest, anxious. The speech very disabled, soundless, unintelligible, afraid of becoming mad. Believes he is persecuted by the Nemesis. His consciousness not disturbed meanwhile... the right pupil is much larger than the left pupil when shining light in his eyes. Understood and answered all questions during the spell.

28 April 1855: He refers the simplest things to the persecutions of the evil daemon. Speaks a lot during visits, but hardly understandable. He expresses the completely unwarranted suspicion that his watch might run too fast.

4 May 1855: During the round writing, unpleasant, said he wrote a letter to a notary in order to sue the doctor.

6 May 1855: Today he spoke of the prosecution of the 'wicked hag'....

8 May 1855: Played almost for two hours the piano, very wild and incoherent... threatened the attendant with a chair... his speech is mumbling similar to a drunken man.

19 May 1855: Yesterday very shy, especially after a visit from Joachim, could stay at the piano only for a short period of time, often his whole body was took by shivers and strong convulsions...

24 May 1855: Played the piano, did not finish the piece, and said this was too tiresome.

16 June 1855: Is completely calm, without any symptoms of madness or of hallucinations.

9 July 1855: …says, he had pain in his belly, he had pain everywhere on touching him, if he liked.

25 July 1855: Was agitated, impulsive, loud, struck the attendant, everything being poisoned, during night-time continuously exited, screaming, angry.

25 August 1855: Yesterday calm, friendly, talking to himself in an exalted good mood, laughing a lot.

8 September 1855: …today sometimes very loud. Not persuaded to write a letter to his wife…

12 September 1855: Wrote down abrupt mentions of melancholic content during the past days and reflections such as for instance: '1831, I was syphilitic and was cured with arsenic'.

10 October 1855: Since yesterday extremely loud, screaming, shouting, also at night: walks in the room during the time of the round, touching several of his works while shouting: this is mine. Appearance very disordered… Poured the wine in the closet, claiming this was urine.

31 October 1855: Yesterday afternoon, during the whole night and today until after breakfast extremely loud, screaming, consequently very hoarse.

19 January 1856: During the evening round friendly, liked to talk coherently and calmly about his compositions.

16 April 1856: Was calm, burned the letters from his wife yesterday evening and today in the morning… disagrees that he burned letters at all…

29 April 1856: Yesterday wrote a clear and coherent letter to his doctors…

9 June 1856: Yesterday received birthday greeting from Mr. Brahms; was unpleasant, speaking of poison when hot chocolate was served.

11 July 1856: …looking sick. His eyes squinting inwardly. His pupils both much widened.

Final Illness and Death

The final entrances in Robert Schumann's record from Endenich read as follows:

29 July 1856: Calm from noon on, at night as well, taking some teaspoons of wine-jelly and some wine from his wife…

30 July 1856: Yesterday at 1pm 60 breaths, pulse almost not palpable. Died yesterday at 4 pm. Begun autopsy on 30 July 3.30 pm.

Clara reports: 'He suffered terribly, although the doctor said differently. His limbs were in continual convulsions. His speaking was often very vehement. Ah! I prayed to God to release him, because I loved him so'. Schumann died in his sleep on 29 July at 4 pm. Clara saw him a half hour later: 'I stood by the corpse of my dearly beloved husband and was at peace; all my feelings went in thanks to God that he was finally free'.[15]

Hugo Wolf also seemingly died in a moment when, besides his attendant, none of his friends who had waited at his deathbed for many days were with him. Pneumonia was considered the final illness, high fever and horrible seizures are reported from his last visitors. Wolf is buried in the Central Cemetery in Vienna, along with many other notable composers.[8,9]

Synopsis of Symptoms and Disease Course

According to the seminal works of Franz Hermann Franken,[2,3,22] the disease courses of both composers are so similar, that one can hardly diagnose different illnesses. A series of cardinal symptoms of general paresis can be derived from the sources (see Table 1 for synopsis). The latency between the infection and the outbreak of general paresis was 22 years in Schumann and 20 years in Wolf. This was preceded by a neurasthenic prodromal phase in both, although part of the 'neurasthenic' behaviour may have been a result of marked personality traits. Wolf died five years after the onset of general paresis, Schumann survived for 30 months. Hallucinations with acoustic, visual, and scenic content were present in both. Paranoid features such as the fear of being poisoned were found in Schumann's patient record at several occasions and were reported in Hugo Wolf's as well. The latter also displayed delusions of grandeur, at times he would claim to be Jupiter and control the weather. Once he wrote to his friends: 'It would be my plan to go on worldwide tours with the Weimar theatrical personnel, much as the Meiningen troupe did, who in their time created a great and

Table 1. Symptoms and disease course in Robert Schumann and Hugo Wolf (modified after Franken[2]).

Symptoms	Schumann	Wolf
Latency between infection and outbreak of general paresis	22 yrs	20 yrs
Neurasthenic prodromal phase	present	present
Latency between outbreak of general paresis and death	2.5–3 yrs	5.5 yrs
Hallucinations and paranoia	marked without delusions of grandeur	marked with delusions of grandeur
Aggressive behaviour	often and markedly expressed	often and markedly expressed
Focal and generalised seizures	present	present
Dysarthria	present, progressing to anarthria	present, progressing to anarthria
Pupillary disturbances	pupils of different size, strabismus	Argyll-Robertson pupils
Stereotypies	marked	present
Fear of being poisoned	permanent	at times
Cachexia, incontinence	marked at final stage	marked at final stage
Pneumonia as final event	present	present
Personality change	present	present
Loss of writing ability	not before 1856	after 1899
Gave up composing	after 1854	after 1899

justified sensation. But the project I have in mind would have an infinitely greater appeal to the public, as works of mine (and exclusively mine) would be performed, which are never to appear in print; therefore they could not be conveyed to the public in any other way than under my direction.' Aggressive behaviour was present in both composers as well as focal seizures. During their final days, generalised seizures were also reported, along with pneumonia as the terminal illness. Already some time before marked cachexia was noted, incontinence was documented. Dysarthria was a very important and marked symptom in both composers, and pupillary abnormalities were observed by medical doctors before the onset of general paresis in Wolf's case and in Schumann's record from Endenich. Schumann showed marked stereotypes in playing dominoes for whole days. Brahms reported from a visit in Endenich, where he found Schumann with an atlas, arranging names of cities and rivers in alphabetical order: 'We sat down, it became increasingly painful for me... He spoke continually, but I understood nothing... Often he only babbled, something like babab — dadada...' Schumann permanently feared being poisoned, and would therefore often refuse his food. On other occasions, he poured the wine on the floor claiming it was urine. Both composers' writing ability deteriorated. Wolf's handwriting became unintelligible from 1899. A late document of his signature from 1899 shows a significantly disabled handwriting. Robert Schumann, in contrast, could write quite well until a year before he died. His last letter to Clara shows a clear handwriting with only minor semantic errors. Regarding their creative output, both composers didn't manage to compose new oeuvres. Schumann's *Geistervariationen* is probably his final original work, although in the early days at Endenich, a piano accompaniment to a Paganini piece for violin and two short very simple chorale settings (the first on the church chorale 'when my last hour is close at hand') are reported and have been published posthumously.

Syphilis and Lied Composition — Two Sides of a Medal

In an interview to a Hugo Wolf programme of the 2003 Salzburg Festival,[30] the famous singer Thomas Hampson compared the composer's work rhythms to a 'light bulb that is about to go out but, before that happens, gives off an especially bright, hot light for one last time'. Hampson assumes that this was surely related to his disease and speaks of the same phenomenon in the work of Franz Schubert and Robert Schumann, all of them known as the three most important German composers of Lieder having suffered from syphilis. In all three composers, periods of 'a frenzy of creativity' were followed by 'sudden fading'.[30] He finally asks the questions, whether the 'illness released in them a heightened sensibility for lyric form, emotional surges, and sensitivity that empowered them to achieve such masterly settings?' And furthermore: 'Would they not have produced Lieder in such quantities if they hadn't been infected?' It is certainly reassuring to the medical pathographer to also have an artist asking these most important and interesting questions.

Certainly the clusters of productivity in Schumann and Wolf are most amazing, since the series of Lied compositions were produced in the shortest periods of time. This is even more remarkable when looking at the total number of songs in their complete oeuvre and the relative short periods of productive life.

The 'heightened sensibility for lyric form' was probably very well nourished in the Romantic era. At a first glance, the personalities of the two composers, however, do not seem to have been susceptible to the 'exaggerated' emotional states of romanticism. From our point of view, the composers' social competence was not perfect in any sense — for instance Wasielewsky reported that Schumann 'lacked the ability to put himself in close rapport with others, and to make his meaning clear to them; this was because he either was silent, or spoke so low that he could not be understood'.[15] Yet, both had definite strengths in expressing their points of view in the written word. Possibly their way to overcome these personal 'shortcomings' was to concentrate on musical expression as a way of showing emotions.

The role of illness is in part quite obvious. Both composers may well have been suspicious about the nature of the primary infection — the medical knowledge of contemporary physicians was good enough to indicate to their patients the possible serious consequence of given sexual adventures. Hence, it may well have been more than the 'typical' anxiety of the period that both Schumann and Wolf expressed their fear of becoming mad. In addition, they were very well aware of the fates of their famous 'co-patients'. Ironically, Wolf in a megalomanic delusion expressed his ambitions to cure Nietzsche, being convinced that he was the director of his asylum in Vienna. Both composers must have had in mind the horrible vision of ending in an asylum; they were by all accounts well aware of the shortness of their remaining productive time. This alone would have justified an intense concentration and condensation of their respective productive periods.

The less obvious role of their illness lies in the potential catalytic effect of the disease with respect to creativity. This is by far the more difficult question to answer. Several authors have speculated about the nature of the very wide mood swings in both composers and have proposed the diagnosis of bipolar affective disorder in both composers to explain the marked mood swings and also fitting with periods of feverish creativity.

However, even when allowing for such a diagnosis as neurosyphilis, it certainly does not explain the final period of illnesses, as this alone is definitely not sufficient explanation for all of the remarkable extremes in creative output in Schumann and Wolf. Hence, we can only speculate about a possible link between neurosyphilis and creativity through the severe organic disturbance of neuronal networks including the limbic brain with its direct links to motivation. Likewise, a disease-mediated disinhibition of frontal-subcortical circuits might interfere with creativity.

In any case, regrettably, neurosyphilis was the cause for the final termination of artistic output in both composers.

References

1. Singh, A. E. and Romanowski, B. (1999). Syphilis: Review with emphasis on clinical, epidemiologic, and some biologic features. *Clin Microbiol Rev* **12**: 187–209.
2. Franken, F. H. (1997). *Die Krankheiten Großer Komponisten Vol. 4.* Wilhelmshafen: Noetzel.
3. Franken, F. H. (1997). Robert Schumann in der Irrenanstalt Endenich. Zum aufgefundenen ärztlichen Verlaufsbericht 1854–1856 von Doktor Franz Richarz. In *Brahms Studien Band 11.* Tutzing: Hans Schneider-Verlag.
4. Mayeda, A., Niemöller, K. W. and Appel, B. R. (Eds.). (2006). *Robert Schumann in Endenich (1854–1856): Krankenakten, Briefzeugnisse und zeitgenössische Berichte.* Mainz: Schott.
5. Hellmer, E. (1921). *Hugo Wolf: Erlebtes und Erlauschtes.* Wien-Leipzig: Wiener Literarische Anstalt.
6. Walker, F. (1960). *Conversations with Hugo Wolf. Music and Letters* **41**: 5–12.
7. Hernried, R. (1945). Hugo Wolf's 'Four Operas': With unpublished letters by Hugo Wolf, Rosa Mayreder, and Oskar Grohe. *The Musical Quarterly* **31**: 89–100.
8. Walker, F. (1968). *Hugo Wolf: A Biography.* New York: Knopf.
9. Werba, E. (1971). *Hugo Wolf Oder Der Zornige Romantiker.* Wien; Munich: Molden.

10. Decsey, E. (1903–1906). *Hugo Wolf.* Leipzig; Berlin: Schuster u. Löffler.
11. Dorschel, A. (1985). *Hugo Wolf.* Reinbek bei Hamburg: Rowohlt.
12. Hanolka, K. (1988). *Hugo Wolf: Sein Leben, sein Werk, seine Zeit.* Stuttgart: Deutsche Verlags-Anstalt.
13. Newman, E. (1966). *Hugo Wolf.* New York: Dover Publications.
14. Boucourechliev, A. (1958). *Robert Schumann.* Reinbek bei Hamburg: Rowohlt.
15. von Wasielewski, W. J. (1906). Robert Schumann. Eine Biographie (Dresden, 1858). In Wasielewski, D. (Ed.). *Vierte, Umgearbeitete und Beträchtlich Vermehrte Auflage.* Leipzig: Breitkopf & Härtel.
16. Mahler, A. (1940). *Gustav Mahler: Erinnerungen und Briefe.* Amsterdam: A. de Lange.
17. Schumann, R. (1971–1987). *Tagebücher, 3 Vols.* Eismann, G. and Nauhaus, G. (Eds.). Leipzig: Deutscher Verlag für Musik.
18. Hopkins Porter, C. (1989). *The Reign of the 'Dilettanti': Dusseldorf from Mendelssohn to Schumann The Musical Quarterly,* **73**: 476–512.
19. Neyses, J. (1927). Robert Schumann als Musikdirektor in Düsseldorf. *Düsseldorfer Almanach* **71**: 73–74.
20. Hernried, R. (1940). Hugo Wolf's Corregidor at Mannheim. *The Musical Quarterly* **26**: 19–30.
21. Grasberger, F. (1991). *Hugo Wolf: Letters to Melanie Köchert.* Translated by Louise McClelland Urban. New York: Schirmer Books.
22. Franken, F. H. (1984). Untersuchungen zur Krankengeschichte Schumanns. In Robert-Schumann-Gesellschaft Düsseldorf (Eds.), *Robert Schumann — Ein romantisches Erbe in neuer Forschung.* Mainz: Schott.
23. Ostwald, P. (1985). Schumann. *The Inner Voices of a Musical Genius.* Boston: Northeastern University Press.
24. Gruhle, H. W. (1906). Brief über Robert Schumann's Krankheit an P. J. Möbius. *Zentralblatt für Nervenheilkunde und Psychiatrie* **26**: 805–810.
25. Kerner, D. (1963). *Krankheiten großer Musiker.* Stuttgart: Schattauer.
26. Möbius, P. J. (1906). *Über Robert Schumanns Krankheit.* Halle: Marhold.
27. Nussbaum, F. (1923). *Der Streit über Schumanns Krankheit.* Dissertation, Köln.
28. Payk, T. R. (1977). Robert Schumann als Patient in Bonn-Endenich. *Confinia Psychiatrica* **20**: 153–161.
29. Slater, E. and Meyer, A. (1959). Contributions to a Pathography of the Musicians: 1. Robert Schumann. *Confinia Psychiatrica* **2**: 65–94.
30. http://www.hampsong.com/blog/images/uploads/Salzburg_Aug04_2003_O.pdf.

Chapter 21

Singing: When It Helps

Gottfried Schlaug

It has been reported that patients with severely non-fluent aphasia are better at singing lyrics than speaking the same words. This observation inspired the development of Melodic Intonation Therapy (MIT), a treatment whose effects have been shown, but whose efficacy has not been proven yet and neural correlates remain unidentified. Because of its potential to engage/unmask language-capable regions in the unaffected right hemisphere, MIT is particularly well suited for non-fluent patients with large left-hemisphere lesions. Proof of principle studies have shown that MIT can lead to significant improvements in propositional speech that generalises to unpracticed words and phrases. Treatment-associated imaging changes indicate that MIT's unique engagement of the right hemisphere, through singing and priming of sensorimotor and premotor cortices for articulation with left-hand-tapping, might account for its effect.

Introduction

Of the estimated 600,000–750,000 new strokes occurring in the US each year, approximately 20% result in some form of aphasia. Non-fluent aphasia generally results from lesions in the left frontal lobe in right-handed individuals including the portion of the left frontal lobe known as Broca's region. Named for Paul Broca (1864), who first associated this area of the brain with non-fluent aphasia, this region is thought to consist of the posterior inferior frontal gyrus (IFG) encompassing Brodmann's areas 44 and 45. However, subsequent reports have shown that a wider array of lesions in the frontal lobes and in subcortical brain structures can also present a clinical picture of a Broca's aphasia (see Kertesz *et al.*, 1977).

Surprisingly, there are no universally accepted methods for the treatment of non-fluent aphasia against which new or existing interventions can be tested, nor have any criteria been established for determining treatment efficacy. Most interventions in the sub-acute stroke phase are conducted by speech therapists who evaluate patients' individual needs, then use a combination of techniques to help recover language and facilitate communication. Despite the lack of specific criteria for success, most therapists would agree that treatment efficacy would be defined by patients' ability to show improvement in speech output that generalises to untrained language structures and/or contexts (Thompson and Shapiro, 2007).

Because the neural processes that underlie post-stroke language recovery remain largely unknown, it has not been possible to effectively target them using specific therapies. To date, functional imaging (mostly positron emission tomography) of language recovery has largely focused on spontaneous recovery, and patients have been imaged mostly after natural recovery has run its course (Weiller *et al.*, 1995; Warburton *et al.*, 1999). Some studies emphasise the role of preserved language function in the left hemisphere (Cappa and Vallar, 1992; Heiss *et al.*, 1999), while others propose that language function is restored when right-hemisphere regions compensate for the loss (Basso, 1989; Moore, 1989; Cappa and Vallar, 1992; Weiller *et al.*, 1995; Cappa *et al.*, 1997; Kinsbourne, 1998; Selnes, 1999; Blasi *et al.*, 2002). Still other studies report evidence for bi-hemispheric language processing (Mimura *et al.*, 1998; Rosen *et al.*, 2000; Winhuisen *et al.*, 2005; Saur *et al.*, 2006; Heiss and Thiel, 2006). Interestingly, only a few studies have examined the neural correlates of an aphasia treatment by contrasting pre- and post-therapy assessments (Small *et al.*, 1998; Musso *et al.*, 1999; Cornelissen *et al.*, 2003; Thompson and Shapiro, 2005; Saur *et al.*, 2006; Schlaug *et al.*, 2008). The general consensus is that there are two routes to recovery. In patients with small lesions in the left hemisphere, there tends to be recruitment of left hemisphere peri-lesional cortex and variable right-hemisphere homolog regions during the recovery process. In patients with large left-hemisphere lesions involving language-capable regions of the fronto-temporal lobes, the only path to recovery might be through recruitment of homolog regions in the right hemisphere.

A 'Speech Therapy' that Targets the Unaffected Hemisphere

Assuming that potential facilitators of language recovery may be either undamaged portions of the left-hemisphere language network, language-capable regions in the right hemisphere, or both, it is necessary to explore treatments that can better engage these regions and ultimately, change the course of natural recovery through neural reorganisation. One therapy capable of engaging language-capable regions preferentially in the right hemisphere, but possibly in both hemispheres is Melodic Intonation Therapy (MIT; Albert *et al.*, 1973; Sparks *et al.*, 1974), a method developed in response to the observation that severely aphasic patients can often produce well articulated, linguistically accurate words while singing, but not during speech (Gerstman, 1964; Geschwind, 1971; Keith and Aronson, 1975; Kinsella *et al.*, 1988; Hebert *et al.*, 2003). MIT is a hierarchically structured treatment that uses intoned (sung) patterns to exaggerate the normal melodic content of speech at three levels of difficulty. The intonation works by translating prosodic speech patterns (spoken phrases) into melodically intoned patterns using just two pitches. The higher pitch represents the syllables that would naturally be stressed (accented) during speech. At the simplest level, patients learn to intone (sing) a series of two-syllable words/phrases (e.g. water', 'ice cream', 'bathroom') or simple, two- or three-syllable social phrases (e.g. 'thank you', 'I love you'). As each level is mastered, patients move to the next, and phrases gradually increase in length (e.g. 'I am thirsty', 'a cup of coffee, please'). Beyond the increased phrase length, the primary differences between the three levels of MIT lie in the way the treatment is administered and the level of support that is provided by the therapist.

MIT contains two unique elements that set it apart from other, non-intonation-based therapies: 1) the melodic intonation (singing) with its inherent continuous voicing, and 2) the rhythmic tapping of each syllable (using the patient's left hand) while phrases are intoned and repeated. Since the initial account of its successful use in three chronic, non-fluent (Broca's) aphasic patients (Albert *et al.*, 1973), reports have outlined a comprehensive program of MIT (Sparks and Holland, 1976; Helm-Estabrooks and Albert, 1991) including strict patient selection criteria (Helm-Estabrooks *et al.*, 1989), and data that showed significant improvement on the Boston Diagnostic Aphasia Examination (BDAE; Goodglass and Kaplan, 1983) after treatment (Sparks *et al.*, 1974; Bonakdarpour *et al.*, 2000). In a case study comparing MIT to a non-melodic control therapy, Wilson *et al.* (2006) found that MIT had a general facilitating effect on articulation, and a longer-term effect on phrase production that they attributed specifically to its melodic component. However, the outcomes of that study were measured by the patient's ability to produce practiced phrases prompted by the therapist, rather than by the transfer of language skills to untrained structures and/or contexts.

Another important characteristic of MIT is that, unlike many therapies administered in the chronic phase that involve one to two short sessions per week, MIT engages patients in intensive treatment up to 1.5 hours a day, five days a week until the patient has mastered all three levels of MIT. In addition to its unique elements, there are several other components that play an important role in MIT, but are also used by other therapies, among them are the slow rate of vocalisation (one syllable/s) and an administration protocol that includes one-on-one sessions with a therapist who introduces and practices words/phrases using picture cues while giving continuous feedback. These shared features must be carefully considered when the efficacy of MIT is tested against a control intervention (Schlaug *et al.*, 2008).

The original interpretation of MIT's path to successful recovery was that it engaged expressive language areas in the right hemisphere (Albert *et al.*, 1973; Sparks *et al.*, 1974), although to date, this has not yet been proven. Alternatively MIT may exert its effect by either unmasking existing music/language connections in both hemispheres, or by engaging preserved language-capable regions in either or both hemispheres. Since MIT incorporates both melodic and rhythmic aspects of music (Albert *et al.*, 1973; Sparks and Holland, 1976; Helm-Estabrooks and Albert, 1991; Cohen and Masse, 1993; Boucher *et al.*, 2001), it may be unique in its potential ability to engage both hemispheres. Belin *et al.* (1996) suggested that MIT-facilitated recovery was associated with the reactivation of left-hemisphere regions, most notably the left prefrontal cortex, just anterior to Broca's region. Although this publication was the first to examine patients treated with an MIT-like intervention using functional neuroimaging, their findings were surprising and somewhat contrary to the hypotheses that had been put forward by the developers of MIT (Albert *et al.*, 1973; Sparks *et al.*, 1974). It is interesting to note that although Belin and colleagues' primary finding was an activation of left prefrontal regions when participants were asked to repeat intoned words, there is an important aspect of their study that is not often reported. In their analysis comparing the repetition of spoken words with the hearing of those words, they found blood flow changes that occurred predominantly in the right hemisphere (including the right temporal lobe and the right central operculum), which concurs with some of our findings (Schlaug *et al.*, 2008).

Assessing Improvements in Speech Production

Most patients undergo a battery of assessments such as the Boston Diagnostic Aphasia Examination (BDAE; Goodglass and Kaplan, 1983) to help classify an aphasic disorder as fluent or non-fluent and to assess the relative impairments in comprehension, naming, repetition, reading and writing. However, these batteries are typically not the best in order to assess improvements in speech production, which would be the desired therapy effect in patient with non-fluent aphasia (Thompson and Shapiro, 2007; Schlaug *et al.*, 2008). Several studies have used investigator-assembled, non-standardised testing batteries to quantitatively measure speech production changes or spontaneous speech improvements in response to an intervention. Among those measures are: 1) *Conversational interviews:* regarding patients' biographical data, medical history, daily activities, description of routine procedures (e.g., cooking a favorite dish or describing favorite hobbies, tools, or routine repair work), etc.; 2) *Descriptions of complex pictures*; and 3) *Naming tasks* (i.e., naming frequent and infrequent picture items) (Snodgrass-Vanderwart, 1980; Kaplan *et al.*, 2001). Patients' responses on the conversational interview and description of complex pictures can be quantified by calculating the average number of Correct Information Units (CIUs) per minute and the average number of syllables/phrase. All meaningless utterances, inappropriate exclamations, incorrect responses (inaccurate information), and/or perseverations need to be excluded prior to scoring.

Measuring Brain Effects of the Therapeutic Intervention

Studies have used a variety of experimental paradigms to examine brain effects of therapeutic interventions ranging from covert to overt naming or word-stem completion tasks. We have used a list of bi-syllabic words/phrases that patients are capable of saying at baseline and repeated that same list of words/phrases at all imaging time points prior to and after the therapeutic intervention. Important is that the rate of speaking/singing does not change comparing pre- to post-therapy assessments, since otherwise the imaging effects would not truly reflect a change in the brain network, but a change in the rate of production. Furthermore, it is important to select appropriate control condition to isolate the neural correlates that change in response to therapy from those that control basic sensorimotor operations of articulation that may not necessarily change. We used a functional magnetic resonance imaging (fMRI) method referred to as sparse temporal sampling that has been previously shown to be ideal for overt tasks in the scanner environment since it uses the natural delay in the neurovascular coupling to isolate the period of greatest movement from the period of scanning the effects of neural activity changes (see Gaab *et al.*, 2003; Ozdemir *et al.*, 2006).

Making Sense of Brain Network Changes

The traditional explanation for the dissociation between speaking and singing in aphasic patients is the presence of two routes for word articulation: one for spoken words through the brain's left hemisphere, and a separate route for sung words that uses either the right or both hemispheres. The small amount of empirical data available supports a bi-hemispheric role in the execution and sensorimotor control of vocal production for both speaking and singing (Guenther *et al.*, 1998; Jeffries *et al.*, 2003; Brown *et al.*, 2004; Bohland and Guenther, 2006; Ozdemir *et al.*, 2006), with a tendency for greater left-lateralisation for speaking under normal physiological conditions (i.e., faster rates of production during speaking than singing). The representation of sensory elements of music and language might be either separate, or in different locations with smaller degrees of overlap (for more details on this see also Koelsch *et al.*, 2002, 2005; Patel, 2003; Peretz, 2003). Nevertheless, if there is a bi-hemispheric representation for speech production, then the question of why an intervention that uses singing or a form of singing such as MIT has the potential to facilitate syllable and word production, still remains. In theory, there are four possible mechanisms by which MIT's facilitating effect may be achieved: 1) *Reduction of speed*: in singing, words can be articulated at a slower rate than in speaking, thereby reducing dependence on the left-hemisphere; 2) *Syllable lengthening*: provides the opportunity to distinguish the individual phonemes that together form words and phrases. Such connected segmentation, coupled with the reduction of speed in singing, can help non-fluent aphasic patients become more fluent, and may receive greater support from right-hemisphere structures; 3) *Syllable 'chunking'*: prosodic features such as intonation, change in pitch, and syllabic stress may help patients group syllables into words and words into phrases, and this 'chunking' may also enlist more right-hemisphere support; and 4) *Hand tapping*: it is likely that MIT engages a right-hemispheric, sensorimotor network through the tapping of the patient's left hand as each syllable is sung (one tap/syllable, one syllable/s), which may in turn provide an impulse for verbal production in much the same way that a metronome has been shown to serve as a 'pacemaker' in other motor activities (rhythmic anticipation, rhythmic entrainment; Thaut *et al.*, 1999). In addition, there may be a set of shared neural correlates that control both hand movements and articulatory movements (Tokimura *et al.*, 1996; Gentilucci *et al.*, 2000; Meister *et al.*, 2003; Uozumi *et al.*, 2004), and further, the sound produced by the tapping may encourage auditory-motor coupling (Lahav *et al.*, 2007).

The two unique elements of MIT most likely to make the strongest contribution to the therapy's beneficial effects are the melodic intonation with its inherent sustained vocalisation, and tapping with the left hand. How might melodic intonation influence recovery? Functional imaging tasks targeting the perception of musical components that require a more global than local processing strategy (e.g., melodic contour, musical phrasing, and/or meter) tend to elicit greater activity in right-hemispheric brain regions than in left-hemispheric regions. It has been shown that tasks that emphasise spectral information over temporal information have shown more right- than left-hemispheric activation (Zatorre and Belin, 2001). Similarly, patients with right-hemisphere lesions have greater difficulty with global processing (e.g., melody and contour processing) than those with left-hemisphere lesions (Peretz, 1990; Schuppert *et al.*, 2000). Thus, it is possible that the melodic element of MIT engages the right hemisphere, particularly the right temporal lobe, more than therapies that do not make use of pitch or melody.

The effects of the left hand tapping should be considered in the same context. Once the right temporal lobe is specifically engaged by the melodic intonation and melodic contour, it is conceivable that the role of the left hand tapping could be the activation and priming of a right-hemispheric sensorimotor network for articulation. Since concurrent speech and hand use occurs in daily life, and gestures are frequently used during speech, hand movements, possibly in synchrony with articulatory movements, may have a facilitating effect on speech production, but the precise role of this facilitation is unknown. We hypothesise that tapping the left hand may engage a right-hemispheric sensorimotor network that coordinates not only hand movements, but orofacial and articulatory movements as well, and may facilitate speech production through rhythmic anticipation, rhythmic entrainment, or auditory-motor coupling (see also Thaut *et al.*, 1999; Lahav *et al.*, 2007).

Acknowledgements

This work was supported in part by grants from the National Institute of Neurological Disease and Stroke (NS045049, DC008796), the Doris Duke Charitable Foundation, the Grammy Foundation, and The Mattina R. Proctor Foundation.

References

1. Albert, M. L., Sparks, R. W. and Helm, N. A. (1973). *Arch Neurol* **29**: 130–131.
2. Aziz-Zadeh, L., Wilson, S. M., Rizzolatti, G. and Iacoboni, M. (2006). *Curr Biol* **16**: 1818–1823.
3. Basso, A., Gardelli, M., Grassi, M. P. and Mairotti, M. (1989). *Cortex* **25**: 555–566.
4. Belin, P., Van Eeckhout, P., Zilbovicius, M., Remy, P., François, C., Guillaume, S., *et al.* (1996). *Neurology* **47**: 1504–1511.
5. Blasi, V., Young, A. C., Tansy, A. P., Petersen, S. E., Snyder, A. Z. and Corbetta, M. (2002). *Neuron* **36**: 159–170.
6. Bohland, J. W. and Guenther, F. H. (2006). *Neuroimage* **32**: 821–841.
7. Bonakdarpour, B., Eftekharzadeh, A. and Ashayeri, H. (2000). *Iran J Med Sci* **25**: 156–160.
8. Boucher, V., Garcia, L. J., Fleurant, J. and Paradis, J. (2001). *Aphasiology* **15**: 131–149.
9. Broca, P. (1864). *Bulletin de la Societe de Chirurgie de Paris* **V**: 51–54.
10. Brown, S., Martinez, M. J., Hodges, D. A., Fox, P. T. and Parsons, L. M. (2004). *Brain Res Cogn Brain Res* **20**: 363–375.
11. Cappa, S. F., Perani, D., Grassi, F., Bressi, S., Alberoni, M., Franceschi, M., *et al.* (1997). *Brain Lang* **56**: 55–67.
12. Cappa, S. and Vallar, G. (1992). *Aphasiology* **6**: 356–372.
13. Cohen, N. S. and Masse, R. (1993). *J Mus Therapy* **30**: 81–99.
14. Cornelissen, K., Laine, M., Tarkiainen, A., Järvensivu, T., Martin, N. and Salmelin, R. (2003). *J Cogn Neurosci* **15**: 444–461.
15. Gaab, N., Gaser, C., Zaehle, T., Chen, Y. and Schlaug, G. (2003). *Neuroimage* **19**: 1417–1426.
16. Gentilucci, M., Benuzzi, F., Bertolani, L., Daprati, E. and Gangitano, M. (2000). *Exp Brain Res* **133**: 468–490.
17. Gerstman, H. L. (1964). *J Speech Hear Dis* **29**: 89–91.

18. Geschwind, N. (1971). *New Eng J Med* **284**: 654–656.
19. Goodglass, H. and Kaplan, E. *Boston Diagnostic Aphasia Examination* (2nd ed.). Philadelphia: Lea and Febiger (1983).
20. Guenther, F. H., Hampson, M. and Johnson, D. (1998). *Psycholog Rev* **105**: 611–633.
21. Hebert, S., Racette, A., Gagnon, L., and Peretz, I. (2003). *Brain* **126**: 1–13.
22. Heiss, W. D., Kessler, J., Thiel, A., Ghaemi, M. and Karbe, H. (1999). *Ann Neurol* **45**: 430–438.
23. Heiss, W. D. and Thiel, A. (2006). *Brain Lang* **98**: 118–123.
24. Helm-Estabrooks, N. and Albert, M. L. (1991). *Manual of Aphasia Therapy.* Austin: Pro-Ed.
25. Helm-Estabrooks, N., Nicholas, M. and Morgan, A. (1989). *Melodic Intonation Therapy.* Austin: Pro-Ed.
26. Jeffries, K. J., Fritz, J. B. and Braun, A. R. (2003). *Neuroreport* **14**: 749–745.
27. Kaplan, E., Goodglass, H. and Weintraub, S. (2001). *Boston Naming Test* (2nd ed.). Baltimore; Philadelpha: Williams & Wilkins.
28. Keith, R. L. and Aronson, A. E. (1975). *Brain Lang* **2**: 483–488.
29. Kertesz, A., Lesk, D. and McCabe, P. (1977). *Arch Neurol* **100**: 1–18.
30. Kinsbourne, M. (1998). In Stemmer, B., and Whitaker, H. A. (Eds.), *Handbook of Neurolinguistics,* pp. 386–393. New York: Academic Press.
31. Kinsella, G., Prior, M. R. and Murray, G. (1988). *Cortex* **24**: 165–169.
32. Koelsch, S., Gunter, T. C., von Cramon, D. Y., Zysset, S., Lohmann, G. and Friederici, A. D. (2002). *Neuroimage* **17**: 956–966.
33. Koelsch, S., Fritz, T., Schulze, K., Alsop, D. and Schlaug, G. (2005). *Neuroimage* **25**: 1068–1076.
34. Lahav, A., Saltzman, E. and Schlaug, G. (2007). *J Neurosci* **27**: 308–314.
35. Meister, I. G., Boroojerdi, B., Foltys, H., Sparing, R., Huber, W. and Topper, R. (2003). *Neuropsychologia* **41**: 401–406.
36. Mimura, M., Kato, M., Sano, Y., Kojima, T., Naeser, M. and Kashima, H. (1998). *Brain* **121**: 2083–2094.
37. Moore, W. H. (1989). *Aphasiology* **3**: 101–110.
38. Musso, M., Weiller, C., Kiebel, S., Muller, S. P., Bulau, P. and Rijntjes, M. (1999). *Brain* **122**: 1781–1790.
39. Ozdemir, E., Norton, A. and Schlaug, G. (2006). *Neuroimage* **33**: 628–635.
40. Patel, A. D. (2003). *Nat Neurosci* **6**: 674–681.
41. Peretz, I. (1990). *Brain* **113**: 1185–1205.
42. Rosen, H. J., Petersen, S. E., Linenweber, M. R., Snyder, A. Z., White, D. A., Chapman, L., *et al.* (2000). *Neurology* **55**: 1883–1894.
43. Saur, D., Lange, R., Baumgaertner, A., Schraknepper, V., Willmes, K., Rijntjes, M. and Weiller, C. (2006). *Brain* **129**: 1371–1384.
44. Schlaug, G., Marchina, S. and Norton, A. (2008). *Music Perception* **25**: 315–323.

45. Schuppert, M., Munte, T. F., Wieringa, B. M. and Altenmuller, E. (2000). *Brain* **123**: 546–559.
46. Selnes, O. A. (1999). Recovery from aphasia: Activating the "right" hemisphere. *Annals of Neurology* **45**: 419–420.
47. Small, S. L., Flores, D. K. and Noll, D. C. (1998). *Brain Lang* **62**: 298–308.
48. Snodgrass, J. G. and Vanderwart, M. (1980). *J Exp Psychol: Hum Learn Mem* **6**: 174–215.
49. Sparks, R., Helm, N. and Albert, M. (1974). *Cortex* **10**: 303–316.
50. Sparks, R. and Holland, A. (1976). *J Speech Hear Dis* **41**: 287–297.
51. Thaut, M. H., Kenyon, G. P., Schauer, M. L. and McIntosh, G. C. (1999). *IEEE Engineering in Medicine and Biology Magazine* **18**: 101–108.
52. Thompson, C. K. and Shapiro, L. P. (2005). *Aphasiology* **19**: 10–11.
53. Thompson, C. K. and Shapiro, L. P. (2007). *Am J Speech Lang Pathol* **16**: 30–42.
54. Tokimura, H., Tokimura, Y., Oliviero, A., Asakura, T. and Rothwell, J. C. (1996). *Ann Neurol* **40**: 628–634.
55. Uozumi, T., Tamagawa, A., Hashimoto, T. and Tsuji, S. (2004). *Neurology* **62**: 757–761.
56. Warburton, E., Price, C. J., Swinburn, K. and Wise, R. J. (1999). *J Neurol Neurosurg Psychiatr* **66**: 155–161.
57. Weiller, C., Isensee, C., Rijntjes, M., Huber, W., Müller, S., Bier, D. *et al.* (1995). *Ann Neurol* **37**: 723–732.
58. Wilson, S. J., Parsons, K. and Reutens, D. C. (2006). *Music Perception* **24**: 23–36.
59. Winhuisen, L., Thiel, A., Schumacher, B., Kessler, J., Rudolf, J., Haupt, W. F. and Heiss, W. D. (2005). *Stroke* **36**: 1759–1763.
60. Zatorre, R. J. and Belin, P. (2001). *Cereb Cor* **11**: 946–953.

Chapter 22

Singing Improves Word Production in Patients with Aphasia

Geir Olve Skeie, Torun Einbu and Johan Aarli

Some patients with expressive aphasia are able to sing with text. Nineteen patients (16 aphasia, three dysarthria) were examined for their ability to resite/sing the text of familiar songs and new material. A short test of musical syntax was also included.

Results: Twelve patients improved their word production by singing the familiar song and five patients improved their reciting ability also by singing the unfamiliar/new song. The improved text output was independent of the musical quality of the song. The intention to sing apparently released the improved word production. All patients with aphasia had problems with the test of musical syntax, while the dysarthria patients managed the test well, indicating a functional overlap between syntax in music and language.

Introduction

Music and language require sound processing for analysis and production and depend on overlapping brain areas for production and comprehension (Cohen and Ford, 1995). Cerebral damage leading to aphasia is therefore often accompanied by amusia (Benton, 1977). Patients with damage in or near the Broca's area have problems with speech and word retrieval (Banich, 1997). This leads to syntactic difficulties such as arranging words together in an ordered structure according to grammatical rules. Patients with Broca's aphasia may also have problems understanding sentences where the syntactic structure is important for the meaning. (Eriksson, 2001). Cerebral damage leading to aphasia is therefore often accompanied by amusia (Benton, 1977).

Intriguingly, some patients with Broca's aphasia retain the ability to sing familiar songs with text. Samson and Zatorre (1992) found that patients with damage to the anterior part of the right hemisphere could sing words but lost the melody, while patients with left-anterior damage retained the melody and lost the words. Transcranial Magnetic Stimulation (TMS) studies found that language was knocked out by stimulation of the inferior frontal dominant hemispheres, while words in songs were retained (Stewart *et al.*, 2001), all indicating a right-hemisphere contribution for text production in songs. Nevertheless, it is intriguing that music can activate words in an aphasic brain. Could this ability be useful in language rehabilitation after stroke? In Music Intonation Therapy (MIT), the intention is to increase interactions between language and music, and thereby get access to language through music. MIT claims to be of some effect in language rehabilitation in about 75% of patients with expressive aphasia (Baker, 2000) but has been criticised. It may be a problem that words become too connected to a melody. MIT can therefore appear as a method with restrictions when recovering words for everyday conversation.

We have examined the singing ability of 16 patients with aphasia and three with dysarthria in the acute phase after stroke. We examined whether singing would help text production in a task of repetition. The musical qualities of the song were quantified (expressive amusia) and musical syntax comprehension examined.

Material and Methods

The study included 19 in-patients with language disturbances treated at the Department of Neurology, Haukeland University Hospital, Bergen, Norway and Department of Neurology, Vestfold Central hospital, Tønsberg, Norway.

Aphasia Screening

The aphasia was classified and graded according to severity by the Norwegian aphasia test (Reinvang and Engvik, 1979). Syntax and semantics of spontaneus speech during the examination was recorded. Sixteen patients had aphasia, three had dysarthria.

Musical Tests

Ability to produce words

None of the patients had received any formal musical training. They were asked to choose between four songs familiar to most Norwegians. Thereafter, the patient was examined for their ability to sing with text according to the following scheme:

1. Examiner humming the melody.
2. Examiner and patient sing the song through together.
3. The patient sings the song alone.
4. The patient repeats the text without singing.

Points 1 to 3 were accompanied by guitar.

The patients' ability to sing short previously unknown phrases (the melody was also new) was also compared with repeating the same phrases without melody.

Syntax in music

The patients' ability to grade musical syntax were examined by asking them to judge completed and uncompleted progression of cords, played on guitar, as finished or unfinished. Patients answered 'yes' or 'no' by pointing at cards or verbally.

Analysis

All patient tests were recorded on a minidisc player and analysed by repeated listening. Only the singing performance when the patient was singing alone was used for the analyses. The study was approved by the local regional ethics committee.

Results

Aphasia and dysarthria

Seven patients had a very severe aphasia (AS 20–63), with no spontaneous speech uttering only one word as yes, no and unarticulated sounds. Five patients had an aphasia of median severity (AS 52–182); they had more mixed aphasias although mainly expressive. Four had minor aphasic problems (AS 146–216) and the last three patients had dysarthria.

Familiar songs

Eleven out of 14 patients with aphasia produced more correct words when singing compared to only reciting the text ($p = 0.01$). The improvement by adding tune to the song was very marked (Fig. 1) in six and moderate in five. Patients with moderate aphasia showed the best effect with singing. Patients with more than a 10% increase in word production had significantly higher AS, 138,65 vs 78,65 than patients who did not increase their word production by singing

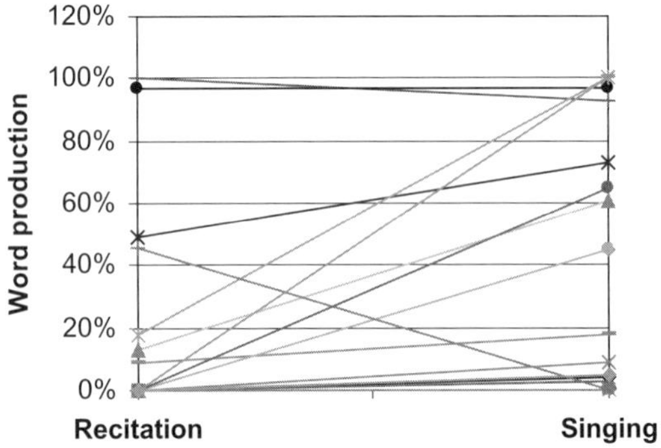

Figure 1. Word production of familiar song.

(p = 0.04). Only one patient with very severe aphasia markedly improved word production by singing, whereas only one of the three dysarthria patients improved word production with singing. There was no correlation between the quality of the patients singing i.e. pitch melody and rhythm, and increased word production by singing. Some patients had very good melody and pitch, but no increase in word production while others increased their word production by singing, although the musical quality of the song was poor. In particular, one patient with very severe aphasia increased his word production markedly although the singing was without any musical quality (the pitch, melody and rhythm). This indicates that the intention to sing increases word production, while the musical quality is of minor importance.

Unfamiliar/new songs

Five out of 14 patients with aphasia improved repetition significantly by singing compared to reciting an unknown/new text (p = 0.02) (Fig. 2). Some patients did not improve, while some had an insignificant worsening. There was no correlation to the severity of aphasia for this task (AS 112,73 for patients with improved word

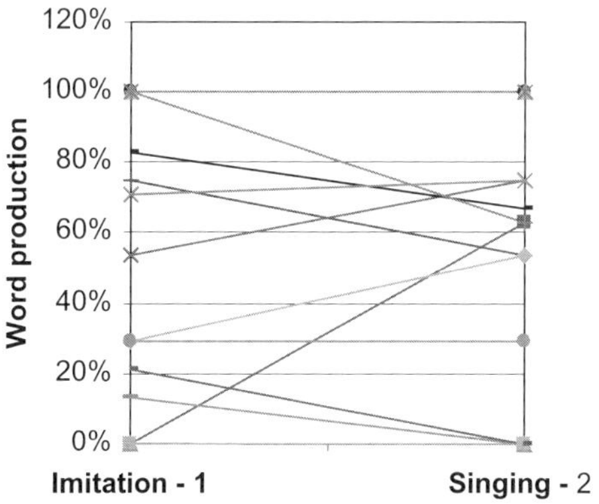

Figure 2. Word production of new song.

production vs 106,61). All patients who improved repetition by singing also improved the word production by singing in the familiar song task. Dysarthria patients did not improve word production by singing in this task.

Twelve out of 17 had severe problems with *pitch* in the unfamiliar text/melody task. Despite this, some patients had marked improvement of repetition/ imitation by singing. Again it seems as if it is the attempt/intention to sing that activates words regardless of the melodic features of the song produced.

Amusia

Most patients had problems with music production — pitch/melody and/or rhythm, but only four had severe amusia. They also had bilateral cerebral damage and had problems with all language functions; all four had poor prosody of speech and two had a monotonous speech without prosody at all. One patient with absent prosody had complete expressive amusia (without pitch, melody, or rhythm); the other patient without prosody had intact rhythm, but no pitch/melody. One patient had deficient rhythm, while melody and pitch were partly preserved. The last patient had some prosody of speech, but lacked melody/rhythm and pitch in the song. Both patients with total expressive amusia had severe stroke with probable bilateral damage, as one had a massive medial infarction with oedema, midline shift and contralateral hydrocephalus requiring urgent craniectomy. The other patient had a severe subarachnoid haemorrhage (SAH) with a haematoma of the left temporal region involving the insula. Severe left-sided damage, in addition to involvement also of the contralateral hemisphere, might explain the severe amusia in these patients.

Overall, in our material, patients with severe language problems also had severe amusia that were accompanied by aprosody. Nevertheless, one patient with amusia/aprosody had a markedly improved text production by singing both the familiar and the new songs.

Patients with particularly well-preserved singing ability

Three patients had particularly good musical quality of their songs in spite of not having a special musical background. Only one of these patients improved in their word production by singing. There were no correlations between good musical abilities i.e. diction, melody, pitch, rhythm and increased word production by singing.

Discussion

Twelve out of 14 patients with aphasia increased their word production by the familiar song chosen. In line with this, Yamadori *et al.*, (1977) found that about 70% of patients with Broca's aphasia could sing familiar songs with text. Some patients with very severe aphasia improved their text production by singing but, in general, patients with moderate to less severe aphasia had the best effect of adding melody to the text. It was probably not only the music's ability to improve rhythm and fluency of speech that increased word production, but rather a speech specific augmentation of language, as adding melody did not facilitate text production in patients with dysarthria.

Several mechanisms have been suggested to explain why some patients with aphasia can sing. The text and melody of a song may be stored together in memory and producing the melody is important for recollection of both (Wallace, 1994). Theoretically, this explains the increased word production in patients with minor or moderate aphasia, but not why patients with very severe speech problems also increased their word production dramatically by singing. Rhymes, song texts, counting and other automatic material are probably stored and reproduced from the right hemisphere and will thus be spared by damage to the language dominant hemisphere (Cohen and Ford, 1995; Kaan and Swaab, 2002; Jeffries *et al.*, 2003). This does not explain why some patients also increased word production in the new/unfamiliar song.

Patients with well-preserved musical abilities did not produce more correct words in either familiar or unfamiliar songs. Indeed, some patients with almost absent melody, pitch, and rhythm of the song, increased their word production by the attempt of singing compared to reciting the text; the attempt or intention to sing therefore seems to facilitate word production in these patients. The attempt of singing may use an alternative route to stored information about text and ways of coding words and this pathway may facilitate the word production even though the musical/melodic output is very poor. This must be an early step in the process and occur before or in parallel with encoding the melody, as the musical output was very bad in most patients.

An interesting new finding in this study is that singing seems to facilitate repetition in 1/3 of patients with aphasia. Selective failure of repetition is thought to be due to damage to the arcuate fascicule or other structures mediating information from the Wernicke to the Broca area, but it will also be problematic for patients with a pure Broca's aphasia. It is more likely that singing improves repetition due to overcoming difficulties with word production rather than facilitating the information stream. All patients with improved word production in the 'new song' also increased word production in the familiar song indicating a similar mechanism for the increased word production.

We found a considerable overlap between aphasia and amusia. Most aphasic patients had some degree of expressive amusia, and problems with pitch, melody or rhythm. This is not surprising, as there are many similarities in the cerebral processing of music and language (Blumstein and Cooper, 1974; Borchgrevink, 1993; Hachinski and Hackinski, 1994; Cohen and Ford, 1995; Cohen, 1995; Altenmuller, 2003; Patel, 2003; Hebert *et al.*, 2003). Only four patients had severe amusia, all had probable bilateral cerebral damage explaining their more severe amusia. Three of these four increased their word production in the attempt of singing showing that the increase in word production by singing is probably an upstream early event occurring before the musical element is added to the process of producing a particular word/sentence in a song. All patients with aphasia had problems with musical syntax, while the patients with dysarthria did well in the simple test for musical syntax in this study. This finding fits well with earlier positron emission tomography (PET) studies (Kaan and Swaab, 2002; Maess *et al.*, 2001; Patel, 2003; Besson and Schøn, 2003) suggesting a similar cortical /processing area (probably overlapping with Broca's area) for syntax in language and music. It fits also with clinical findings as patients with Broca's aphasia have problems with producing sentences with intact syntax and in understanding sentences where syntax is of importance for the meaning (Haarmann and Kolk, 1991; Caplan and Waters, 1999).

Some patients with aphasia may receive language therapy without any improvement in verbal output. Music intonation therapy (MIT) was grounded in search for getting more effective methods in recovering verbal communication. Through four progressive levels, the melodic line, the rhythm and point of stress are important elements in the transition from song intonation to spoken prosody, and the patients should at the end be able to mainly use the spoken prosody in uttering sentences. Familiar songs are not used, because the lyrics are stored in a close connection to the melody (Cohen, 1992; Cohen and Ford, 1995). Better auditory comprehension than verbal expression is important for patients to succeed (Sparks *et al.*, 1974). It remains still a question if MIT is functional and individual enough to improve the verbal communication ability at a satisfactory level. The melodic phrases must be internalised to facilitate words (Baker, 2000), and the challenge is to let the patients themselves know how melodies can release words.

Acknowledgements

Trygve Aasgaard, PhD, associate professor, MusikkHøyskolen i Oslo. Are Brean, Department of Neurology, Vestfold Central Hospital, Tønsberg, Norway.

References

1. Altenmuller, E. O. (2003). How many music centres are in the brain? In Peretz, I. and Zatorre, R. (Eds.), *The Cognitive Neuroscience of Music*. Oxford: Oxford University Press.

2. Baker, F. A. (2000). Modifying the melodic intonation therapy program for adults with severe non-fluent aphasia. *Music Therapy Perspectives, Vol. 18*. PT ISS/2.
3. Banich, M. (1997). *Neuropsychology — The Neural Basis of Mental Function*. Boston: Houghton Mifflin Company.
4. Benton, A. L. (1977). The amusias. In Critchley, M. and Henson, R. A. (Eds.), *Music and the Brain. Studies in the Neurology of Music*. London: William Heinemann Medical Books Limited.
5. Besson, M. and Schøn, D. (2001). Comparison between language and music. In Zatorre, R. and Peretz, I. (Eds.), *The Biological Foundations of Music*. Annals of The New York Academy of Sciences, Volume 930.
6. Blumstein, S. and Cooper, W. E. (1974). Hemispheric processing of intonation contours. *Cortex* **10**: 146–158.
7. Borchgrevink, H. M. (1993). Musikk, hjerne og medisin. *Tidsskr Nor Lægefor nr 30* **113**: 3743–3747.
8. Caplan, D. and Waters, G. S. (1999). Verbal working memory and sentence comprehension. *Behavioral and Brain Sciences* **22**: 77–126.
9. Cohen, N. S. (1992). The effect of singing instruction on the speech production of neurologically impaired persons. *Journal of Music Therapy* **29**: 87–102.
10. Cohen, N. S. and Ford, J. (1995). The effect of musical cues on the nonpurposive speech of persons with aphasia. *Journal of Music Therapy* **32**: 46–57.
11. Eriksson, H. (2001). *Neuropsykologi. Normalfunktion, demensformer og afgrænsede hjerneskader*. København: Hans Reitzels Forlag.
12. Hachinski, K.V. and Hachinski, V. (1994). Music and the brain. *Can Med Assoc J* **151**: 293–296.
13. Haarmann, H. J. and Kolk, H. H. J. (1991). Syntactic priming in Broca's aphasics: Evidence for slow activation. *Aphasiology* **5**: 247–263.
14. Hébert, S., Racette, A., Gagnon, L. and Peretz, I. (2003). Revisiting the dissociation between singing and speaking in expressive aphasia. *Brain* **126**: 1838–1850.
15. Jeffries, K. J., Fritz, J. B. and Braun, A. R. (2003). Words in melody: An H2 15 O PET study of brain activation during singing and speaking. *NeuroReport*. Philadelphia: Lippincott Williams & Wilkins.
16. Kaan, E. and Swaab, T. Y. (2002). The brain circuitry of syntactic comprehension. *Trends in Cognitive Sciences* **6**: 350–356.
17. Maess, B., Koelsch, S., Gunter, T. C. and Friederici, A. D. (2001). Musical syntax is processed in Broca's area: An MEG study. *Nature Neuroscience* **4**: 540–545.
18. Patel, A. D. (2003). Language, music, syntax and the brain. *Nature Neuroscience* **6**: 674–681.
19. Reinvang, I. and Engvik, H. (1979). *Handbok Norsk grunntest for afasi*. Oslo: Universitetsforlaget.
20. Samson, S. and Zatorre, R. J. (1992). Learning and retention of melodic and verbal information after unilateral temporal lobotomy. *Neuropsychologia* **30**: 815–826.

21. Sparks, R.W., Helm, N. A. and Albert, M. L. (1974). Aphasia rehabilitation resulting from Melodic Intonation Therapy. *Cortex* **10**: 313–316.
22. Stewart, L., Walsh, V., Frith, U. and Rothwell, J. (2001). Transcranial magnetic stimulation produces speech arrest but not song arrest. In Zatorre, R. and Peretz, I. (Eds.), *The Biological Foundations of Music*. Annals of The New York Academy of Sciences, Volume 930.
23. Wallace, W. T. (1994). Memory for music: Effect of melody on recall of text. *Journal of Experimental Psychology: Learning, Memory and Cognition* **20**: 1471–1485.
24. Yamadori, A., Osumi, Y., Masuhara, S. and Okubo, M. (1977). Preservation of singing in Broca's aphasia. *Journal of Neurology, Neurosurgery, and Psychiatry* **40**: 221–224.

Chapter 23

Nerve Compression Syndromes in Musicians — A Surgeon's View

Ian Winspur

Not all that tingles is a compressed or stretched nerve and nowhere is this more true than in musicians. Many symptoms suggestive of nerve compression in fact relate to the musician's technique or contact with the instrument. Carpal tunnel syndrome (CTS) is the most common form of nerve compression, but it is less common in this group, surprisingly, than in the general population. The indications for surgery are different in musicians and the surgical techniques may need to be modified.

Introduction

I hardly need to remind my readers that all that tingles is not necessarily a compressed or stretched peripheral nerve and nowhere is this more important than when dealing with musicians. I first started seeing musician patients in my practice in Santa Barbara in the early 1980s. I was immediately impressed by a number of younger classical pianists who presented after periods of intense practice of difficult repertoire — Shostakovich seemed to be the most incriminated — with classical carpal tunnel syndrome (CTS)-like symptoms, but normal nerve conduction tests. At that time, I had become very friendly with Professor Buchtal, then retired in California but previously Professor of Neurophysiology in Denmark and Sweden, and a very early pioneer of nerve-conduction testing and EMGs. Giving my observations on these young musicians, I asked him what the pathophysiology of the median neuropathy might be, and he replied instantly: 'Oedema'. Never forgetting that, these young musicians seen in California and many subsequent younger musicians presenting CTS-like symptoms do not have mechanical compression of the median nerve and certainly do not require surgical release of their carpal tunnel; in fact they fare badly after such operations. They have flexor tenosynovitis of the wrist manifest by the boggy swelling seen on the flexor aspect of the wrist and their treatment is conservative including steroid injection.

Carpal Tunnel-Like Symptoms in Musicians

There is in fact a further sub-group of musicians with carpal-tunnel-like symptoms who certainly do not have anatomical compression of the nerve. These are young guitarists presenting with carpal tunnel-like symptoms after playing, seldom having nocturnal symptoms and having normal nerve conduction tests (NCT). On analysis, their playing position is usually found to be incorrect, with the guitar rotated excessively and either the left or the right wrist, or both wrists excessively flexed when playing. They have what I have termed acute positional CTS, and do not need surgical release, instead supervision by a guitar teacher. The third group of musicians presenting with carpal tunnel-like symptoms tend to be older, have nocturnal symptoms, and good technique on the instrument, but abnormal nerve conduction tests and, of course, have classical CTS as do so many of the general public. If in doubt on the diagnosis, nerve-conduction testing can be extremely helpful in these sub-groups. Those with acute positional CTS have normal NCTs as do those with flexor tenosynovitis, but those with classical CTS have abnormal tests. I would even state that if a musician is suffering from carpal tunnel-like symptoms, he or she should not have carpal-tunnel release surgery unless the nerve-conduction tests are positive. When injecting the carpal tunnel, one should be particularly careful in neither injuring the median nerve nor injecting the nerve, and the key landmarks are the axis of the ring finger in the longitudinal plane and at a point 1.5 cm proximal to the wrist crease in the transverse plane; a very short needle should be used.

Traumatic Neuromas

Traumatic neuromas can occur on peripheral nerves as the result of external compression and these can occur in musicians, presenting as painful nodules on a peripheral or digital nerve with associated tingling and numbness. Two classical sites exist in musicians — hornplayer's thumb where the ulnar digital nerve of the left thumb is compressed in the ring one uses to stabilise a French horn, and in a second situation where the radial digital nerve of the left index finger is compressed when holding a flute, flutist's finger. In both situations, modification of the instrument and the provision of an additional support, avoiding the critical area will usually solve the problem, not surgery. A young female French horn player developed a neuroma on her left thumb ulnar digital nerve after purchasing a new French horn. Fortunately, she brought the instrument with her to the consultation and I found there was an adjustable support incorporated into the ring which had not been highlighted when she bought it, and had not been adjusted to suit her thumb. The simple release of a grub screw and repositioning of the support within the ring solved the problem, but not all digital neuromas are so simple to deal with. Cellists are particularly susceptible to tiny distal neuromas on the terminal branches of the digital nerves in their fingertips. The tension on the modern steel cello strings is 30 times that of a violin and one sees traumatic neuromas presenting small isolated areas of tenderness and possibly altered feeling at the very tip of the fingers. Changing temporarily to gut strings, which are not strung too tightly or to a Baroque instrument may help. A steroid injection given with a tiny needle around the neuroma also may rapidly reduce symptoms, since surgery is not an option for these neuromas.

Non-Specific Arm Pain

Dr Wynn Parry has analysed over 1,000 cases of musicians' upper-limb problems presenting to the BAPAM Clinic in London over the last 20 years, and confirms that less than 50% have an identifiable medically definable problem. The remainder, with painful arms or hands, are suffering from non-specific symptoms of fatigue, fatigue beyond the point of tiredness, poor positioning, and possible hypermobility. A higher proportion of younger musicians and music students are so afflicted. Many of these will complain of intermittent numbness and tingling, occasionally in a recognised anatomical distribution, but nevertheless flitting, inconsistent and changing from examination to examination. Some will have symptoms suggestive of thoracic outlet syndrome (TOS). All investigations, however, will be normal, including MRI scans, CT scans and computerised angiography. Those with symptoms suggestive of TOS, will be improved markedly by physiotherapy focused on strengthening the shoulder girdle and improving positioning. Indeed, most musicians with TOS appear to have 'dynamic' TOS in relation to dropped-shoulder girdles. Surgical release is seldom required and indeed there are documented cases of disasters following first rib resections in professional musicians.

The Role of Surgery

Of the 45% or so musicians in whom a recognised medical diagnosis is possible, only 4% were found to have a surgically amenable condition. Of these, the greatest number are musicians involved in coincidental trauma — car accidents, DIY and kitchen accidents — and of the remainder, the most common group of conditions presenting for surgery are nerve compression syndromes (NCSs). In my experience, over the last 27 years, this represents 25% of all operations on professional musicians, the majority being for CTS. This may seem a large proportion, but CTS is so common in the general public it usually represents well over 25% of elective-hand surgical operations performed. In my own practice over the last seven months, of a total number of 236 operations performed, 104 were for NCTs of which 90% were for CTS. This represents 44% of the surgical load in elective non-emergency surgical practice. Clearly, the 25% in musicians is lower then this. Repetition has traditionally been associated with the development of, and linked causally to, the development of NCSs. The fact that they are less common in the professional musician, who is subject to repetition by a factor of many tens of thousands greater than the general population, calls into serious question the role of repetition in the development of NCSs, particularly CTS.

In assessing a musician medically, one has to look at the whole musician, their posture and technique, the instrument played, and the interface between the instrument and that musician. The source of the problem, and indeed the solution, may lie in the interface. Indeed, even if the source is not found here, the solution may be. One is obligated, as a treating physician dealing with the musician, to seek solutions in this area before even considering surgery. If no conservative solution is available and the interface cannot be adjusted, then in certain focal situations surgery may be a solution. But the indications for surgery and indications for surgical release of a peripheral nerve may be different in a professional musician from those in the general population — can they no longer play at the required high level, and is there no solution by adjusting the instrument or the interface? Hence, in the care of CTS, one may be forced to operate earlier than one would in the general public. Again, I would make the further strong assertion that one should only release a peripheral nerve in a musician if the NCTs — assuming the nerve in question can be tested — are positive. Using this criterion, one will not be tempted to release the carpal tunnel of younger musicians with flexor tenosynovitis who will in fact be harmed by surgery, as previously discussed. Also one will not be tempted to release the ulnar nerve of musicians who are in fact suffering from dynamic TOS or non-specific arm pain without clear physical entrapment of the ulnar nerve at the elbow. One clinical situation exists where the decision to operate has to be clinical, as NCTs are not reliable, and this is in radial tunnel syndrome (RadTS). This is an uncommon but nevertheless recognised cause of fairly non-specific arm pain and can occur in musicians as well as in the general public. This is due to dynamic entrapment of the radial nerve in the radial tunnel, by ECRB, by abnormal muscles or by the fibrous arch of the supinator. Many musicians with RadTS will settle spontaneously over six to nine months with conservative care. Some will not, and dramatic relief can be obtained by surgical release of the nerve. I have personally released one professional musician with radial tunnel syndrome and he returned to full playing.

Surgical Techniques

As regards surgical techniques, the same techniques are used as in the general public, but with a few nuances:

(a) The endoscopic techniques for nerve release, with their documented slightly higher serious complication rate of direct injury to the nerve is to be avoided in professional musicians. Clear visualisation of the entire segment of the nerve being released is mandatory in musicians and there is little place for keyhole surgery.

(b) When releasing the ulnar nerve at the elbow, the decision whether or not to transpose the nerve has to be made at the time of surgery, bearing in mind the musician's normal playing position. If the normal playing position is with the elbow flexed, then the nerve should be transposed unless there is good reason for it not to be. If the musician's playing position is with the elbow extended, and following decompression the nerve appears to be lying tension-free in the condylar groove, the nerve should not be transposed and a simple release will prove effective. If, however, in these patients, when the elbow is flexed at the time of surgery the nerve is clearly stretched, even these should be transposed. When transposing the nerve, I prefer a simple sub-cutaneous transposition, rather then the more complicated sub-muscular transposition or those involving bony excision of the epicondyle. One important factor in treating a musician is to perform surgery which allows the earliest possible return to limited playing on the instrument. The simple releases and subcutaneous transpositions meet this criterion as opposed to sub-muscular transposition. A young cellist who developed symptoms of left ulnar nerve entrapment at the elbow some ten years previously had a simple release which relieved her symptoms for five years, but her symptoms redeveloped. The left elbow of the cellist is held for long periods in flexion, and therefore I re-released the nerve, transposing it at this time, and since then she has been trouble-free. I have personally released the ulnar nerve in a number of professional pianists and organists and have restricted the operation to simple release, the nerve not being visibly tight when the elbow was flexed at the time of surgery, and they have all done extremely well.

(c) It is important that post-operatively, the patient be allowed to return to the instrument as soon as is physically possible. The only exception to this is following carpal tunnel release in musicians where the playing position is with the wrist flexed. Playing should be delayed in this group at least 12 days after the operation to prevent the uncommon, but potentially disastrous, painful bow-stringing of the flexor tendons. In over 4,000 carpal tunnel-release operations I have performed over the last 30 years, post-operatively keeping the wrist extended by an external splint for at least 10 days, I have never encountered this most troublesome post-operative complication, and similarly in 22 carpal tunnel releases in professional musicians, there have been no complications including bowstringing.

Surgical Results

What are the results of surgery on NCTs in musicians? Provided the surgery is performed for the correct reasons (indications) and the various technical aspects I have discussed are adhered to, the musician patient will recover as well as, if not better than, the non-musician patient. They are motivated, highly co-ordinated, and will compulsively follow post-operative instructions. Of the 26 professional musicians undergoing operations for NCTs (10 cases were bilateral), all but one returned to full-time playing at their previous levels. This is the advantage of monitoring professional musicians as there is a definable end point — did they return to their previous high levels of playing and earning level, or not? The average time off the instrument was eight days (prolonged by the CTS releases in violinists and guitar players as explained) and the average time to return to full playing was three weeks. The one failure concerned an older pianist who presented with classical right CTS and whose NCTs also showed significant delay. He made only minor progress after carpal tunnel release and was subsequently shown to have fairly severe three-level disc disease in the neck, which also requires decompression and stabilisation. He returned to teaching, but not to performing.

This again reminds us that all that tingles (particularly in musicians) is not a compressed peripheral nerve. But if such a lesion exists in a musician and there is no doubt about the diagnosis, the surgical release can be expected to give as great, if not better results, than in the general public and can on occasion salvage a musical career.

References

1. Birch, R., Bonnie, G. and Wynn Parry, C. (1998). *Surgical Disorders of the Peripheral Nerves.* Edinburgh: Churchill Livingstone, Chapters 12 and 19.
2. Allieu, Y. and Mackinnon, S. (2002). *Nerve Compression Syndromes of the Upper Extremity.* London: Martin Dunitz, Chapters 7, 8, 11 and 14.
3. Winspur, I. and Wynn Parry, C. (1998). *The Musician's Hand.* London: Taylor & Francis, Chapters 4, 5, 6, 9 and 13.

Chapter 24

Focal Hand Dystonia Affecting Musicians

Katherine Butler

In 1911, Oppenheim coined the term dystonia to describe disordered motor control, characterised by an association of hypotonia and tonic muscle spasm. Focal hand dystonia is one form of this disorder, in which symptoms are often task-specific and occur during skilled movements such as writing (writer's cramp) or playing a musical instrument (musician's cramp). Much research has been conducted on the pathophysiology of dystonia, but the underlying mechanisms still remain unclear. Hypotheses about functional central nervous system alterations continue to gain more support. Scientific treatment-based publications on focal dystonia are sparse, and progress in evidence-based treatment options is necessary in order to assist this patient group. This chapter will review the literature, documenting dystonia classification criteria, manifestations, pathophysiology and treatment options. Current treatment techniques that will be discussed include oral medications, botulinum toxin (BTX) injections, rehabilitative therapies and supportive approaches. Future research areas relating to focal hand dystonia and the musician are highlighted.

Introduction

Dystonia is a syndrome characterised by involuntary prolonged muscle contractions that can lead to sustained twisting postures (Fahn, 1998; Fahn *et al.*, 1987, 1998). It is a set of disorders that are characterised by abnormal postures and unwanted spasms that interfere with motor performance (Hallett, 2004). Three criteria can be utilised to assist in classifying this syndrome: age of onset, etiology and distribution of symptoms (Fahn *et al.*, 1987, 1998). Onset before 28 years of age is classified as early, and after this age is classified as late onset dystonia. Etiology can be divided into primary/idiopathic (no obvious affects on the brain) or secondary/symptomatic (often the basal ganglia are affected, resulting in more generalised symptoms) (Fahn, 1998). Distribution can be:

general — symptoms manifest in all extremities including the trunk;
hemi — symptoms are focussed on one side of the body;
segmental — a segment of the body is affected or
focal — a single body part is affected.

Any part of the body can be affected by focal dystonia including the neck, eyelids, vocal cords or hand (Berardelli *et al.*, 1998; Deuschl and Hallett, 1998). This chapter will focus on focal hand dystonia (FHD), and late onset, mostly primary dystonia, that is often task specific and includes writer's cramp and musician's cramp (Figs. 1a and 1b).

Focal Hand Dystonia in Musicians

FHD in musicians is a primary dystonia that is painless and tends to be task specific, focal and of late onset. Symptoms can include lack of co-ordination, cramping and tremor (Jankovic and Shale, 1989) and tend to be specific for each individual and related to the instrument played rather than hand dominance (Figs. 2a and 2b).

Patients can respond to sensory tricks and, if they do, this is usually a good indicator of successful hand therapy. **Sensory tricks** can be used to 'fool' the brain and give a 'nonsense' feedback loop that breaks the fixed link in the sensory motor loop for a short moment (Berardelli *et al.*, 1998; Hallett, 1995; Lederman, 1991). Often the novelty will only be effective for a short time until the brain recalibrates to an automatic pattern which is the dystonic one. Coban, Blue Tac, latex gloves and splints can all be used as sensory tricks (Figs. 3a and 3b).

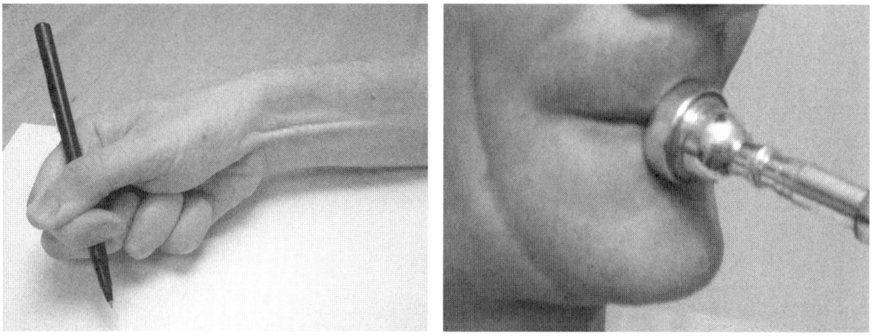

(a) (b)

Figures 1a and 1b. Task-specific action-induced FHD has different forms, including writer's cramp and musician's dystonia (embouchure and hand).

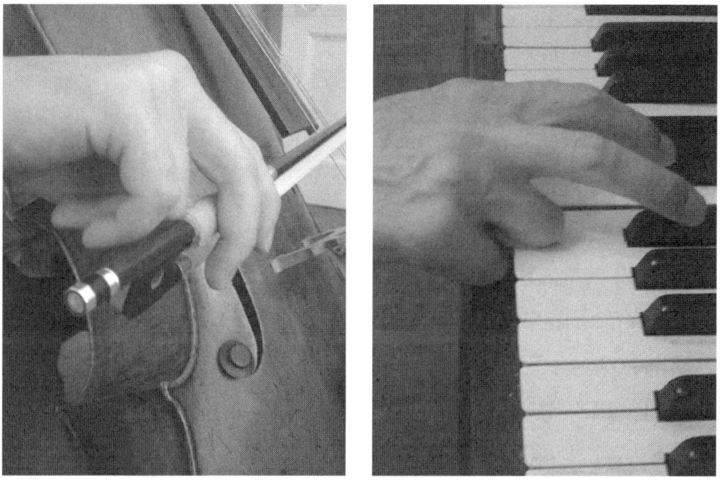

(a) (b)

Figures 2a and 2b. FHD in musicians showing lack of motor co-ordination or loss of voluntary control in cellist's and pianist's right small and ring fingers.

Epidemiology

Jabusch *et al.* (2004) has presented a study where he outlines the largest epidemiology series to date of musicians affected by FHD; he analysed 144 cases and the results show that this condition occurs more frequently in males than females and the average age of onset is

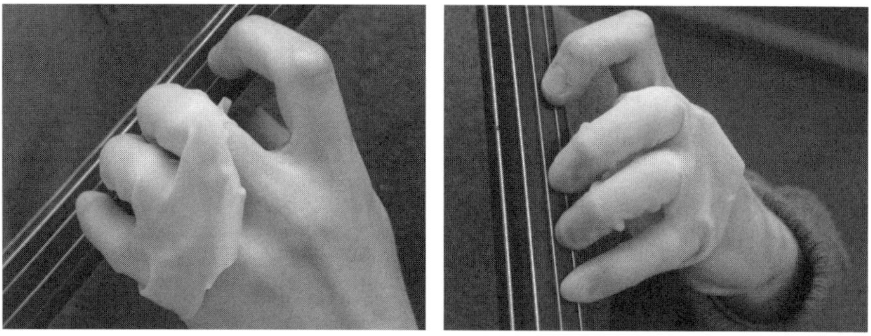

(a) (b)

Figures 3a and 3b. Cellist utilising latex glove as a sensory trick. Figures 1a, 1b, 2a, 2b, 3a and 3b are all originally published in Butler K, Norris R. Assessment and treatment principles for the upper extremities of instrumental musicians. Rehabilitation of the Hand. 6th edn. Philadelphia: Elsevier. In print.

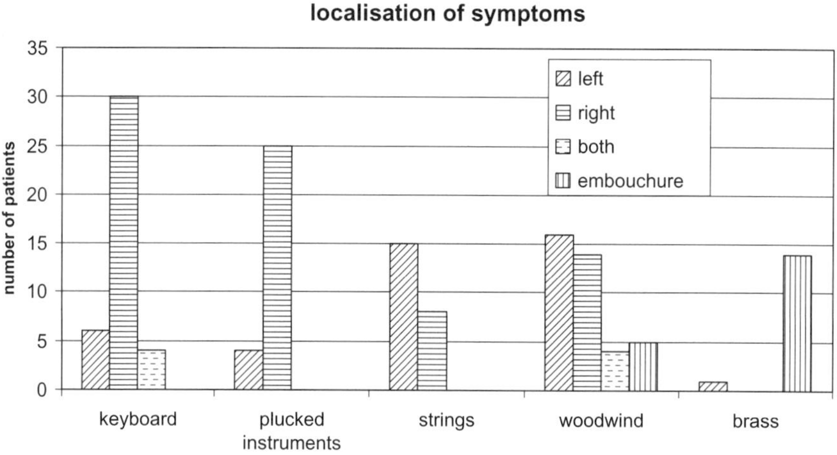

Figure 4. Numbers of patients affected by FHD according to instrument played and affected body part.

33 years. Dystonia mainly affects brilliant professional classical soloists, who play the keyboard, are male, work exceptionally hard and have high-performance constraints.

Figure 4 (Jabusch, 2006), displays that FHD predominately affects the right hand of musicians who play the keyboard or plucked instruments, the symptoms predominately occur while playing and affect one or more fingers.

The estimated prevalence of FHD among professional musicians is about two to ten per cent (Jabusch, 2004; Brandfonbrener, 1995; Lim *et al.*, 2001), which is higher than writer's cramp (0.1%) in the general population (Nutt *et al.*, 1998). This high percentage reflects the specific demands made upon musicians.

Pathophysiology

There are few studies specifically relating to the pathophysiology of musicians' dystonia.

The general concepts as to why this complex sensorimotor network problem occurs are presented by Rosenkranz *et al.* (2005):

- Lack of inhibition on many levels of the nervous system, for example, at the spinal-cord level there is a lack of reciprocal inhibition and at the level of the motor cortex there is a reduction in intracortical inhibition
- Impairment/failure of sensory integration results from impaired discrimination of the temporal/spatial ability and sensory input activates and leads to abnormal motor activation.
- Impairment/loss of somatotopic organisation in sensory and motor areas.

Hallett (2006) raises the idea of patients who suffer from writer's cramp having a disordered neuroplasticity. There may be a genetic base in developing task specific focal hand dystonia, perhaps with a decrease in inhibition, an increase in plasticity or impairment in sensory functions. The idea that writer's cramp develops due to excessive writing in people who are genetically predisposed further supports this hypothesis with respect to musicians developing focal hand dystonia. Thus musicians who develop FHD may be genetically predisposed to developing this condition and then with excessive playing or overtraining may 'bring' the condition on. Ongoing work to identify abnormal genes in patients with focal dystonia continues and results of further studies are eagerly awaited.

Many factors can 'trigger' the manifestation of FHD in musicians such as a sudden increase in playing time, change in technique, return to playing after a long break, trauma, history of nerve entrapment, psychological aspects or change of instrument.

Byl (1996a, 1996b) proposes a model to explain why FHD develops. Two monkeys were trained to repeatedly squeeze an object for five to six days a week for a reward of a nut. The first monkey did this for 12 weeks and the second for 25 weeks. After five and eight weeks respectively, each monkey presented with difficulty releasing the object and 'dystonic' like symptoms were evident. The monkeys were sacrificed and the sensory area 3b was somatographically mapped and displayed smearing and overlapping of the finger representation areas. In 'healthy' brains, there is a strict somatotopic organisation, and in the 'dystonic' brain there is a lack of clarity and organisation with overlapping and deformity in the map. You can imagine if this patterning was transferred to the motor cortex of a human, there would be a lot of confusion in the resulting movement patterns.

If we make this model relate to a human being, then repetitive movements induce stereotypical feedback messages, which lead to disorganisation in the sensory cortex and a failure in sensorimotor integration, which in turn induces lack of motor co-ordination (Fig. 5). However, if this is the case, why do more musicians not develop this condition, seeing as so many of them practice for long lengths of time? The issue of whether there are preventative measures that can decrease the occurrence of FHD is raised.

An increase in cortical representation of the left hand in string players (Elbert *et al.*, 1995), and a blurring/overlapping of cortical representation in musicians with focal hand dystonia (Elbert *et al.*, 1998), may reflect the animal study findings.

Treatments

At present there is no 'cure' for dystonia, and many of the treatments available have significant limitations. Current treatments include oral medication, botulinum toxin injections, surgery, rehabilitative therapies and supportive approaches. In 2006, Butler and Rosenkranz published

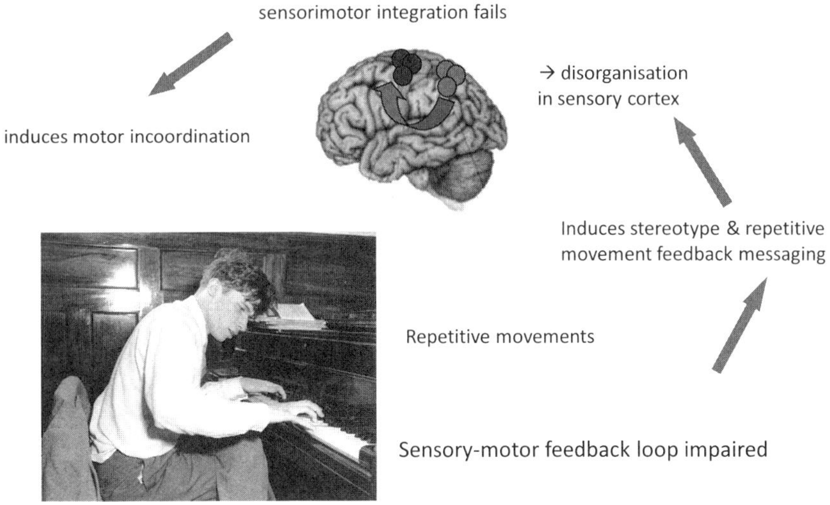

Figure 5. Byl model of focal hand dystonia in humans.

two papers that clearly outline many of the treatments that have been researched and trialled with patients affected by FHD. In this chapter some of the more commonly used treatment techniques and current research findings will be explored.

Oral medications

These do not cure FHD, but can be used as palliative treatments, although side effects limit their long term use (Altenmüller, 1998).

Botulinum toxin injections

Botulinum toxin type-A is the most widely studied and utilised for treatment of FHD. Injections are limited due to the associated weakness of non-dystonic muscles (Priori *et al.*, 2001) and the fact that the injection treats the muscle over-activity but not the affected motor co-ordination (Cole *et al.*, 1991). This treatment technique is not the first that would be tried with musicians, but is often resorted to when other more conservative techniques are not assisting the patient.

Scheuele *et al.* (2005) found that out of 84 musicians with task specific FHD, treatment with EMG-guided botulinum toxin injections, 69% experienced improvement and 36% reported long-term benefit in their performance.

Sheean (2007) comments that botulinum toxin injections appear to be effective with writer's cramp and musician's dystonia patients. They state that there is evidence to suggest that botulinum toxin injections can produce transient improvement in some cortical abnormalities, possibly through altering sensory input from the periphery by direct and indirect means. Thus the idea of using the effects of botulinum toxin injections in combination with a sensory motor retraining program is raised.

Surgery

Some clinicians believe surgery in contraindicated (Winspur, 1998), others comment that peripheral surgical techniques can alleviate symptoms in severe cases (Lozano and Linazasoro, 2000) as long as it is carried out by an experienced surgeon and is thoroughly planned (Singer and Weiner, 1995).

Rehabilitative therapies

The aim is to re-establish integration of the sensory-motor feedback loop, so that appropriate controlled motor commands can be executed. There are many interlinking techniques that can be utilised to assist in achieving this goal. **Sensory re-education** has a focus on sensory discrimination. Treatments that combine both the sensory and motor aspects of FHD include: **sensory motor retuning** and the **multi-disciplinary approach**, which includes hand therapy. **Limb immobilisation** is another treatment option and this interrupts motor performance and decreases afferents from the limb. **Supportive approaches** can be useful and can include: **assistive devices, instrument modification, Alexander technique and psychotherapy**.

Sensory re-education

Byl (1996a, 1996b) suggests through her studies with primates that repetitive motions can induce plasticity changes in the sensory cortex, which may degrade the hand representation and interfere with motor control. Through this research, the possibility of utilising sensory training to successfully treat patients with FHD is raised. Sensory re-education programmes are used in rehabilitation to facilitate and positively influence the re-learning process with improvement of functional sensibility. Sensory discrimination training is thus emphasised and is part of the home exercise programme (Byl and McKenzie, 2000; Byl *et al.*, 2000; Byl and Topp, 1998). However, the same number of repetitions that lead to degradation may be required to restore the hand representation, so patient co-operation is essential for successful treatment (Byl *et al.*, 1996a, 1996b). To facilitate normal sensation and perception, and reinforce hand function, patients are asked to visualise healing, imagine normal sensory processing, motor control, and task execution. Byl expects patients to complete one to two hours a day of sensory discrimination activities at home (Byl and McKenzie, 2000). These activities can include: matching objects/shapes or textures, Braille reading, or identifying and manipulating common household objects with vision occluded (Figures 6a, 6b and 6c).

Byl *et al.* (2000) present a case report of a flautist which shows cortical topographic changes in hand representation, with increases in

(a) (b) (c)

Figure 6a, 6b and 6c. Manipulating embossed items such as dominoes (a), identifying sensory stimulation (b) and discriminating and matching common household items (c).

fine motor control and somatosensory discrimination following a rehabilitation programme that emphasised sensory retraining. Byl *et al.* (2003) and Byl (2000) conclude that therapy with an aggressive sensory re-education element, accompanied by exercises that facilitate fitness and musculoskeletal health, can improve sensory processing and motor control of the hand consistent with the principles of neural adaptation.

Zeuner *et al.* (2002) report on their studies of the efficacy of learning to read Braille as a method of sensory training for patients with FHD. Ten patients with FHD were compared with age- and gender-matched controls; the evaluation tools included the Fahn dystonia scale and time taken to write a standard paragraph. Each individual was trained in Grade 1 Braille reading for eight weeks and monitored to see that they were regular in their performance of this activity for at least 30 to 60 minutes per day. Both controls and patients demonstrated improvement on the spatial acuity test, with the latter showing a significant mean difference on the Fahn dystonia scale from baseline to eight weeks, that improvements correlated positively with improved sensory perception. Of the patients, 60% displayed a decrease in time taken to write the standard paragraph, following Braille-reading training. The authors conclude that training in Braille reading improves spatial discrimination and decreases the level of disability in patients with FHD. Zeuner *et al.* (2003) present the results of three patients who continued the Braille training for a year. They show that further improvements in the Grating Orientation Discrimination Task (GOT), writing the standard paragraph and self-rating scales were gained by these three patients. The authors conclude that sensory training lasting longer than eight weeks may lead to continued improvement.

Rosenkranz *et al.* (2008) present proprioceptive training as a sensory intervention in order to assist in increasing control of movements while playing the piano. The research assessed whether 15 minutes of proprioceptive training can improve the sensorimotor organisation (SMO) of the motorcortical hand area and the motor control in musicians suffering from focal hand dystonia. It is a well-known fact that SMO is abnormal in musicians affected by focal hand dystonia and this research set out to see if it can be restored by a 15-minute intervention with proprioceptive stimulation. It is unclear whether this neurophysiological effect is associated with an improvement in motor control. Thus, five healthy pianists, six healthy non-musicians and six musicians that showed a task-specific, ring-finger flexion pattern while playing the piano, were recruited. Proprioceptive training lasted 15 minutes and consisted of attended muscle vibration applied discontinuously to one of three hand muscles at random. Before and after proprioceptive training, the SMO was explored by measuring changes in short-interval intracortical inhibition (SICI) during short periods of hand-muscle vibration, the performance of a five-finger exercise was objectively evaluated by a MIDI piano, and subjects rated their performance subjectively on visual-analogue scales. The expression of dystonic symptoms were scored on the BMF and Tubiana-Chamagne scales (TCS).

At baseline, the SMO in healthy subjects was spatially differentiated: SICI is reduced in projections to the vibrated, but enhanced to the non-vibrated muscles. In musician's dystonia, this pattern was completely abolished. Proprioceptive training strengthened the spatial differentiation of SMO in all groups; particularly in musician's dystonia, the inhibitory effects of vibration on projections to non-vibrated muscles were restored. This was associated with a significant improvement in motor control during piano playing, as objectively shown in the MIDI data and the BMF and TCS scores. Patients perceived this improvement for up to 24 hours.

Thus the authors conclude that proprioceptive training applied for only 15 minutes significantly and immediately restored a differential pattern of SMO in musician's dystonia and improved motor performance on the piano objectively and subjectively for up to 24 hours. This intervention is a highly promising tool for rehabilitation and it is hoped that further investigations can continue into this exciting treatment technique.

Sensory re-education can increase definition of the hand-representation area in the somatosensory cortex, and assist in treating patients affected with FHD. This treatment technique is certainly utilised heavily by the author within the clinic setting and many patients do respond very well to it. Careful explanation and massive amounts of encouragement are required to make sure this treatment technique is carried out long enough for sensory changes to occur and the effects noted in the act of playing their instrument.

Sensory Motor Retuning (SMR)/Constraint Induced Movement Therapy (CIMT)

CIMT is a behavioural therapy that has been proven beneficial with stroke patients (Taub *et al.*, 1993, 1999). The non-affected limb is immobilised in a cast, thus encouraging the affected limb to be used. These principles have been trialled for treatment of patients with FHD and the treatment has been labelled SMR (Candia *et al.*, 1999). The 'compensating' finger is fixed in a splint, while the 'dystonic' finger carries out exercises, which are completed under supervision and involve one or more of the other digits to exercise up to 2.5 hours per day for eight consecutive days, see Fig. 7 (Candia *et al.*, 2002, 2003).

It is believed that SMR produces functional improvement associated with neuronal reorganisation. Candia *et al.* (2002) present results of 11 professional musicians who took part in a prospective case series that had a follow up comparison group of three to 25 months for piano and guitar subjects and zero to four months for oboe and flute subjects. SMR was seen as being a valuable treatment technique for pianists and guitarists since each patient displayed improved performance without the splint. The wind players did not display any improvement, and the authors state that perhaps this is because wind players exert a fairly constant and firm force while playing, or finger-mouth co-ordination affects brain mechanisms. fMRI demonstrated an increase in activation in the primary motor cortex compared with pre therapy baselines. Thus, treatments that alter movement patterns may provide remediation for patients with FHD.

Zeuner *et al.* (2005) wanted to evaluate if improvement in dystonic symptoms was possible with motor training, and developed a motor-retraining program for individual finger movements. Ten patients with writer's cramp participated in the program. The evaluation tools that were utilised included the Fahn dystonia

The 'focal dystonic digit' and the digit(s) that perform compensatory movements for the dystonic one are identified. A splint that immobilises the main compensatory finger and in turn permits independent movement of the dystonic finger is then made.

Sequential exercises are performed in which the subject makes movements of two or three digits in extension, including the focal dystonic digit. These exercises are performed for a 10-minute period, in a continuous ascending and descending order (e.g. D2, D3, D4, D3, D2 etc. with D4 being the focal dystonic finger and D5 the immobilised main compensatory finger). The patient then rests for two minutes, following which a different sequence of movements of two or three fingers, including the focal dystonic finger, are completed. Five blocks of exercises are performed in an hour.

Initially, the exercise task is paced by a metronome and begins at a medium tempo (60 bpm). The tempo is then increased and gradually decreased, as some musicians with dystonia find slow, controlled movements more difficult than fast ones.

After completing the first five blocks of exercises, the splint is removed and patients can rest for 10 minutes. Following this, four more 10-minute blocks of exercise with two-minute rest breaks between the blocks are completed. A variety of possible finger movements are performed in the different exercise blocks.

Subjects then have a rest of about 40 minutes, often which they are encouraged to play their instrument without the splint. They are invited to play a piece of music of their choice for 15–30 seconds. If they cannot do this, they are encouraged to try a second time. After two successful repetitions, they are asked to play a new, longer segment of the piece, until they have played for 15 minutes (excluding rest breaks). After a five-minute break, if the patient is not too fatigued, the splint is reapplied and a second series of alternating digital manoeuvres, each of five-minute duration, is performed. This regimen continues for eight consecutive days.

Figure 7. Exercises and exercise schedule for SMR.

scale, kinematic analysis of handwriting, transcranial magnetic stimulation (TMS), and electroencephalography (EEG). The Fahn dystonia scale displayed significant clinical improvements, and six patients reported an improvement in writing. After training in simple exercises it was noted that the handwriting analysis showed a trend for improvement. No changes were evident in cortical excitability measured with TMS and EEG. It does not appear that sufficient improvements were made to reverse motor cortex abnormalities measured by TMS and EEG, but motor training for four weeks did lead to mild subjective improvements as well as improvement in handwriting.

Limb immobilisation

Priori *et al.* (2001, 2004) designed a study (N = 8) to investigate if immobilisation for four to five weeks of the forearm and hand is an effective treatment for upper limb dystonia (Figs. 8a and 8b). At 24 weeks post-splintage, the improvement in symptoms remained marked for four patients, but decreased to moderate for three patients and had disappeared for one patient. The severity of the dystonia was measured using the Arm Dystonia Disability Scale, Tubiana and Chamagne Score and a subjective self rating score. Improvements included: regaining voluntary control of the hand, decreased severity of dystonic symptoms and improved motor-task performance. They concluded that limb immobilisation can be a simple, effective and safe treatment for focal dystonia and the immobilisation allows for plastic changes to occur at the cortical level (Pesenti *et al.*, 2001). A larger sample group is required to validate this treatment for general use and information regarding ideal length if immobilisation, number of joints and post-splintage therapy regimes need to be investigated.

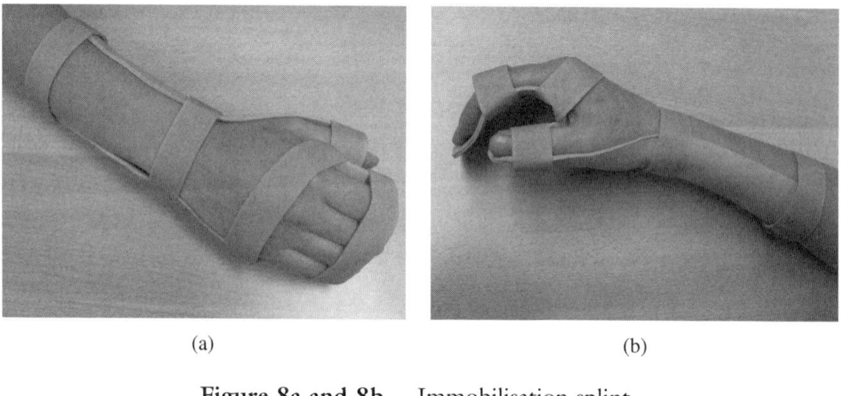

(a) (b)

Figure 8a and 8b. Immobilisation splint.

Another question has to be raised: are the muscles just weakened as a result of being immobilised, and thus will this be an effective long term treatment approach?

Slow-down exercise therapy

Sakai (2006) reported 20 pianists with FHD who were successfully treated through slow-down exercise (SDE) therapy. This treatment has five steps:

1. The patient chooses a piece of music that causes a dystonic hand movement.
2. The performance speed is reduced until there are no dystonic movements evident and the metronome marking is noted.
3. At this slow tempo the patient rehearses the piece for 30 minutes per day for two weeks, and is allowed to play other pieces freely.
4. After two weeks, the speed is increased by 10%. If symptoms do not appear with this increase in speed, the patient proceeds to practice for an additional two weeks at this tempi. If the dystonic movement does reappear then the speed is decreased to the prior one.
5. After two weeks the speed is gradually increased again by another 10%, and the programme continues.

Prior to the SDE therapy, four patients reported moderate difficulty and 16 marked difficulty but, after treatment, symptoms improved to normal for 12 patients and to mild for eight. There was a significant clinical result ($p < 0.05$) evaluated by the Tubiana and Chamagne Score. It is concluded that SDE therapy allowed patients to reduce the speed of movement below the level that memories associated with dystonia exist and allows for retraining. This treatment technique is utilised in the clinic setting with great effect and many patients find it encouraging to be able to play their instrument freely when not doing the slow-down exercises. It is also encouraging to patients to see the improvement in metronome markings, as they progress and this acts as an objective marker for them.

Hand therapy and the MDT

No single treatment modality seems to be effective for the treatment of FHD. When treating musicians, MDT approach can be very helpful and necessary. This team will frequently include: the musician, neurologist, hand therapist, music teacher, instrument maker and psychologist.

MDT treatments can include: traditional hand-therapy modalities, encouraging rest, psychological support, modifications to the instrument, Alexander Technique or Feldenkrais therapy, retraining the whole body and associated movement patterns and mirror treatment techniques. A brief outline of each of the above treatments will follow.

Traditional hand-therapy modalities, such as splinting (Figs. 9a, 9b, 9c and 9d), adaptive devices, heat, ice, exercise, strengthening, rehabilitation and preventative measures for development of FHD. Soft-tissue massage may be required to decrease muscular tension and education and liaison with the teachers and other members of the MDT is integral.

Lederman (1998) reports on four bagpipers with FD, one of whom found splints that prevented flexion of the affected fingers useful. This patient also reported a reduction in involuntary flexion in the non-dystonic fingers on the affected and to a lesser degree the unaffected side. Splints can block the dystonic movements or those of the compensatory finger and can be viewed as a sensory trick.

Gowers (1893) and Poore (1887) over 100 years ago identified that dystonia was difficult to treat, but advocated that rest from the activity that induces dystonia was an effective treatment. Gowers claimed that the earlier the rest period occurred after symptom onset, the shorter and more beneficial it would be. Hoppmann in 2001 also advocates that rest is advantageous when treating FHD. Tubiana (2000) states that after a period of not playing, musicians must return with slow graded progression in duration, tempo, and complexity of playing, often with psychological support. An example of a graded return to play programme can be seen in Fig. 10 (Warrington, 2003).

Dystonic movements occur predominately while performing perceptual motor tasks involving emotion. There is difficulty changing

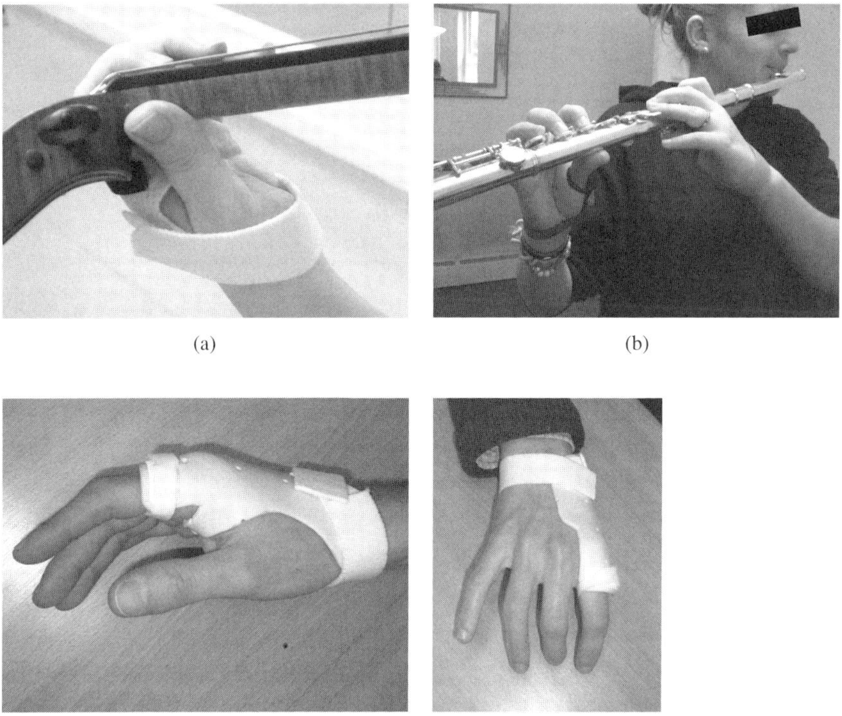

(a) (b)

(c) (d)

Figure 9a, 9b, 9c and 9d. Splints that can assist in retraining movement patterns and that can act as sensory tricks, or blocks for the hyperextension of the 'compensatory' finger shows a splint originally presented by Butler and Svens (2005). Figure 9b originally published in Butler K, Norris R. Assessment and treatment principles for the upper extremities of instrumental musicians. Rehabilitation of the Hand. 6th edn. Philadelphia: Elsevier. In print.

emotional and motor traces that have become associated, and this may lead to preservation of dystonic symptoms. Emotional supports or referral for professional help may be necessary for some patients.

Modifications to the instrument or instrument support may be able to decrease or improve symptoms through eliminating postural triggers (Figs. 11a and 11b). Modifications could include: changing to a smaller instrument, using a neck support, altering location of thumb supports, altering bridge height and extending or altering finger supports or replacing ring keys with plateau keys in woodwind instruments

Practice Sessions Per Day	Minutes of Playing
Two sessions Shadow Playing	3–5 minutes
Two sessions On Instrument	3–5 minutes
Two sessions	5–10 minutes
Two sessions	15 minutes
Two sessions	20 minutes
Three sessions	15 minutes
Three sessions	20 minutes

Practice Sessions Per Day	Minutes of Playing
Four sessions	20 minutes
Four sessions	30 minutes
Three sessions	45 minutes

Start with **Simple, Slow** and **Soft** music.
Double minutes of playing every few days.
Drop back a level if pain is elicited.
A 5 minute break is encouraged every 20 minutes at the higher levels of playing.

Figure 10. Return to play schedule.

(Altenmüller, 2003). Also altering playing positions can be beneficial in decreasing symptoms, for example, standing up, kneeling or lying down to play the instrument.

Feldenkrais and the Alexander Technique can be useful to help patients gain awareness of control, with simple movements being practiced and then more complex patterns being introduced once muscle activity and relaxation have been learnt (Jabusch, 2004). Biofeedback can be used to help patients eliminate muscle co-contraction.

Tubiana (2000) and colleagues propose a four stage treatment programme for patients with FHD. Retraining of the whole body is emphasised and the stages are:

- Reconstruction of the patient's body image
- Relaxation training and muscle differentiation
- Individual muscle retaining, and
- Technical retraining on the instrument.

Chamagne (2003) comments that exercises, stretching and removable orthoses are utilised in their treatment approach.

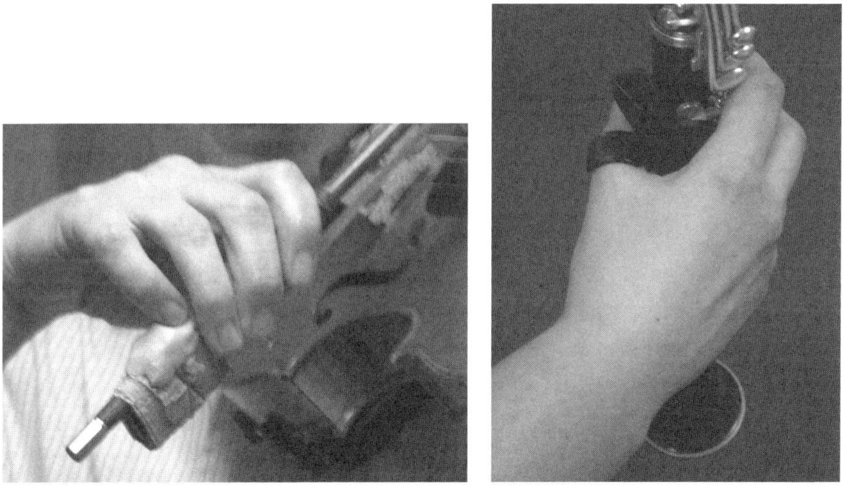

(a) (b)

Figure 11a and 11b. Instrument modifications can include wrapping the end of a bow in Coban and adding a Blu Tac bow 'build-up' or using supports that alter or maintain joint positioning. Figure 11b originally published in Butler K, Norris R. Assessment and treatment principles for the upper extremities of instrumental musicians. Rehabilitation of the Hand. 6th edn. Philadelphia: Elsevier. In print.

Chamagne particularly focuses on the control of the scapula and humerus and thus emphasises arm-swinging exercises as a major part of the home-exercise programme. Tubiana *et al.* (1983) gave the results of 438 patients utilising their treatment programme and report that 22% returned to concert performance, 65% had partial improvement, and 13% reported no improvement.

Jabusch *et al.* (2004) presented long-term outcome measurements of 144 musicians with focal dystonia after treatment with botulinum toxin (n = 71), trihexiphenidyl (n = 69), pedagogical retraining (n = 24), ergonomic changes (n = 51), or non-specific exercises on the instrument (n = 78). Subjective rating of cumulative treatments response and response to individual therapies were utilised as the outcome measures. Alleviation of symptoms was evident for 54% (n = 77), and of this, 33% reported the change was due to the trihexiphenidyl, 49% with the botulinum toxin, 50% with pedagogical retraining, 56% with unmonitored technical exercises, and 63% with ergonomic changes.

Ramachandran (1998) first described the mirror technique to restore the disruption of normal interaction between intention to move the limb and the absence of appropriate sensory feedback in patients suffering from phantom limb. This technique has also been described as successful with patients who suffer from CRPS Type 1 by assisting in re-establishing the pain-free relationship between sensory feedback and motor execution. This technique was effective for early CRPS Type 1, but not for chronic disease, as it was felt in the latter that there were fewer plastic neural pathways and that trophic changes had occurred in the muscles (McCabe *et al.*, 2003). Instant visual feedback with mirrors can help patients recognise dystonic and non-dystonic movements, with the mirror positioned so the musician can see the uninjured hand looking like the injured one. The illusion is created and the brain thinks there is activity in the injured hand (Rosen *et al.*, 2005); it is very useful to use this technique with pianists in particular, but this cannot be used with many musicians due to the different tasks both hands have to perform.

Prevention is better than cure!

Animal studies show that highly repetitive motor movements contribute to degradation in the somotosensory cortex. However, when speed and force of the repetitive motor tasks are varied and interspersed with other regular activities, the degradation of hand cortical representation and loss of motor control can be minimised. Thus, it is important to maintain instruments in top-playing condition, with the hope of decreasing excessive energy outlay for the desired level of performance. Practice must be interspersed with other activities, stress and anxiety before a performance and practice schedules must be controlled to minimise the chances of the musician developing FHD. Freedom in musicians who improvise, and decreased external pressures in amateur musicians, may be preventative factors to the development of FD.

Research and the Future

Byl is working on a computerised system of sensory re-education in order to provide intense, goal-directed, suitable, motivating, repetitive, and discriminatory sensory stimulation for patients (personal communication). The challenge is to negotiate with compensatory bodies and maintain motivation levels with patients, so that sensory discriminative task performance continues until enough cortical remapping has occurred to allow normal motor control. Computerised equipment may increase the intensity of repetition and the gradation of sensory decisions, and may make tasks more interesting and shorten retraining times.

Certain questions still remain: why are only some musicians affected and not all? Are there risk and preventative factors? Could the adaptational mechanisms of the brain be used to 'reverse' this process? Rosenkranz *et al.* (2005, 2008) have published a study that tests sensory-motor integration by using muscle vibrations and transcranial magnetic stimulation. This study shows that sensory motor integration in healthy subjects is somatotopically organised and that the somatotopic representation is mildly changed in healthy musicians, but is completely lost in musicians with FHD. There are differences between writer's cramp and musician's dystonia and the way the brain deals with sensory input. Further research is being performed analysing the use of proprioceptive stimulation to specific muscles that relate to the dystonic finger, to retrain the somatotopic organisation of the somatosensory cortex in musicians with FHD.

Conclusions

The mechanisms by which FD develops in musicians need to be identified. Treatment must assist in re-establishing sensory-motor control. A comprehensive therapy programme with an aggressive sensory re-education element can improve sensory processing and motor control of the hand. SMR is of value for treating FD in pianists and guitarists. Scientific research investigating preventative measures and appropriate treatments for FHD is essential. Collaboration and a MDT approach to prevention, treatment and research are imperative and will be of benefit to all.

Acknowledgements

I am indebted to Dr Karin Rosenkranz, Dr Christopher Wynn Parry, Mr Ian Winspur and Graeme and Ruth Butler for their constant support, professional inspiration, and for proof reading this article.

References

1. Altenmüller, E. (2003). Focal dystonia: Advances in brain imaging and understanding of fine motor control in musicians. *Hand Clinics* **19**: 523–538.
2. Altenmüller, E. (1998). Causes and cures of focal limb dystonia in musicians. *International Society for Study of Tension in Performance* **9**: 13–17.
3. Berardelli, A., Rothwell, J. C., Hallett, M., Thompson, P. D., Manfredi, M. and Marsden, C. D. (1998). The Pathophysiology of primary dystonia. *Brain* **121**: 1195–1212.
4. Brandfonbrener, A. G. (1995). Musicians with focal dystonia: A report of 58 cases seen during a ten-year period at a performing arts medicine clinic. *Medical Problems of Performing Artists* **10**: 121–127.
5. Butler, K. and Svens, B. (2005). A functional thumb metacarpal extension blocking splint. *J Hand Ther* **18**: 375–377.
6. Butler, K. and Rosenkranz, K. (2006). Focal hand dystonia affecting musicians. Part I: An overview of epidemiology, pathophysiology and medical treatments. *Br J Hand Ther* **11**: 72–78.
7. Butler, K. and Rosenkranz, K. (2006). Ibid.: 79–87.
8. Byl, N. N., Nagajaran, S. and McKenzie, A. L. (2003). Effect of sensory discrimination training on structure and function in patients with focal hand dystonia: A case series. *Arch Phys Med Rehabil* **84**: 1505–1514.
9. Byl, N. N., McKenzie, A. and Nagarajan, S. S. (2000). Differences in somatosensory hand organization in a healthy flutist and a flutist with focal hand dystonia: A case report. *J Hand Ther* **13**: 302–309.
10. Byl, N. N. and McKenzie, A. (2000). Treatment effectiveness for patients with a history of repetitive hand use and focal hand dystonia: A planned, prospective follow-up study. *J Hand Ther* **13**: 289–299.
11. Byl, N. N., Nagarajan, S. S., Newton, N. and McKenzie, A. (2000). Effect of sensory discrimination training on structure and function in a musician with focal hand dystonia. *Phys Ther Case Reports* **3**: 94–113.

12. Byl, N. and Topp, K. S. (1998). Focal hand dystonia. *Phys Ther Case Reports* **1**: 39–52.
13. Byl, N., Merzenich, M. M. and Jenkins, W. M. (1996a). A primate genesis model of focal dystonia and repetitive strain injury: Learning-induced dedifferentiation of the representation of the hand in the primary somatosensory cortex in adult monkeys. *Neurology* **47**: 508–520.
14. Byl, N., Wilson, F., Merzenich, M., Melnick, M., Scott, P., Oakes, A. and McKenzie, A. (1996b). Sensory dysfunction associated with repetitive strain injuries of tendonitis and focal hand dystonia: A comparative study. *J Orthop Sports Phys Ther* **23**: 234–244.
15. Candia, V., Wienbruch, C., Elbert, T., Rockstroh, B. and Ray, W. (2003). Effective behavioral treatment of focal hand dystonia in musicians alters somatosensory cortical organisation. *Proc Natl Acad Sci USA* **100**: 7942–7946.
16. Candia, V., Schafer, T., Taub, E., Rau, H., Altenmüller, E., Rockstroh, B. and Elbert, T. (2002). Sensory motor retuning: A behavioural treatment for focal hand dystonia of pianists and guitarists. *Arch Phys Med Rehabil* **83**: 1342–1348.
17. Candia, V., Elbert, T., Altenmüller, E., Rau, H., Schafer, T. and Taub, E. (1999). Constraint-induced movement therapy for focal hand dystonia in musicians. *Lancet* **353**: 42.
18. Chamagne, P. (2003). Functional dystonia in musicians: Rehabilitation. *Hand Clinics* 19: 309–316.
19. Cole, R. A., Cohen, L. G. and Hallett, M. (1991). Treatment of musician's cramp with botulinum toxin. *Medical Problems of Performing Artists* **6**: 137–143.
20. Deuschl, G. and Hallett, M. (1998). Focal dystonias: From occupational cramp to sensorimotor disease that can be treated. *Aktuelle Neurologie* **25**: 320–328.
21. Elbert, T., Candia, V., Altenmüller, E., Rau, H., Sterr, A., Rockstroh, B., Pantev, C. and Taub, E. (1998). Alteration of digital representations in somatosensory cortex in focal hand dystonia. *Clin Neurosci* **9**: 3571–3575.
22. Elbert, T., Pantev, C., Wienbruch, C., Rockstroh, B. and Taub, E. (1995). Increased cortical representation of the fingers of the left hand in string players. *Science* **270**: 305–307.
23. Fahn, S. (1998). Concept and classification of dystonia. *Clin Neuropharmacol* **9**: S37–S48.
24. Fahn, S., Bressman, S. B. and Marsden, C. D. (1998). Classification of dystonia. *Adv Neurol* **78**: 1–10.
25. Fahn, S., Marsden, C. D. and Calne, D. B. (1987). Classification and investigation of dystonia. In Marsden, C. D. and Fahn, S. (Eds.), *Movement Disorders 2*. London: Butterworths.
26. Gowers, W. R. (1893). *A Manual of Diseases of the Nervous System* (2nd ed., vol. II). Reprinted by Hafner Publishing Company, Darien, Connecticut.

27. Hallett, M. (2006). Pathophysiology of writer's cramp. *Hum Movement Sci* **25**: 454–463.
28. Hallett, M. (2004). Dystonia: Abnormal movements result from loss of inhibition. *Adv Neurol* **94**: 1–9.
29. Hallett, M. (1995). Is dystonia a sensory disorder? *Annals of Neurology* **38**: 139–140.
30. Hoppman, R. A. (2001). Instrumental musicians' hazards. *Occupational Medicine* **16**: 619–631.
31. Jabusch, H. C. (2006). Epidemiology, phenomenology and therapy of musician's cramp. In Altenmüller, E. (Ed.), *Music, Motor Control and the Brain*. Oxford: Oxford University Press.
32. Jabusch, H. C., Zschucke, D., Schmidt, A., Schuele, S. and Altenmüller, E. (2005). Focal dystonia in musicians: Treatment strategies and long-term outcome in 144 patients. *Movement Disorders* **20**: 1623–1626.
33. Jabusch, H. C., Muller, S. V. and Altenmüller, E. (2004). Anxiety in musicians with focal dystonia and those with chronic pain. *Movement Disorders* **19**: 1169–1175.
34. Jankovic, J. and Shale, H. (1989). Dystonia in musicians. *Semin Neurol* **9**: 131–135.
35. Lederman, R. J. (1998). Piper's palsy: A focal dystonia. *Medical Problems of Performing Artists* **13**: 14–18.
36. Lederman, R. J. (1991). Focal dystonia in instrumentalists: Clinical features. *Medical Problems of Performing Artists* **6**: 132–136.
37. Lim, V. K., Altenmüller, E. and Bradshaw, J. L. (2001). Focal dystonia: Current theories. *Hum Movement Sci* **20**: 875–914; Clinical lecture on certain conditions of the hand and arm which interfere with the performances of professional acts, especially piano-playing. *BMJ* **1**: 441–444.
38. Lozano, A. and Linazasoro, G. (2000). Tratamiento quirurgigo de la distonia. *Revista de Neurologia* **30**: 1073–1076.
39. McCabe, C. S., Haigh, R. C., Ring, E. F., Halligan, P. W., Wall, P. D. and Blake, D. R. (2003). A controlled pilot study of the utilty of mirror visual feedback in the treatment of complex regional pain syndrome (type 1). *Rheumatology* **42**: 97–101.
40. Nutt, J. G., Muenter, M. D. and Melton, I. J. (1988). Epidemiology of dystonia in Rochester, Minnesota. *Adv Neurol* **50**: 361–365.
41. Oppenheim, H. (1911). Über eine eingenartige Krampfkrankheit des Kindlichen und jugenlichen Alters (Dystonia musculorum deformans). *Neurologisches Zentralblatt* **30**: 1090–1107.
42. Pesenti, A., Priori, A., Scarlato, G. and Barbieri, S. (2001). Transient improvement induced by motor fatigue in focal occupational dystonia: The handgrip test. *Movement Disorders* **16**: 1143–1147.

43. Poore, G. V. (1887). Clinical lecture on certain conditions of the hand and arm which interfere with the performances of professional acts, especially piano-playing. *BMJ* **1**: 441–444.
44. Priori, A., Pesenti, A., Cappellari, A., Scarlato, G. and Barbieri, S. (2001). Limb immobilization for the treatment of focal occupational dystonia. *Neurology* **57**: 405–409.
45. Ramachandran, V. S. and Hirstein, W. (1998). The perception of phantom limbs. The D.O. Hebb lecture. *Brain* **121**: 1603–1630.
46. Rosen, B. and Lundborg, G. (2005). Training with a mirror in rehabilitation of the hand. *Scand J Plast Reconstr Surg* **39**: 104–108.
47. Rosenkranz, K., Butler, K., Cordivari, C., Lees, A., Williamon, A. and Rothwell, J. C. (2008). Behavioural and neurophysiological effects of proprioceptive training in musician's dystonia. *Movement Disorders* **23**: S165.
48. Rosenkranz, K., Butler, K., Williamon, A., Cordivari, C., Lees, A. J. and Rothwell, J. C. (2008). Sensorimotor reorganisation by proprioceptive training in musician's dystonia and writer's cramp. *Neurology* **70**: 304–315.
49. Rosenkranz, K., Williamon, A., Butler, K., Cordivari, C., Lees, A. J. and Rothwell, J. C. (2005). Pathophysiological differences between musician's dystonia and writer's cramp. *Brain* **128**: 918–931.
50. Sakai, N. (2006). Slow-down exercise for the treatment of focal hand dystonia in pianists. *Medical Problems of Performing Artists* **21**: 25–28.
51. Schuele, S., Jabusch, H. C., Lederman, R. J. and Altenmüller, E. (2005). Botulinum toxin injections in the treatment of musician's dystonia. *Neurology* 64: 186–187.
52. Sheean, G. (2007). Restoring balance in focal limb dystonia with botulinum toxin. *Disability Rehabilitation* **29**: 1778–1788.
53. Singer, C. and Weiner, W. J. (1995). Primary dystonias current therapeutic recommendations. *CNS Drugs* **3**: 186–193.
54. Taub, E., Uswatte, G. and Pidikiti, R. (1999). Constraint-induced movement therapy: A new family of techniques with broad application to physical rehabilitation — A clinical review. *J Rehabil Res Develop* **36**: 1–21.
55. Taub, E., Miller, N. E., Novack, T. A., Cook, E. W. and Fleming, W. C. Nepomuceno, C. S., Connell, J. S. and Crago, J. E. (1993). Technique to improve chronic deficit after stroke. *Arch Phys Med Rehabil* **74**: 347–354.
56. Tubiana, R. (2000). Musician's focal dystonia. In Tubiana, R. and Amadio, P. C. (Eds.), *Medical Problems of the Instrumentalist Musician* (1st ed.), pp. 329–342. London: Martin Dunitz.
57. Tubiana, R. and Chamagne, P. (1983). Occupational 'cramps' of the upper limb. *Annales de Chirugie de la Main* **2**: 134–142.
58. Warrington, J. (2003). Hand therapy for the musician: Instrument-focused rehabilitation. *Hand Clinics* **19**: 287–301.

59. Winspur, I. (1998). Surgical indications, planning and technique. In Winspur, I. and Wynn Parry, C. B. (Eds.), *The Musician's Hand: A Clinical Guide*, pp. 41–52. London: Martin Dunitz.
60. Zeuner, K. E., Shill, H. A., Sohn, Y. H., Molloy, F. M., Thornton, B. C., Dambrosia, J. M. and Hallett, M. (2005). Motor training as treatment in focal hand dystonia. *Movement Disorders* **20**: 335–341.
61. Zeuner, K. E. and Hallett, M. (2003). Sensory training as treatment for focal hand dystonia: A 1-year follow-up. *Movement Disorders* **18**: 1044–1047.
62. Zeuner, K. E., Bara-Jimenez, W., Noguchi, P. S., Goldstein, S. R., Dambrosia, J. M. and Hallett, M. (2002). Sensory training for patients with focal hand dystonia. *Annals of Neurology* **51**: 593–598.

Index